Quantum Mechanics

Ali A. Abdulla
Professor of Physics
Physics Department
College of Science, University of Baghdad
Iraq

Contents

Introduction

The theory of quantum physics was developed after the discovery of the duality property of a moving particle. This property was discovered by Louis de Broglie in 1924. He proposed that any moving particle is associated with wave. Its wavelength is given by $\lambda = \frac{h}{P}$, where λ is the wavelength, *and h* is Planck constant. It is a universal constant; its value is 6.63×10^{-34} joules second (energy flow), and P is the momentum of the moving particle (mv, where m is the mass of the particle and v is the particle velocity). This Planck constant was introduced by Planck during his work on the properties of the black body radiations for different temperatures. In this experiment, which was performed in 1900, it was found that the radiation energy is discrete, not continuous as was thought before in Maxwell equations 1964. The radiation is consisted of quanta; the energy of each quanta is given by E=hv, where v is the frequency of the radiation wave. So h is called the proportional constant of the energy with the frequency, named after Planck. Sometimes the reduced h is used; it is $\hbar = \frac{h}{2\pi}$. This experiment result was a breakthrough in the theory of physics. Therefore, 1900 was considered the birth year of the quantum physics.

In 1905, Einstein interpreted the photoelectric effect on the bases that the incident light consists of quanta called photons. Then was the discovery of the duality behavior of matter or particles where they are in moving state, as mentioned before, in 1924 by de-Broglie. That was the starting moment of finding a mechanics other than the known classical mechanics to deal with a microsystem in a wavy state. Thus, the period from 1926 to 1927, a quantum, or wave mechanics, was developed to describe the character of moving, interacting, and the behaving of the quanta or the microscopic systems. These are, in fact, the building blocks of the microscopic systems.

Therefore, the behavior of the microscopic system is, indeed, the average behavior of these blocks. Hence, quantum mechanics is the general mechanics corresponding to the classical mechanics, which is the limit to the quantum mechanics as $h \rightarrow 0$. So in general, the classical physics is a limit to the quantum physics when $h \rightarrow 0$ and the principal quantum number $n \rightarrow \infty$.

From 1926 to 20014, the theory of quantum physics have been developed extensively in all the scientific branches. But in physics, it can be noticed that the great role of the quantum mechanics in different branches of physics are atomic physics, nuclear physics, molecular physics, solid-state physics, astrophysics, cosmology, and field theory.

This book is based on lectures given by the author to graduate students for twenty-five years at the physics department, college of science, Baghdad University, Iraq. It contains ten chapters starting with the basics principles up to the briefing of the quantum field theory. Also, there are appendixes that contain some solved problems. These chapters are

Chapter One: Basics Principles of Quantum Mechanics
Chapter Two: Operators, Concepts, and Properties
Chapter Three: General Formalism of Quantum Mechanics
Chapter Four: Schrödinger and Heisenberg Pictures
    (Representations) with Some Applications
Chapter Five: Angular Momentum in Quantum Mechanics
    (Properties and Applications)
Chapter Six: Identical Particles in Quantum Mechanics
Chapter Seven: Scattering Theory in Quantum Mechanics.
Chapter Eight: Approximation Methods, Theory, and
    Applications
Chapter Nine: Basic Elements of Relativistic Quantum
    Mechanics

Chapter One
Basic Principles

1.1. *Discreteness*

Before (1900), it was thought that the classical physics (mechanics and electromagnetic), had reached a stage that it can deal with any physical phenomena met. From Maxwell equations (1864), it was thought the emitted energy has a continuous range of values. In 1900, there was a breakthrough in the theory of physics. Planck found, by working on black body radiations with different temperature, that energy is discrete; it is consisted of quanta. The energy is given by E=hv, for each quantum. E is the energy, v the frequency of the radiation, and h is Planck constant; its value is found to be 6.63x10^{-34} joules. Second. It is a universal constant. Therefore, the energy of the quanta is proportional to its frequency (v). Hence, the energy in the microsystems is discrete, not continuous. These systems cannot be described by classical mechanics.

Quantum mechanics was worked out to deal with the microscopic systems. This discovery represents a new era in science. In quantum theory, all physical quantities in these systems are quantized, such as energy, angular momentum, spin, etc. Hence, the concept of quantization is a quantum feature of the physical dynamical variables in quantum theory. Therefore, the classical physics is the limit of the quantum physics as $h \to 0$, and the principal quantum number $n \to \infty$. This was clearly demonstrated by the corresponding principle proposed by Niles Bohr. The energy spectrum of the hydrogen atom quietly well clarify this discrete phenomenon. The following schematic diagrams show the idea of discreteness:

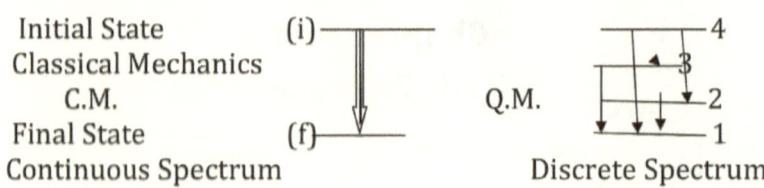

Initial State (i)

Classical Mechanics

C.M.

Final State (f)

Continuous Spectrum

Q.M.

Discrete Spectrum

In conclusion, classical mechanical is dealing with the continuous quantities of macroscopic systems, while quantum mechanics is dealing with the discrete quantities of microscopic systems, such as atoms, nuclei, and molecules.

1.2. *Wave-Particle Duality*

As mentioned in the introduction, the dual behavior of the particle as a particle and as a wave was of a serious discussion between scientists. In 1924, de- Broglie proposed that every moving mass is associated with a wave. Its wavelength is given by $\lambda = \frac{h}{p}$. One can notice that if m is a mass of macroscopic system. The expected wavelength will be so small, not possible to be detected. But if it is a mass of a microscopic (atomic and below), the wavelength might be detectable. This behavior of light and radiations is observed in the diffraction, the interference, and the photoelectric effect. The first two phenomena represent the waving behavior of the light and the radiations. But the photoelectric effect represents the particle behavior of the light and radiations. For example, in the light incident on a metal, an electron is ejected with energy $E_e = E_{light} - \omega$, where E_{light} is the energy of the incident light, and ω is the work function of the metal. It is different for different metals. But if the light passes through narrow slits, it shows a diffraction and interference on the screen behind the slits as shown in the plots below:

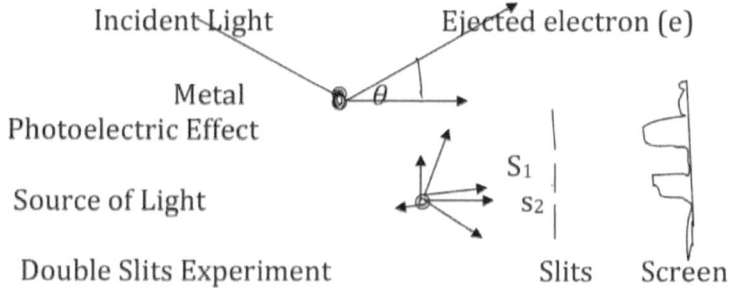

Incident Light   Ejected electron (e)

Metal
Photoelectric Effect

Source of Light S_1
                                   S_2

Double Slits Experiment   Slits   Screen

These two physical phenomena demonstrate the light behavior subjected to two different physical phenomena as corpuscular (particle) and wave respectively. This behavior is called a wave-particle duality. It has been puzzling scientists for about twenty-five years.

In 1924, as mentioned before, de- Broglie proposed that any moving system is associated with wave of wavelength given by $\lambda = \frac{h}{p}$, where h is Planck constant, and p is the momentum of the moving particle of mass (m) and velocity (v), so p = mv, so λ = h/mv; λ represents the wave behavior, and the mass m represents the particle behavior.

Therefore, each moving particle behaves as particle and as wave at the same time. The type of the physical situation subjected to imposes the behavior of a particle or a wave, as shown in the above diagrams.

1.3. *Hiesenberg Uncertainty Principle*

In classical physics, the canonically related dynamical variable values can be measured quite accurately and simultaneously, such as position (x) and momentum (p), energy (E) and time (t).

In quantum mechanics, this measurement is not possible. This physical fact is due to the intrinsic behavior of the canonical relation between the measured values of the dynamical variables and the exchange effect between the canonically related physical quantities during the measurement of any of them. Thus, there is a restriction on the measurement such that the inaccuracy in measuring the momentum is (Δp) and that in measuring the position (Δx) are related as $\Delta x \Delta p \geq \hbar$,

where $\hbar = 1.04 \times 10^{-34}$joule.sec.

The same is valid for the energy with time, that is to say $\Delta E \Delta t \geq \hbar$,. Take the following example for moreillustration:

$$\text{x} \xrightarrow{\quad L \quad} \text{x+dx}$$

$$(\frac{\Delta p}{p})(\Delta x / x) \geq \hbar / L$$

Now notice, if $L \to \infty$, then $\frac{\hbar}{L} = 0$, so classical physics is a limit of quantum physics. Thus, for macroscopic system, no such error is expected in the measurements. It is an accurate measurement which implies the determinism philosophy as classical concept. In quantum physics, the dimensions are in the range $(10^{-16}-10^{-8})$ cm, so $\Delta p \Delta x \geq \hbar / L \neq 0$.

Therefore, the intrinsic relation is basic and very important in studying the microscopic systems. It is a mathematical concept and intrinsic character in quantum mechanics only. This mathematical physical concept is called Heisenberg uncertainty principle.

1.4. *Probability Concept*

In classical physics, as mentioned earlier, the measurements can be performed for two canonically related physical dynamical variables, such as p_i and q_i, E_i and t accurately and simultaneously within the ordinary statistical deviation errors. But in quantum physics, any accurate measurements are impossible.

In 1896–1898, the radioactivity of some heavy elements was discovered. Since these radioactive elements consist of so many atoms, it is not possible to determine which atom is the radiating one. The concept of the probability was developed to find the probability that a certain atom is emitting the radiations. This development of physical thoughts have led to the law of the radioactivity. It is given by $A = A_0 e^{-\lambda t}$. It is a probabilistic law. Here, A is the activity of the radioactive element at time(t) (life time), (A_0) is the activity at $(t = 0)$, (initial activity), and λ is the so-called disintegration constant. Therefore, if we deal with the microscopic systems, there is no place to the concept of the determinism but the probability.

Hence, the probability concept is the one describes the physical behavior of microscopic systems. Now, how can one deal with the description of the physical situation of such microscopic systems as the atoms, nuclei, molecules, nucleons, leptons, mesons, and quarks? As mentioned earlier, these physical identities behave as corpuscular and wave simultaneously; therefore, the wave mechanics, developed by Schrödinger(1926), was to describe these microscopic systems; at the same time, Heisenberg developed the quantum mechanics using the matrix concept, but both came to the same results.

Then came Max Born to give the wave function (describes the system) the statistical concept. It contains the full information about the described system. This wave function denoted as $\psi(x, t)$. So this wave function described by Born as a statistical nature or behavior contains full information of the described system. Hence, the wave function describes the state of the system at the position point (x, y, z); but for static system (t =

0), $\psi(x, y, t = 0) = \psi(x, y, z)$, then $|\psi|^2$ is the probability amplitude to find the particle described by this wave function at a certain position. Suppose there is a radioactive source emitting radiations placed in front of a screen, as shown below:

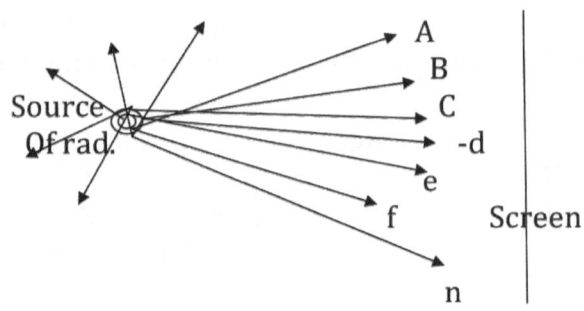

Here, it is not possible to know exactly which path the radiation is following toward the screen, among so many possible paths. Is it *a, b, c, d, e, f*... or *n?* Hence, the problem here is probabilistic one; its value is within the range (0–1). If the wave function is describing a dynamical system, then it will be time dependent, i.e.,it contains a maximum information about the described system.

1.5. *Operator Concept*

In classical physics, the position (*x*) and the momentum (*p*), as examples, are measured directly using the well-known classical way of measurements. But in quantum physics, where quantum mechanics is the tool to deal with describing a microscopic system, this classical type of measurement is not workable. This is because these dynamical variables are canonically related; their space wave functions are mathematically related through Fourier transformation. SO *x* behaves differently in momentum space and vice versa.

Therefore, each dynamical variable is associated with an operator operating on the wave function describes the system to produce the eigenvalue (measured value) of the dynamical variable through the probability of finding the system somewhere in its defined space. These physical concepts and their quantumtreatment will be dealt with in the next chapters. For simple illustration, consider a dynamical variable A; its operator is A_O, and then the measured value (average value) of A is given by, $<A> = \int \psi^* A_O \psi\, dv$, also the eigenvalue equation is $A_O = a$; a is the eigenvalue.

The operators for the position and the momentum in position and momentum spaces are given as follows:

Space (Position)   *Operator*

X   X

P $-i\hbar \frac{y}{\partial x}$

Space (Momentum)

P   P

X $-i\hbar \frac{\partial}{\partial p}$

Energy Operator $(E = -i \frac{\partial}{\partial t} \hbar)$

1.6. *Expected Values*

In quantum physics, there is no exact measured value, as pointed out previously. This is because the method used for the measurement, the probability, where each dynamical variable is associated with an operator, it operates on the eigenfunction, describing this (DV) to find the eigenvalue (measured value=expected value of the DV). This measured value is an average value, not the exact one. This is expected, or the average value is designated as <>. For example, to measure the average value of the momentum (P) in position space, where the wave function is $\psi(x, t)$, we should write,

$<P_x> = \int \psi^*(-i\hbar \frac{\partial}{\partial x})\psi dv$, ψ^* is the complex conjugate of ψ.

But in momentum space, $<p_x>$ given by,

$$<p_x>=\int \phi^*(p,t)\ p\ \phi(p,t)\ dv$$

Note: The operator of a dynamical variable in the same space is the dynamical variable itself.

1.7. Eigenvalue Equation

Its equation connects mathematically between the wave functionψ and describes the dynamical variable called eigenfunction, and its measured value is called eigenvalue. So since the wave function is specified, it should be written as$A_0\ \psi = a\psi$. Therefore, an eigenvalue equation for a physical dynamical variable A associated with an operator A_0 is given by

$A_0\psi = a\psi$; (a) is the eigenvalue representing the average measured value of A.

1.8. Some Important Postulates

In view of the previous features of quantum mechanics, the following postulates are proposed:

(a) *Postulate I*: Matter behaves partly as a mechanical mass and partly as a wave motion. This leads to:

1. Uncertainty Principle, i.e., $\Delta x \Delta p \geq \hbar$
2. Statistical interpretation of the wave function describing the wave property of matter.
3. As long as $\hbar/L \cong 1$, the system is quantum mechanical one, where L is the angular momentum. Now looking through a microscope (as shown

below) at an object, the following information will be obtained:

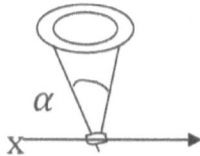

$\Delta p_x = \left(\frac{v}{c}\right)\hbar\sin\alpha = (\hbar/\lambda)\sin\alpha$, errors in measuring P. And ΔX is $\frac{\lambda}{\sin\alpha}$. Hence, $\Delta p_x \Delta x = \hbar$

Therefore, the uncertainty principle is verified.

(b) *Postulate II:* Radiations behave partly as a wave motion and partly as a mechanical mass. This is related to the proposal of de- Broglie where the wave-matter duality is a property of moving system. This is canonically relating the wave property with the corpuscular property by the law, $\lambda = \frac{\hbar}{p} = \frac{\hbar}{m}\frac{1}{V}$. Here, λ is represnting a wave property, while m is representing a corpuscular property; so if m is constant (nonrelativistic), then it is known that the velocity V of a moving microscopic system is close to the light velocity; but for moving macroscopic system, the velocity is too far below the light velocity. Then λ is too small to be detected with the available technology, if V is constant, and the mass is variable, the same argument is valid.

As mentioned before, this is for the nonrelativistic motion only.

In conclusion, the mass of microscopic system is compared to 10^{-31} kgm. The wavelength will be about angstrom ($1A^0$); it is easy to detect. For macroscopic system, the mass very large compared to 10^{-31}kgm may be 10^{31} larger. In this case, λ will be about $10^{-28}A^0$. It is undetectable with the available technology.

(c) *Postulate III: Operational Postulate*

In classical physics, as pointed out earlier, the macroscopic system can be found in a definite state if its initial conditions are known because it is not quietly affected by observation. Conversely, in quantum physics, the observation clearly affects the observed microscopic system. Thus, it is nonsense to say the system is in a definite position after a certain lapsed time.

As an example, consider the double-slit experiment as shown in the diagram below:

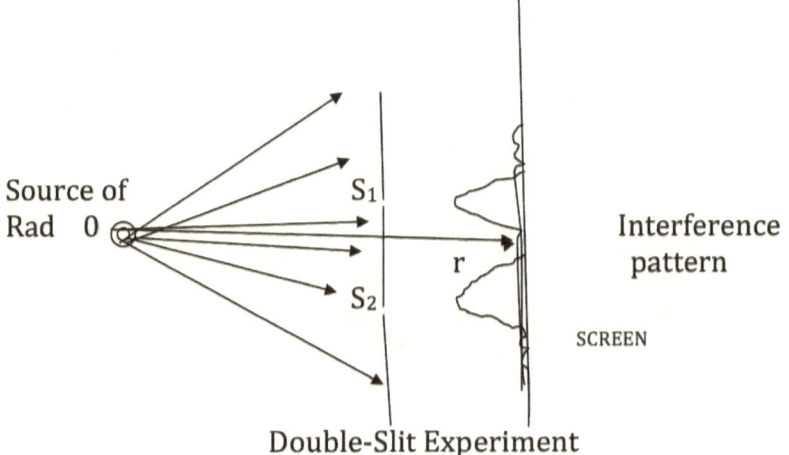

Double-Slit Experiment

The passing of radiation or electron will incident on the screen.

The observation of the event on the screen will show interference pattern as shown above. This pattern does not show where the electron is situated. Therefore, the concept of probability density of finding the particle within the neighborhood of r was developed. It is defined as [$P(r, t)d^3r$] in position space and [$Q(p, t)d^3p$] in momentum space. These two probabilities density are related by Fourier transformation, as will be shown later on. Therefore, one can use either one of

24

them if the wave function is known. They have the following characters:

1. Both are real and positive.
2. They are normalized to one, i.e., = 1.

1.9. Super Position Principle

This principle is well known in the physical optics, where the interference phenomenon and its applications take place in many physical fields. In quantum physics, the wave mechanics play a big role in describing the wave equations of motion. The super position principle is of great importance here. Let ψ_1 and ψ_2 are wave functions for system one and system two, respectively. If the two systems are observed, then the wave function describing the new system (1+2) designated as $\Psi = c_1\psi_1 + c_2\psi_2$ which is a linear combination of ψ_1 and ψ_2,

c_1, and c_2 are arbitrary coefficients. The probability density accordingly is P(r, t) = l$\Psi(r, t)$l^2 = l$c_1\psi_1 + c_2\psi_2$l$^2 \geq 0$, the same for Q(p, t) = l$\phi(p, t)$ l$^2 \geq 0$.

To find $\Psi(r, t,)$ in terms of $\phi(p, t)$ and conversely use the Fourier transformation such as,

$$\Psi(r, t) = \frac{1}{(2\pi)^{3/2}} \int A(k) \, e^{i(k.r - \omega t)} \, dk \text{ -----------(1).}$$

Where $P = \hbar k$, k is a wave number, $\omega = E/\hbar$, E = energy.

Equation 1 is a plane wave. Also, E = $p^2/2m$ for free system (V(r) = 0).

1.10. Hamiltonian and Energy Operator

In classical mechanics, Hamiltonian is given by
H = T + V = Kinetic Energy + Potential Energy. ------------(2)

If the system is free, there is no external force, $V(r) = 0$. Then $H = P^2/2m$ and $H_0 = P^2/2m = -\hbar^2/2m.\nabla^2$, where $\nabla^2 = \frac{\partial^2}{\partial x^2} + \frac{\partial^2}{\partial x^2} + \frac{\partial^2}{\partial x^2}$. The energy is canonically related to the time of an action because it is a product of event, and the time starts as action begins. Therefore, $E_0 = -i\hbar\frac{\partial}{\partial t}$, since T_0 (kinetic energy operator) can be written as, $(H_0 = -\frac{\hbar^2}{2m}\nabla^2)$. Hence, Hamiltonian operator for $V(r) = 0$. Now $(H_0 + E_0) = (T_0 + V(r) + E_0) = T_0 + E_0 = -V(r)$.

Then $(T_0 + E_0) = 0$
$H_0\Psi_i(r,t) = E_0\Psi_i(r,t)$ --------------------------------(3).

Equation 3 is an eigenvalue equation, where E_0 is the eigenvalue, and Ψ_i is the eigenfunction. H_0 is Hamiltonian.

In case the system is subjected to an external or internal force, $V(r)$, then $H_0 = T_0 + V$, $V_o = V(r)$ (same space).

Hence, $H_0 = (-\hbar^2\nabla^2/2m) + V(r)$; therefore, $H_0 = (-\hbar^2\nabla^2/2m) + V(r)$. Hence, $H_0\Psi_i(r,t) = \varepsilon_i\Psi_i$--------------(4).

Equation 4 again is eigenvalue equation.

And $H_0\Psi(r,t) = E\Psi(r,t)$------------------------(4*) is Schrödinger time-dependent equation.

1.11. *Continuity Equation and Probability Current Density*

For free particle, $V(r) = 0$, Schrödinger equation is

$$(-i\hbar^2\nabla^2/2m + \frac{\hbar}{i}\frac{\partial}{\partial t})\Psi(r,t) = 0 \text{ ---------------------------- (5)}$$

Using Fourier transformation, where
$$\Psi(r,t) = \frac{1}{(2\pi)^{3/2}} \int A(k)e^{i(k.r-\omega t)}d^3 k$$

Equation(5) can be written as,

$$[\frac{\hbar^2}{2m}\nabla^2 + \frac{\hbar}{i}\frac{\partial}{\partial t}] \Psi(r,t) = 0 =$$
$$\frac{1}{(2\pi)^{3/2}} \int A(k)e^{i(k.r-\omega t)} [\frac{\hbar^2k^2}{2m} - \hbar\omega] d^3 k$$

Notice that p²/2m = = K.E=E(total energy)$=-\hbar^2\nabla^2/2m$, for free system(V(r)=0), hence,

$$(-\hbar^2\nabla^2/2m + \frac{\hbar}{i}\frac{\partial}{\partial t})\Psi(r,t) =$$
$$\frac{1}{(2\pi)^{3/2}} \int A(k) e^{i(k.r-\omega t)}[\frac{\hbar^2k^2}{2m} - \hbar\omega]d^3k$$

Notice, $E = \hbar\omega$ = total energy. $V(r) = 0$
The normalization of Ψ and $P(r, t)d^3r$ gives
$\int |\Psi(r,t)|^2d^3r = \int P(r,t)d^3r = 1$, so $\frac{\partial}{\partial t}\int P(r,t)d^3r = 0$. This implies that the probability is time independent; hence,

$$\frac{\partial}{\partial t}\int \Psi^* \Psi d^3r = 0 \Longrightarrow \int (\Psi\frac{\partial\Psi^*}{\partial t} + \Psi^*\frac{\partial\Psi}{\partial t})d^3r = 0 \text{ --------(6).}$$

Also, for free system, $-\frac{\hbar^2}{2m}\nabla^2 = \hbar i\frac{\partial}{\partial t} = E(r, t) = k.E.$

Taking this in consideration, one gets

$(i\hbar/2m)\, \nabla \cdot [\Psi^*\nabla\Psi -\nabla\Psi\Psi^*] =0$. Using Green theorem,
$\int (\,\Psi^*\nabla\Psi -\Psi\nabla\Psi^* \cdot ds = 0$ (over surface).

Define the probability density as $\rho = \Psi^* \,\Psi$
and the probability current density as

$J = \dfrac{\hbar}{2mi}\,(\Psi^*\nabla\Psi -\Psi\nabla\Psi^*)$–current density.

These leads to, $\dfrac{\partial \rho}{\partial t} + \nabla \cdot J = 0$--------------------(7).

Equation (7) is the continuity equation.
From the so far formulated equations and stated concepts, the following can be concluded:

(1) $\dfrac{k^2\hbar^2}{2m}\nabla^2\psi = i\hbar\dfrac{\partial}{\partial t}\psi$, Schrödinger equation for free system V(r)=0.

(2) $j = \dfrac{\hbar}{2mi}[\Psi^*\nabla\Psi -\Psi\nabla\Psi^*]$, the probability current density.

(3) $\dfrac{\partial \rho}{\partial t} +\nabla \cdot J = 0$, the continuity equation.

(4) $\dfrac{\partial}{\partial t}\int P(r,t)d^3\,r = 0$, the probability is constant (0 - 1)

(5) $\int \nabla \cdot j\, d\,v = \int j.\,ds =0, \psi \to 0$ as r$\to \infty$, i.e.,

$\psi \to 0$ Faster than 1/r.

These remarks are important to be understood.
Consider:

(i) $\psi \to 0$ as r$\to \infty$
(ii) Boundary conditions (BC). They are very necessary to satisfy that the surface integral is $\int j.\,ds = 0$.

These BC are necessary to satisfying the surface integral.

For BC, take the case

-L | B.C | +L

Where $\psi(+L=\psi(-L))$ is a periodic boundary condition.
Consider this for one dimensional space, then $\psi \sim e^{ipx}$.

Thus, by (ii)$e^{ipL} = e^{-ipl} \rightarrow e^{2ipL} = 1 \rightarrow 2PL = 2n\pi \rightarrow p = n\pi/L$,
where n is an integer number, n = 0, 1, 2, 3 . . .n. This indicates
the spectrum is discrete. As it is clear that it is due to boundary
conditions. Therefore, the spectrum of a bounded system, such
as atoms, nucleus, molecules is discrete.

1.12. Parsival Equality of $\int P(r,t)d^3r$ and $\int Q(p,t)\,d^3p$

By definition, as stated before, P(r, t)d³r is the probability
density in spatial coordinate, and Q(p, t) is the probability
density in momentum space (p, t). The eigenfunctions in these
spaces are mathematically related by the Fourier
transformation, as illustrated below:

$$\psi(r,t) = \frac{1}{(2\pi)^{3/2}} \int A(k)e^{i(\vec{k}.\vec{r}-\omega t)}d^3k =$$
$$\frac{1}{(2\pi)^{3/2}} \int d^3k\, \phi(k,t)\, e^{i\vec{k}.\vec{r}}, \text{ where,}$$
$$\phi(k,t) = \frac{1}{(2\pi\hbar)^{3/2}} \int \psi(r,t)\, e^{-i\vec{k}.\vec{r}}d^3r, \; p=\hbar k$$

Now let us find P(r, t)d³r and Q(p, t)d³k and the relation
between them. It is defined that $P(r, t) = |\psi(r,t)|^2$, and
$Q(p, t) = |\phi(r,t)|^2$ then.

Now, Schrödinger equation in momentum space is

$$\frac{\hbar^2}{i}\frac{\partial\phi(p,t)}{\partial t} = \frac{p^2}{2m}\phi(p,t) \text{ for free system.}$$

This can be rewritten as

$(-i/\hbar^2 \frac{p^2}{2m})dt = \frac{\partial\phi(p,t)}{\phi(p,t)}$. The solution is,

$\phi(p,t)=\phi(p,0) \exp(\frac{-i}{\hbar})\frac{p^2}{2m}t.$, $V(r) = 0$.

Now $\int \psi^* \psi \, d^3r = \int \phi^* \phi d^3p = 1$, normalized.

Take $\frac{1}{(2\pi)^{3/2}} \iint d^3pd^3p^- \phi^*\phi e^{\frac{-i(p-p^-)}{\hbar}} = \int \psi^* \psi d^3r$.

Using Dirac delta function properties, it is obtained that

$= \frac{1}{(2\pi)^{3/2}} \int d^3p\phi^*(p,t)\phi \, (p,t) = \int Q(p,t) \, d^3p = \int \psi^*(r,t)\psi d^3 r$
$= \int P(r,t) \, d^3r$. Hence, $\int P(r,t)d^3 r = \int Q(p,t) \, d^3p$.

So Parsival equality is proved.

1.13. *Expected Value (Average Value) with Some Details*

As mentioned before, in quantum mechanics, it is not possible to measure exact value of any physical dynamical variable.Because here, the problem is dealing with a microscopic systems of dimensions within 10-16-10-8 cm and a mass of 10-28 gm. Therefore, the observation of such systems, by any means, will affect their situation. Hence, the concept of the determinism in classical mechanics is no longer valid in quantum mechanics, where the world of probability takes place. The probability is the important tool in quantum mechanics. Thus, looking for a particle in the neighborhood of the position (r), the probability is the method to be used.

The same will be for the momentum space (ϕ). Since the probability is depending on the probability density to find a particle somewhere in the neighborhood of the position(r) or the position(p). Therefore, for measuring the expected value of any physical dynamical variable, the probability density has to be used. For example, to measure the expected value of the position (X) described by the wave function $\psi(r,t)$.

The expected value of (X) *is* designated by

$$<X> = \frac{\Sigma_r(X|\psi|^2}{\Sigma|\psi(r,t)|^2} = \frac{\int d^3 r X\psi\,\psi^*}{\int \psi^*\psi d^3\,r}$$
$$= \int X\,\psi^*\psi d^3 r \ , \quad \int d^3\,r\,\psi^*\psi = 1, \text{normalized}$$

So $<X> = \int X|\psi|^2 d^3 r$----------------------- (8).

Where $X_0 = X$ in x-position space. Let us show graphically the position for a symmetric and nonsymmetric distribution.

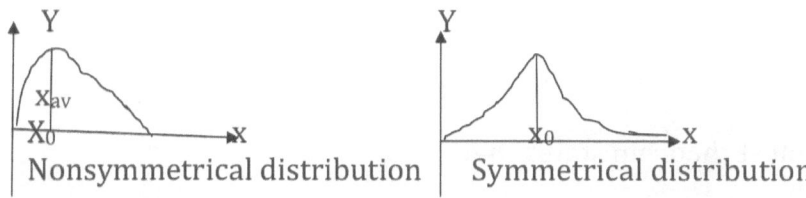

| Nonsymmetrical distribution | Symmetrical distribution |

1.14. *The Continuity Equation*

Let us write Shrodinger equation under the effect of V(r)as,

$$[\frac{\hbar^2\,\nabla^2}{2m} +V(r\,)]\psi(r,t) = i\hbar\frac{\partial}{\partial t}\psi(r,t)$$ ----------------(9)

The probability charge density is given by $\rho = \psi^*\psi$, as defined before for free system. The probability current density, as defined before, for free space, is given by

$$J = \frac{\hbar}{2mi}[\psi^*\nabla\psi-\psi\nabla\psi^*].$$ Also, the continuity equation is

$\frac{\partial\rho}{\partial t} +\nabla.J = 0$; for free system, V(r) = 0.

Now if $V(r) \neq 0$, then $\frac{\partial \rho}{\partial t} = \psi^* \frac{\partial \psi}{\partial t} + \frac{\partial \psi^*}{\partial t} \psi$---------------------(10).

From equation 9 and equation10, one obtains

$$\frac{\partial \rho}{\partial t} = \frac{\psi^*}{i\hbar}[-\frac{\hbar^2}{2m}\nabla^2 + V(r)] - \{\frac{1}{i\hbar}[-\frac{\hbar^2}{2m}\nabla^2 + V^*(r)]\psi^*\}\psi)$$

$$= -\nabla \cdot J + \frac{1}{i\hbar}\psi^* \psi (V-V^*)$$ ----------------------------(11).

Now if *V*, the potential, is not real, then the charge is not conserved, and the continuity equation is not valid. But if V = V*, that means *V* is Hermitian; its value is real. Consequently, charge is conserved, and the continuity equation is valid. In the case the charge is not conserved, this indicates a certain particle is injected into the system.

1.15. *Ehrenfest Theorem*

Ehrenfest theorem states that

$<P_X> = m\, d<X(X,t)>/dt$
$d<P_X>/dt = -<V>$.

To prove that, from section I-13, the expected value of the position and the momentum is given by

$<X(t)>$ and $<P> = m<V> = m\, dX(t)/dt$.

From classical mechanics, it is known too that $dp/dt = m(dV/dt) = ma$, for nonrelativistic system. Hence, $dp/dt = F$(force), so $d<P>/dt = <F>$.

These are in classical mechanics. What is it in quantum mechanics? Take the average value of *X* in quantum mechanics:
$<X(t)> = \int \psi^0 X_0 \psi\, d^3r = \int \psi^* X \frac{\partial}{\partial t} \psi(r,t) d^3r$

Now $\frac{d<X(t)>}{dt}=\frac{d}{dt}\int \psi^* \, X \, \psi d^3r=\int [\frac{\partial \psi^*}{\partial t}X\psi +\psi^*X\frac{\partial \psi}{\partial t}+\psi^*]d^3r,$

but since X is an operator which can be at any position on the range shown in the figure below, the$n$ $\frac{\partial x}{\partial t}$=v is not considered.

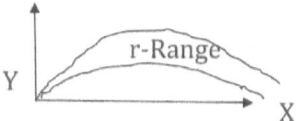

Therefore,

$$\frac{d<X>}{dt} = \int [\frac{\partial \psi^*}{\partial t}X\psi + \psi^*X\frac{\partial \psi}{\partial t}]d^3r = \int X \frac{\partial}{\partial t}(\psi^*\psi) \; d^3r$$

$$=\int X \frac{\partial \rho}{\partial t} \, d^3r = -\int X(\nabla.j) \, d^3r = \int \nabla.(Xj) \; d^3r = \int X\nabla.J d^3$$

$$+ \int j_x d^3r \qquad \underset{\nearrow}{0}$$

$$= \int Xj ds \text{ (over surface)} + \int j_x d^3r = \int j_x d^3r.$$

Using continuity equation, this was obtained.
Now substituting for j_x (equation of the current density), one obtains

$$\frac{d<X>}{dt} = \int \frac{\hbar}{2mi}(\psi^*\nabla_x\psi-\psi\nabla_x\psi^*)d^3r = \frac{\hbar}{2mi}\int 2(\psi^* \frac{\partial \psi}{\partial t})d^3r =$$

$$\frac{1}{m}\int \psi^*(\frac{\hbar}{i}\frac{\partial}{\partial x})\psi d^3r = \frac{1}{m}\int \psi^* \, P_o\psi d^3r, \text{ this implies that,}$$

$$<P_x> = m\frac{d<x(t)>}{dt}\text{-}\text{-}\text{-}\text{-}\text{-}\text{-}\text{-}\text{-}\text{-}\text{-}(12).$$

By the same way, it is easy to find that

$$\frac{d<P_x>}{dt} = \text{-}<\nabla V > = <F>. \; V \text{ is the potential, and } F \text{ is the force. To}$$

show that, take $<P_x> = m\frac{d<X>}{dt} = \frac{d}{dt}(\hbar/i)\int \psi^* \frac{\partial}{\partial x}\psi \; d^3r.$

By carrying the calculations and using Schrödinger equation, the following result is obtained:

$$\frac{d<P_x>}{dt} = -\int \psi^* \frac{\partial V(r)}{\partial x} \psi d^3r = -(\psi, \nabla_x V(r)\psi) = -<V(r)>.$$

Hence, $\frac{d<P_x>}{dt} = -<V(r)> = <F>$ (force) ------------------------ (13).
Equations 12 and 13 are the proof of Ehrenfest theorem.

1.16. The Conservation of Total Energy of Isolated System

The total energy of an isolated system is given by E=k.E+P.E, the operator of E, is given by $E_0 = i\hbar \frac{\partial}{\partial t}$.

Therefore, $<E> = <P^2/2m> + <V(r)> = <(-\hbar^2/2m)> + <V(r)>$
$<E> = \int \psi^* E_0 \psi d^3r = \int \psi^* (i\hbar \frac{\partial}{\partial t}) \psi d^3r$
$= \int \psi^*, (-\frac{\hbar^2}{2m}\nabla^2)\psi \, d^3r + \int \psi^* V(r)\psi d^3r$, since $V_0 = V$

Since KE is time independent and V(r) is independent of time as well, therefore, $\frac{d<E>}{dt} = 0$, which implies that E is conserved for isolated physical system.

Now is the energy a real quantity? Let us find the answer.

Take the Schrödinger, $[-\frac{\hbar^2}{2m}\nabla^2 + V(r)]\psi = i\frac{\partial}{\partial t}\psi(r,t) \Longrightarrow$

$H_0 \psi_i = E_i \psi_i$ represents eigenvalue equation.
$\psi(r,t)$ can be written as $\psi(r,t) = \psi(r)\phi(t)$, so Schrödinger equation takes the form $\frac{1}{\phi(t)}[-\frac{\hbar^2}{2m}\nabla^2 + V]\psi(r) = \frac{\partial\phi\, i\hbar\frac{\partial}{\partial t}}{\phi(t)} = \frac{\partial\phi}{\phi}E.$

This is written so (they are independent of each other); hence, Schrödinger equation will be as the solution of the right side given as $\phi(t) = \phi_0 e^{\frac{-iEt}{\hbar}}$, time-dependent solution.

The solution of the time-independent Schrödinger equation is well known according to the described system.

But the probability density $\rho = (\psi^*(r,t)\,\psi\,) = |\psi\,(r)\,|^2, |\phi(t)|^2 = 1.$

Now take $\frac{\partial <\rho>}{\partial t} = \frac{\partial}{\partial t}\int \psi^*(r,t)\psi(r,t)d^3r = \int \frac{\partial}{\partial t}(\psi^*\psi)d^3r =$

$i/\hbar[E^* - E]e^{\frac{i[E^*-E]t}{\hbar}}[\int \psi^*\psi\,d^3 = 1] = (i/)\hbar(E^*-E)e^{\frac{i(E^*-E)t}{\hbar}}$

But $\frac{\partial \rho}{\partial t} + \nabla . j = 0$; hence, $\frac{\partial <\rho>}{\partial t} = v\int -\nabla . j\,d^3r = s\int j.ds = 0$ (Gaussian theorem). This implies that
$i/\hbar\,[E^* - E]e^k = 0, k = i/\hbar\,[E^* - E] \neq 0.$

Therefore, $[E^* - E] = 0$, so $E = E^*$, i.e., E is real.

1.17. *Wave Packet and Particle Concept*

It has already been seen that the boundary conditions imposed on the wave function describing abounded system leads to discrete spectra of energy, such as $E_1, E_2, E_3, \ldots E_n$, where $E_1 < E_2 < E_3 \ldots < E_n$. These energies are corresponding to wave functions, $\phi_1, \phi_2, \phi_3 \ldots \phi_n$; therefore, ψ_i, i = 1, 2, 3 ... n is an n-fold degenerate wave function. It is a complete set of basis vectors.

Hence, it is quite possible to construct an eigenfunction, such as $\psi(r,t) = \sum C_k\,\phi_k e^{\frac{iE_k t}{\hbar}}$. So for many states corresponding to the same energy (E) level (degeneracy) will be

$$\psi(r,t) = \sum_{nk} C_k^n \, \phi_k^n e^{\frac{-iE_k t}{\hbar}} \text{---------------(14).}$$

This idea leads to the concept of wave packet. In quantum mechanics, the wave packet is quite important because finding a particle somewhere in the neighborhood of (r) or (p), the probability is used. In other words, the path of the particles are not well defined; therefore, all possible paths to be considered is the basis on which Feynman (1947–1948) built his method of path integral. Hence, the wave packet is the suitable wave function to describe the system under study; it contains all possible particle paths.

In wave packet, the energy $E_k = \frac{\hbar^2}{2M}(k_x^2 + k_y^2 + k_z^2)$, where $E_k = \hbar\omega_k$. The corresponding wave function is given by $\psi(r,t) = \int C_k \exp.i(\vec{k}.\vec{r} - \omega_k t)\text{------(15).}$

Where, $\psi(r,t)$ can be represented graphically as

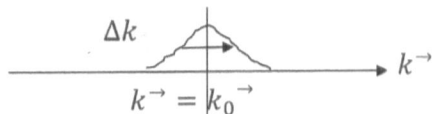

For detailed description to wave packet, one might go to some texts (I. Schiff, Saxon, Merezbacker, etc.).

The parity concept is an intrinsic characteristic of quantum mechanics. It describes the inversion of the coordinates of a physical system through the origin. The figure bellow illustrates that.

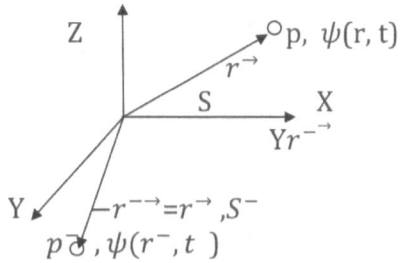

Notice $\psi(r, t)$ is in the S-System and $\psi(r^-, t)$ in the S^- -System. The Schrödinger equation in the primed system(S^-) is given by,

$$[-\frac{\hbar^2}{2m}\nabla^2 + V(r^-)]\psi(r^-, t) = i\hbar\frac{\partial}{\partial t}\psi(r^-, t)\text{-----------} (16),$$

but V(r) = V(-r) because V depends on the magnitude of $r^{\rightarrow}$ only.

Schrödinger equation formally is the same; hence, it is the solution of the equation for the same energy (E). This is only true if V(r) = V(-r), i.e., V(r) is symmetric. Therefore, these two solutions can be combined together as symmetric and antisymmetric solutions, such as

$\phi_+ = 1/2[\psi(r^{\rightarrow}) + \psi(-r^{\rightarrow})]$ for symmetric case.

$\phi_- = 1/2[\psi(r^{\rightarrow}) - \psi(-r^{\rightarrow})]$ for antisymmetric case. $\Big\}$---- (17).

The parity is denoted by π as a physical dynamical variable. It is associated also with operator denoted as π_0. It describes the inversion of the coordinates through the origin as stated earlier. It is either even (denoted by +) or odd (denoted by -), such as,

$\pi_0\phi_+ = +\phi_+$ ----- even parity.

$\pi_0\phi_- = -\phi_-$ ------ odd parity.

Now if $E = E_0 + iE_1$ is total energy, then $E^* = E_0 - iE_1$ is a complex conjugate of E. Therefore, $E^* - E = -2iE_1$.

So now the probability density $\rho = \psi^*\psi = \exp[i/\hbar(-2E_1)t\ \phi^*\phi$, if $E_1>0$, $\rho \to 0$, for $t \to \infty$, which indicates that the particle density leaks out of the system. But $E_1<0$, $\rho \to \infty$, for $t \to \infty$, ρ gets larger and larger. Also, $\phi \to 0$ as $r \to \infty$.

This implies the system is bounded; its energy (E) is discrete with values $E_1, E_2, E_3, \ldots E_n$, as shown in the following diagram,

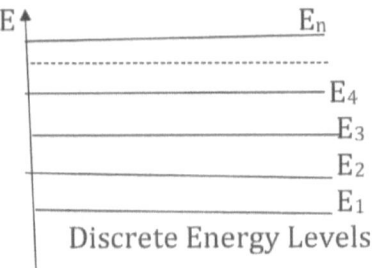

Discrete Energy Levels

If two or more wave functions correspond to the same energy, it is called degenerate level. Its rank is defined by the number of wave functions sharing this level. As mentioned previously, if $V(r) = V(-r)$, it is symmetric, then $\phi(r)$ and $\phi(-r)$ correspond to the same energy (E); hence

$\phi_+ = 1/2[\phi(x) + \phi(-x)]$ --------even function
$\phi_- = 1/2[\phi(x) - \phi(-x)]$ ---------odd function

It is important to notice it. It may happen there is an even state or odd state because of the reflection through the origin corresponding to the same energy. This implies to a degeneracy of rank two. This case might occur in the ground state too.

For more clarification, take the case of free system (particle) where $V(r) = 0$; its energy is given by $E = P^2/2m$. It is independent of J_z, the Z -component of total angular momentum (J), but it is dependent of J. It is also known that

$J_z|njm>$ = $m|njm>$, where m takes the values $(2J+1)$. This plays the role of lifting of the degeneracy.

1.18.1. *Conservation of Parity*

In general, parity is conserved. It means going from state A to state B by a reflection or inversion through the origin. The energy in state A is equal to the energy in state B, also the same for the total angular momentum. It follows that the parity in state A is the same as in state B.

In nuclear weak interaction, Beta-decay and Mu-decay, parity is not conserved. The symmetric potential V(r) is invariant under inversion and parity, as consequence of the symmetrical character of the V(r). One of the known transformations is t→ −t (time reversal). The solution of Schrödinger equation for this case has two values for the energy.

For the wave function $\psi(r,t)$, the energy is E>0, positive. But E<0, negative, for the wave function $\psi(r,-t)$. If the complex conjugate of Schrödinger equation is considered, the equation can be written as:

$$[(-\hbar^2/2m)\nabla^2 + V(r)] \, \psi(r,t) = \text{-}i\hbar\frac{\partial}{\partial t}\psi(r,t). \text{ --- non-time reversal}$$

$$[(-\hbar^2/2m) + V(r)]\psi^*(r,\text{-}t) = \text{-}i\hbar\frac{\partial}{\partial t}\psi^* (r,\text{-}t). \text{ --- time reversal}$$

Since V(r) is real, time-independent = V*. If the parity is conserved, its transformation is linear, i.e., $\pi(a\psi) = a\pi\psi$.

But for time reversal, the transformation is anti-linear, that is $T(a\psi) = a^*\psi^*$.

1.19. *Charge Conjugate (C) Concept*

Charge conjugation is a physical phenomenon concerned with the charged particles and their antiparticles, like electron and positron, proton and antiproton, neutron and antineutron, etc. For example, the charge (+) is conjugate to the charge (-) and vice versa; the property of conservation of (CPT) together proposed to be valid!

These basic principles are introduced to be an entrance to the rest chapters of the book.

Chapter Two

Operators, Concept, and Properties

2.1. *Operator Concept*

In quantum mechanics, as pointed out in chapter one, each physical dynamical variable is associated with an operator. The operator either used in its own space or in another canonically related space to measure the value of the physical dynamical variable described by a wave function either in its own space or in a space related canonically to its space, like spatial and momentum spaces. As examples of these operators are the position r(x, y, z) and the momentum P (p_x, p_y, p_z) operators; they were clarified in chapter one. They are,

$P_O = -i\hbar \frac{\partial}{\partial x}$, $X = X$, in spatial space;

$P = p$, $X_O = -i\hbar \frac{\partial}{\partial p}$,   in momentum space.

For the measurement or finding the value of a physical dynamical variable, such as *P* or *X*, its operator will operate on the wave function that describes the dynamical variable called the eigenfunction. The measured value is called the eigenvalue (proper value). The Schrödinger equation here is called the eigenvalue equation as stated before in the previous chapter. To find the expected value (or the average value), as example, of the momentum (P) in its own space, it is given by,

$<P> = \int \phi^*(p,t)p\ \phi(p,t)d^3p$, in its own space (momentum).

But if the spatial wave function is used, then it is given by,

$<P> = \int \psi^*(r, t)(-i\hbar \frac{\partial}{\partial x})\psi d^3r$, in spatial space (position).

It is known in classical mechanics that the mean square deviation occurs in the measurement, denoted by $<X>^2 = (X-<X>)^2$, but in quantum mechanics, this is calculated as,

$$<x> = \int \psi^*(r, t)(X_0-<X>)^2 \, \psi \, dx, \qquad X = X_0$$

$$\text{or } <p^2> = \int \psi^*(r, t) \, (P_0-<p>)^2 \psi \, d^3r, \qquad P_0 = -i\hbar \frac{\partial}{\partial x}.$$

This indicates that there is a range of measurement, say $P \pm \Delta p$, which is quite important in the measurement of the expected value, where the possible errors can be calculated. This is considered in the analysis of the results.

As a conclusion out of what so far stated in chapter one and in the above notes, the following remarks can be stated:

1. Any physical dynamical variable is associated with an operator, which operates on the describing function to find the value of the measured dynamical variable (DV).
2. The measured value for (DV) represents the eigenvalue of its associated operator.
3. The eigenfunctions of this operator is the wave function describing the (DV). It contains a full information about the (DV).

2.1.2. Some Important Properties of the Operators

The properties of operators are very important in quantum physics. Therefore, the acquainting with these properties is the key of dealing well with quantum mechanical problems. Therefore, in the following, these properties will be clarified.

2.1.2.1. Hermiticity Property

This property has important applications in quantum mechanics; it means the eigenvalue is a real quantity. For example, if the operator of a dynamical variable A is a Hermitian, then its expected value can be written as <A> = $\int \psi^* A\psi d\tau$, as the expected value of A, where A should be real, i.e., $A_0 = A_0^*$. This means A_0 is Hermitian. The Hermiticity is a mathematical terminology in the name of the mathematician Hermit. It means any dynamical variable its operator is Hermitian; its value is real.

2.1.–2.2. Hermitian Adjoint

Take an operator such as L_0, then its Hermitian adjoint is L_0^*, such that $\int f^* L_0^* g d\tau = \int (L_0 f)^* g d\tau$ where f and g are mathematical functions. Here, L_0 acts on f, and in $\int f^* L_0 g d\tau$, L_0 acts on g. If $L_0 = {}^*L_0$, then L_0 is called self-adjoint.

2.1–2.3. Some Important Theorems

Theorem 1: If L is self-adjoint, it is Hermitian.

The proof: Take g = f or ψ, then $\int \psi^* L\psi \ d\tau = \int (L\psi)^* \psi d = \int L^* \psi^* \psi d\tau$; hence, L = L*. Therefore, L is self-adjoint and Hermitian.

Theorem 2: If L is Hermitian, it is self-adjoint.

The proof: If L is Hermitian, then $\int \psi^* L\psi d\tau = \int (L\psi)^* \psi d\tau = \int L^* \psi^* \psi d\tau$.

Now let $\psi = \phi + \lambda g$, where ϕ and g are arbitrary functions. Then

$\int \psi^* L\psi \ d\tau = \int (\phi + g)^* L(\phi + g) d\tau$
$= \int \phi^* L \phi d\tau + l\lambda \ l^2 \int g^* Lg d\tau + \lambda \int \phi^* Lg\phi d\tau + \lambda^* \int g^* L\phi d\tau$
$= \int (L\phi)^* \phi d\tau + l\lambda \ l^2 \int (Lg)^* g d\tau + \lambda \int (L\phi)^* g \ d\tau + \lambda^* \int (Lg)^* \phi d\tau.$

Taking the coefficients of λ and λ^* from both sides, it is found that $L = L^*$, so L is self-adjoint operator.

Theorem 3: If h_1 and h_2 are Hermitian operators, then $(h_1 + h_2)$ is Hermitian operator too.

The proof: Consider:

$$\int ((h_1 + h_2), \psi)^* \psi d\tau = \int (h_1 \psi)^* \psi d\tau + \int (h_2 \psi)^* \psi d\tau$$
$$= \int \psi^* h_1^* \psi d\tau + \int \psi^* h_2^* \psi d\tau.$$

By proposal, $h_1^* = h_1$ and $h_2 = h_2^*$; hence,

$$\int \psi^* h_1 \psi d\tau + \int \psi^* h_2 \psi \, d\tau$$
$$= \int \psi^* (h_1 + h_2) \psi \, d\tau.$$

Therefore, $(h_1 + h_2) = (h_1 + h_2)^*$. Thus, $(h_1 + h_2)$ is also Hermitian.

Theorem 4.
$(B^*)^* = B$

The proof: Take $\int \psi^* B^* \psi d\tau = \int (B\psi)^* \psi d\tau$

Write the complex conjugate of both sides, such as
$\int (B^* \psi^*)^* \psi d\tau = \int (B\psi) \psi^* d\tau = \int \psi^* B \psi d\tau \rightarrow ((B^*)^* = B$.

Theorem 5
If A is not a Hermitian operator, it is possible to construct $(A + A^*)$ and $i(A - A^*)$ as Hermitian operators.

The proof: $(A + A^*)$ is Hermitian operator:
Consider, $\int ((A + A^*)\psi)^*, \psi \, d\tau =$
$= \int (A\psi)^* \psi d\tau + \int (A^* \psi)^* \psi d\tau = \int \psi^* A^* \psi d\tau + \int \psi^* (A^*)^* \psi d\tau$
$= \int \psi^* A^* \psi d\tau + \int \psi^* A \psi d\tau = \int \psi^* (A + A^*) \psi d\tau.$

Therefore, A + A* is Hermitian, i.e., (A + A*)* = A + A*
As an exercise, by the same way, show that i(A - A*) is Hermitian operator.

Theorem 6:
$(CD)^* = D^*C^*$

The proof: Write $\int \psi^*(CD)^*\psi d\tau = \int (CD\psi^*)^* \psi d\tau = \int (D\psi)^*C^*\psi d\tau$
$= \int \psi^*(D^*C^*)\psi d\tau$

Comparing LHS with RHS, it is clear that
$(CD)^* = D^*C^*$, as required.

Theorem 7:
As an exercise: If *A* and *B* are Hermitian, then $(A, B)^* = BA$.

2.1.2.4. *Hermitian Operators and the Real Eigenvalues*

In quantum physics, where the quantum mechanics is the tool of handling the problems, the operators associated with physical dynamical variables are quite important in finding the expected (average) value of these dynamical variables. Let F_0 is a Hermitian operator, associated with the physical dynamical variable (F) described by the wave function (ψ_i) called eigenfunction, the eigenvalue equation is $F_0\psi_i = f_i\psi$; f_i is the eigenvalue of the operator F_0. It is the expected value of *F*. The question is, is f_0 a real quantity? The answer is followed from the following theorems:-

Theorem 1: If F_0 is a Hermitian operator, then the eigenvalue f_i is real.

The proof: Take $\int \psi_i^* f_0 \psi_i d\tau = \int \psi_i^* f_i\psi_i \, d\tau = f_i\int \psi_i^* \psi_i d\tau$ --- (a)
or $\int (f_0 \psi_i^*)^*\psi_i \, d\tau = f_i^* \int \psi_i^* \psi_i \, d\tau$ ------------ (b).

Subtract equation b from equation a; the result is $[f_i - f_i^*] \int \psi_i^* \psi_i d\tau = 0$

since $\int \psi^* \psi \, d\tau = 1$, normalized. Hence, $[f_i - f_i^*] = 0$, so the eigenvalue of a Hermitian operator is real.

Theorem 2: If the eigenvalues f_i and f_j for ψ_i and ψ_j, respectively, are not equal, the eigenfunctions are orthogonal, then

$\int \psi_i^* \psi_j d\tau = \delta_{ij} = 0$, $i \neq j$, orthogonal.

The proof: Consider $\int \psi_i^* f_0 \psi_j \, d\tau = f_i \int \psi_i^* \psi_j \, d\tau$ ----------- (c) or $\int (\psi_i f_0)^* \psi_j \, d\tau = f_j^* \int \psi_i^* \psi_j d\tau$ -------- (d)

Subtract equation d from equation c; the result is, $(f_i - f_j) \int \psi_i^* \psi_j d\tau = 0$, $f_i - f_j \neq 0$.

$\int \psi_i^* \psi_j d\tau = \delta_{ij} = 0$, since $i \neq j$, orthogonal.
δ_{ij} is Kronecker delta function.

2.1.2.5. *Linear Properties of the Operator*

The linearity feature of the operator in quantum mechanics is so well useful to deal with a mathematical formalism. For example. take the operator A_0, let it operate on a given eigenfunction ψ_a. It transforms ψ_a to $\psi_{\bar{a}}$, i.e., $A_0 \psi_a \rightarrow \psi_{\bar{a}}$. Remembering the eigenvalue equation, $A_0 \psi = a\psi \equiv \Psi$, where it is new eigenfunction, $\psi_{\bar{a}}$.

Now the following are the more general properties of the operator A_0:

1. $A_0(\psi_a + \psi_b) = A_0\psi_a + A_0\psi_b$
2. $A_0(\lambda\psi_a) = \lambda(A_0 \psi_a)$, λ is acomplex quantity.

3. If (1) and (2) are satisfied, then A_o is a linear operator.
4. If $A_o\psi = B_o\psi$, then $A_o = B_o$, for any and all vectors.
5. $A_oB_o\psi = A_o(B_o\psi)$, $A_o B_o \neq B_oA_o$, they do not commute.
6. $= (A_0 + B_0)\psi = A_0\psi + B_0\psi$ is identity operator.

These properties will be clarified in more details through their applications in the next chapters.

Projection Operators

If ψ_i is a complete set of basis vectors, then it is possible to construct a wave function $\Psi = \sum_i a_i \psi_i$, where a_i is the amplitude of the coefficient. Multiplying both sides by ψ_i, it is obtained that $(\psi_i, \Psi) = \sum_i a_i (\psi_i, \psi_i) = \sum_i a_i$, since, $(\psi_i, \psi_i) = 1$, normalized.

Hence, $\sum_i a_i = (\psi_i, \Psi)$. Taking the ith component $\psi_i (\psi_i, \Psi)$, *(using the projection method such as, $Pi (\Psi) = \psi_i(\psi_i, \Psi) = \sum_i a_i \psi_i$)*

This is the definition of the projection operator (P_i).

It is so important to demonstrate some properties of this projection operator, because it plays main roles in the mathematical formalism and the formal theory of quantum mechanics. These properties might be summarized as follows:-

1. It is a linear operator, i.e., $P_i(\psi_a + \psi_b) = P_i\psi_a + P_i\psi_b$.
2. $P_I(\lambda\psi_a) = \lambda (P_i\psi_a)$, λ is a complex quantity.
3. (1) and (2) can be shown by the definition of the projection operator P_i (as mentioned before) as follows:

$P_i(\psi_a + \psi_a) = \psi_i(\psi_i, \psi_a + \psi_b) = \psi_i(\psi_i, \psi_a) + \psi_i(\psi_i\psi_b) = P_i\psi_a + P_i\psi_b$. Then, $\psi_i(\psi_i, \lambda\psi_a) = \lambda\psi_i(\psi_i, \psi_a) = \lambda(P_i, \psi_a)$ as in (2).

4. $P_i \psi_i = \psi_i (\psi_i, \psi_i) = \psi_i$.
5. If i≠j, then $P_i\psi_j = 0$, because $P_i\psi_j = \psi_i(\psi_i, \psi_j) = 0$, i ≠ j.
6. $P_i^2 = P_i \Rightarrow (P_i^2 - P) = 0 \rightarrow P_i(P_i - 1) = 0, \rightarrow$ P=1 or 0.

To verify, this take,

$$P_i^2\psi = P_i\psi \text{ or } P_i(P_i\psi) = P_i\psi = \psi_i(\psi_i, \psi) = (\psi_i, \psi_i)P_i \psi = P_i\psi$$

This is called idempotent or idemfactor.

7. $P_i P_j = P_j P_i = 0$ if i≠j
8. $\sum_i^\infty P_i = 1$ (unit operator)

Now let $A_0\psi = \psi^-$, assuming$\Psi = \sum a_i \psi_i$, where ψ_i is a complete set of basis vectors, as defined before. Then $\Psi^- = \sum_i a_i^- \psi_i$

Therefore, $A_0\Psi = \sum a_i (A_0\psi_i) = \sum a_i a_i^- \psi_i$ --------------- (1)

Also, it is possible to expand $(A\psi_i)$ to obtain a new vector in the same space; hence, $A_0\psi_i = \sum A_{ij} \psi_j$, where

$A_{ij} = (\psi_i, A\psi_j)$ is the matrix elements of A, then
$A\Psi = \sum a_i A_{ij}\psi_j$---------------------------------- (2).

Comparing equations1and 2, one obtains $a_j^- = \sum_i A_{ij} a_i$ --- (3).
A_{ij} depends on the operator A_0 only. Therefore, since A_{ij} is known, a_j^- can be found from the matrix relation below:-

$$
\begin{pmatrix} a_i^- \\ a_2^- \\ a_3^- \\ \cdot \\ \cdot \\ a_n^- \end{pmatrix}
=
\begin{pmatrix} A_{11} & A_{12} & A_{13} \text{------} A_{1n} \\ A_{21} & A_{22} & A_{23} \text{------} A_{2n} \\ A_{31} & A_{32} & A_{33} \text{-----} A_{3n} \\ \text{---------------------} \\ \text{-----------------------} \\ A_{n1} \text{-------------------------} A_{nn} \end{pmatrix}
\begin{pmatrix} a_1 \\ a_2 \\ a_3 \\ a_4 \\ \text{-------} \\ \text{-------} \\ a_n \end{pmatrix}
\quad \text{---- (4)}
$$

This is infinite matrix elements, unless a subspace is taken in Hilbert space, where a finite matrix element is to be found.

Take $(\psi_b, A\psi_a) = (b_i\, \psi_i, A_{jk}^- a_k \psi_j) = b_i^* \hat{A}_{jk} a_k (\psi_i, \psi_j)$, number.

Here, (ψ_i, ψ_j) = number either 0 or 1. It depends on $i \neq j$ or $i = j$.
$= b_i^* A_{jk} a_k =$

$$
[b_1^*, b_2^*, ---- b_n^*]
\begin{pmatrix} A_{11} \ A_{12} \ ----A_{1n} \\ -\ -\ -\ \qquad - \\ \text{--------------------} \\ \text{--------------------} \\ A_{n1} A_{n2} ----A_{nn} \end{pmatrix}
\begin{pmatrix} a_1 \\ a_2 \\ a_3 \\ \cdot \\ a_n \end{pmatrix}
\quad ----(5)
$$

The Self-Adjoint Operator:

The self-adjoint operator as defined before is
$\int (Af)^* g d\tau = \int f^* A^* g d \rightarrow (A\psi_a, \psi_b) = (\psi_a, A^*\psi_b) =$
$(\psi_b, A\,\psi_a)^*$ or $(\psi_i, A^*\psi_j) = (\psi_j, A\,\psi_i)^*$.

This implies that $A_{ij}^+ = A_{ji}^+ \Rightarrow [A^+] = [(A_{ji})^*]^T$. T means transposed.

If A⁺=A, it means that the operator A is self-adjoint or Hermitian operator. Therefore, $A_{ij}^+=A_{ij}$, but if $A_{ji}^*=A_{ij}$, the diagonal elements of the matrix is real, as shown in the following matrix elements:

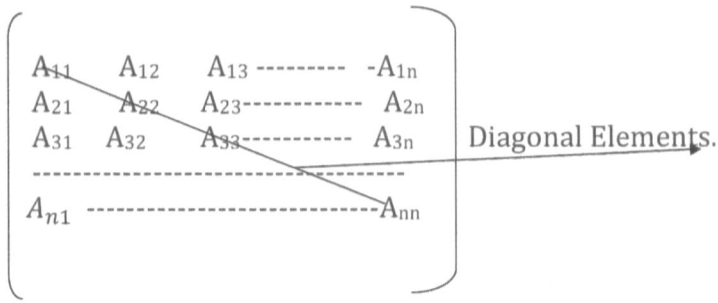

Now defining an operator, such as $A_0 B_0=B_0 A_0=1$, where $B_0= A_0^{-1}$, i.e., B_0 is inverse of A_0. It is called inverse operator. If $A^+A=A$ $A^+=1$, it indicates $A^+ = A^{-1}$, and AA^+ is a unitary operator.

Now take $(\psi_a, A^+A\psi_b)=(\psi_a,\psi_b)$ or $(A^+\psi_a, A\psi_b)=(\psi_a^-,\psi_b^-)$.

This clearly shows that under the transformation $(\psi^- = A\psi)$, the inner product is conserved. It implies the norm is conserved too.

The question now is, is the projection operator has inverse? Why? The answer is no; it has no inverse. The reason is the eigenvalue of the projection operator is one or zero.

In physics and in mathematical physics, in general, it is dealing with dynamical variables and operators described by their own eigenfunctions. Naturally, these physical systems are subjected to changes of their positions in space and their frame of references. Therefore, their vectors basis is subjected to changes too. That requires certain mathematical formalism, taking in consideration the possible changes under the effect of the motion or the influence of internal or external forces.

Accordingly, suppose there is a basis vector related to another one designated by ψ_i, propose another ϕ_k, related to it such as $\phi_k = S_{ik}\psi_i$. Multiply both sides by ψ_i; it is obtained that $S_{ik} = (\psi_i, \phi_k)$. Here, ψ_i is the basis vector in the old frame of reference, while ϕ_k is the basis vector in the new frame of reference.

Now let $\Psi = a_i\psi_i = a_k^-\phi_k$, $\Rightarrow \Psi = a_k^-\phi_k = a_k^-S_{ik}\psi_i$. So in general, $[a = Sa^-]$.

Let us study this effect on the operator A_0 then. $A_{ij} = (\psi_i, A_0\psi_j)$ is in the old frame.

Let $A_{ij}^\sim = (\phi_i, A_0\phi_j)$ be in the new frame.

Hence, it can be written that

$(\phi_i, A_0\phi_j) = (\psi_\ell S_{\ell i}, A_0\psi_m S_{mj}) = S_{\ell i}^* S_{mj} A_{\ell m}$;
therefore, $A_{ij}^\sim = S_{\ell i}^* A_{\ell i} S_{mj}$, but $S_{\ell i}^* = S_{i\ell}^+$,
then $A_{ij}^\sim = S_{i\ell}^+ A_{\ell m} S_{mj} \rightarrow$
$[A^\sim = S^+ A\, S]$ similarity transformation---------------- (6).

Now the question is, are the eigenfunctions $\phi_k = S_{ik}\psi_i$ normalized, i.e., does the transformation affect the normalization? This can be checked as follows:

$(\psi_i S_{ik}, \psi_\ell S_{\ell k^-}) = S_{ik} S_{\ell k^-} = S_{kk^-} = 1$, where $k = k^-$, $= 0$ if $k \neq k^-$. Also, it is known that $S_{ik}^* = S_{ki}^+$; this leads to $SS^{-1} = 1$. It is a unitary operator.

Here, the use of a unit matrix is considered to conserve the normalization (or the orthonormality) from the old frame to the new frame. Hence, it can be written as $[A^\sim] = [S^+ A\, S]$ so that

$S^+S=1$, so S is a unitary operator. Hence, the answer is yes, the normalization is applicable to $\phi_k =S_{ik} \psi_i$.

Infinitesimal Transformation:-

Sometimes the physical system in the frame S rotates infinitesimally the system coordinates through a very small angle.

$S_\theta=n\delta\theta$

$n\to \infty, \delta_\theta \to 0$

Then $S_{\delta_\theta}=1+ i\epsilon f$

Where $\epsilon =\dfrac{\theta}{n}=\delta_\theta$

So it can be written as $S_\theta =(1+i\epsilon f)^n$; taking the limit of S_θ, it is obtained that $(1+i\dfrac{\theta}{n} f)^n=e^{i\theta f} =1+i\theta f+\dfrac{(i\theta)^2}{2!}+$ higher terms. Using $S^+S =1$, it is obtained that

$(1-i\epsilon f^+)(1+i\epsilon f)=[1+i\epsilon(f-f^+)+\epsilon^2 \pm - -0.\longrightarrow$

$S^+S=1=1+i\epsilon(f-f^+) \longrightarrow (f-f^+)=0 \longrightarrow f=f^+.$

Hence, f is Hermitian operator. Now take $A^\sim =S^+AS=(1-i\epsilon f)A(1+i\epsilon f)=A—i[f, A]+$ neglecting higher order in ϵ, $=$ $=-i[f, A]$. If f commutes with A, then $[f, A]=0$, which leads to $A^\sim =A$; therefore, A is invariant under the infinitesimal transformation $(1+i\epsilon f)$. As an example, take the kinetic energy $P^2/2m$ and the potential $V(r)$, where $V(r)$ has the same value in all kind of frames. Why?. Find the answer.

Since f is infinitesimal generator of transformation, it is possible to show that f is the angular momentum L_x for the rotation about the X-axis. Also, a linear momentum (P) is a generator of translation, such as

$$A(x) \qquad\qquad A(\,X+\delta h)$$
$$X\!\circ\!\xrightarrow{\hspace{4cm}}\!\circ(X+\delta h)$$

So $A(X+\delta h) = A(X)+\delta h\,\frac{\partial}{\partial x}A(x)) = (1+\delta h\frac{\partial}{\partial x})A(X)$. The potential $V(r)$ is known to be independent of the lrl direction; therefore, it commutes with the generator of the transformation, i.e., [V, G]=0. G is the generator of the transformation.

The Hamiltonian (H) is kinetic energy plus potential energy, i.e., $H=T+V(r)=P^2/2m+V(r)$. It is invariant under transformation, but it is a generator for infinitesimal-time transformation, such that $i\hbar\frac{\partial}{\partial t}\psi = H\psi$. The transformation of a wave function $\psi(r,t)$ under infinitesimal change in time can be written as

$$\psi(r,t)\longrightarrow\psi\,(r,t+\delta t)= \psi(r,t)+\delta t\frac{\partial\psi}{\partial t}=(1+\delta t\frac{\partial}{\partial t})\psi(r,t).$$

These will be shown in details in the next chapters and sections. It will be noticed also that any operator commutes with Hamiltonian H is a constant of the motion, a very important physical concept.

Some Notes on the Commutation Algebra

1. *Uncertainty Principle and Commutation:-*

Consider A and B as canonically conjugate dynamical variables. The uncertainty relation is written as [A, B]=AB-BA=ic, where c is Hermitian $(c^*=c)$ and [ ] is a commutator symbol. For example, $[X,P_0] = [X,-i\hbar\frac{\partial}{\partial t}]=i\hbar$, where $\hbar$ is comparable to c.

Now if [A, B] = 0, it means A commutes with B, which physically implies that A and B can be measured simultaneously, and there is no effect for the uncertainty principle.

Now from the fact that $\int |(A-<A>)\,\psi +i\lambda\,(B-<B>)\,\psi|^2 dr^3 \geq 0$.

Here, $\Delta A \Delta B = 0$ (no effect to uncertainty principle), this indicates that A-<A> = 0 and B-<B> = 0, so the integral above is 0.

Hence, (A-<A>) = (B-<B>) =0. This implies two operators can have common eigenfunctions. This will be met in details in the mathematical formalism of operators' applications in quantum mechanics. The following notes might be noticed:

(a.) If A and B do not commute, then their measurements are affected by the uncertainty principle, so[A, B]=ic≠0.
(b.) The uncertainty principle is a quantum mechanical characteristic. It is a very important principle in quantum physics theory. It states for two canonically related dynamical variables, it is impossible to measure their values simultaneously and accurately.

{As an exercise, start from the forms, $(\Delta A)^2 = \int \psi^*(A-<A>)^2 \psi d^3 r$
$=(\psi, (A-<A>)^2 \psi)$.

And $(\Delta B)^2 = \int \psi^* (B-<\Delta B>)^2 \psi d^3 r =(\psi,(B-<\Delta B>)^2 \psi)$
show that [A, B]=ic.

2. *Properties of the Commutator [ ]*

[ ] is called a commutator in quantum mechanics corresponding to { } as passion bracket in classical mechanics.

The properties of the commutator can be summarized in the following:

1. $[A, B]=-[B, A]$ ---antisymmetric
2. $[A, B+C]=[A, B]+[A, C]$
3. $[A, BC]=[A, B]C+B[A, C]$
4. $[A, [B,C]]+[A, B]C+B[A, C]=?$ Find it.
5. $[A, B^n]=\sum_{s=0}^{n-1} B^s[A, B]B^{n-s-1}$

As an exercise, prove these properties, and also show that,

$\vec{L}=\vec{r} \times \vec{p}$, where $l_x=(yp_z-zp_y)$, $L_y=(xp_z-zp_x)$, $L_z=(xp_y-yp_x)$

Exercise 2: Show the following:

1. $[L_x, L_y]=iL_z$, $[L_y, L_z]=iL_x$, $[L_z, L_x]=iL_y$
2. $[\vec{L}, \vec{L}]=i\hbar\vec{L}$, $[L^2, L]=0$, $L^2=L_x^2 +L_y^2 +L_z^2$

These commutation relations are the basic principles of the quantization which in itself is an intrinsic physical concept in quantum physics.

10. *Energy-Time Uncertainty Relation*

Philosophically the time is clue of an event, but the event is due to energy, which is the ability to perform work in unit time. Hence, time and energy are canonically related. So they obey the effect of the uncertainty principle. Therefore:
$\Delta E \Delta t \geq \hbar$, where t is the time parameter. Also, $\Delta x \Delta p \geq \hbar$, x, and p are operators. This can be shown from considering a particle moving toward a detector, as shown below:

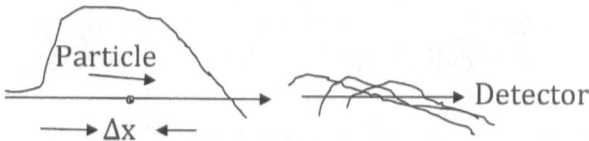

Particle

Δx

Detector

So $\Delta t = \Delta x / V_x$, $\Delta p_x \geq \hbar \frac{1}{\Delta x}$. V_x is the velocity of the detected particle. Since E is a function of p, i.e., $\partial E = \frac{\partial E}{\partial p_x} \Delta p_x \rightarrow \partial E = \frac{p_x}{m} \Delta p_x = V_x \cdot \Delta p_x = (\frac{\Delta x}{\Delta t}) \cdot \Delta p_x = \frac{\geq \hbar}{\Delta t} \implies \Delta E \Delta t \geq \hbar$, as required.

Dirac Notation:

So far, the wave function, in general, is designated by the symbol $\psi(r,t)$. This symbol was used by Schrödinger and Max Born as wave function of statistical character. It carries the whole information of the physical system described in the microscopic scale. But Dirac, as a mathematician, introduced a new symbol; it is a mathematical representation to the wave function. Hence, for the complex conjugate wave function assigned the bra <l, as ψ^* and the ket l> assigned to the wave function $\psi(r,t)$. Therefore, the scalar or the inner product (ψ, ψ) is <l> which is a symbol of the bra-ket. This new representation to the wave function might be summarized as follows:

1. $\psi^* \rightarrow$ bra (state vector)
2. $\psi \rightarrow$ ket (state vector)
3. $(\psi, \psi) \rightarrow$ <alb >----Scalar (inner) product
4. $(\psi_a, \psi_b) = (\psi_b, \psi_a)^* \rightarrow$ <alb> = <bla>*
5. $(\psi_a, \psi_b) \geq 0 \rightarrow$ <alb> ≥ 0
6. <ilj>=δ_{ij} =1, if i=j, =0, if I≠j, orthonormalization
7. Projection operator $P_k \psi_k = \psi_k (\psi_k, \psi) \rightarrow P_k$ la>=lk><kla>

So as stated before, the projection operator is given by, P_k=lk><kl, where $\sum$ li><il=1. It is a unitary operator (closure relation). It is also called the completeness relation.

8. Hermitian conjugation: <alH⁺lb>=<blHla>*. The eigenvalue equation is A_0la>=ala>.

A_0 is the operator, as is its eigenvalue a; la> is its eigenfunction.

Chapter Three
General Formalism of Quantum Mechanics

3.1. *State Vector:* It is a state of a physical system. It is a complex linear vector space of infinite dimension. It is called Hilbert space. The following is a simple illustration to this state vector. Take a point P in Hilbert space, and then define a state vector in this space.

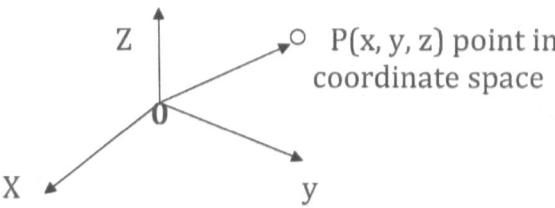

Now a vector space (as basis vector) can be taken as State Vector (V)

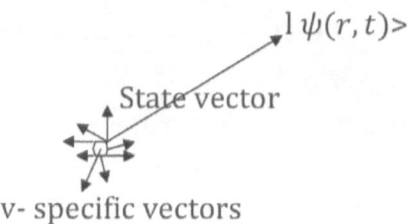

$|\psi_i>$ is any vector space as a basis vector in Hilbert infinite space. Let $|\psi>$ be a state vector with respect to specific state vectors (v), with the following properties:

1. $|\psi_a> + |\psi_b> = |\psi_b> + |\psi_a>$.
2. $(|\psi_a> + |\psi_b> + |\psi_c>) = |\psi_a> + (|\psi_b> + |\psi_c>)$.
3. $\lambda (\mu|\psi_a>) = (\lambda\mu|\psi_a>)$.
4. $\lambda(|\psi_a> + |\psi_a>) = \lambda |\psi_a> + \lambda|\psi_b>$.

3.1.1. *Linear Vector Space:*

It is defined as $|\Psi> = \lambda_1|\psi_1> + \lambda_2|\psi_2>$.

Now $|\Psi> = \int_{\xi_1}^{\xi_2} a(\xi)|\psi(\xi)>d\xi$, where $|\Psi>$ and $|\psi>$ belong to the same space. If $|\psi_1>$ and $|\psi_2>$ are linearly independent, then the condition $\lambda_1|\psi_1> + \lambda_2|\psi_2> + \text{----} = 0$ is implied.

1.1. The dimensionality of each vector is defined such as $N = \sum_i^N C_i |\psi_i>$.

Where N is the dimension of the vector space.

3.1.3. *Basis Vectors:* They are defined as a set of linearly independent vectors representing a complete set. Therefore, $|\psi>$ cab be written such as,

$$|\psi> = \sum_i a_i|\psi_i> + \int_{\xi_1}^{\xi_2} a(\xi)|\psi(\xi)>d\xi \text{-------------} (1).$$

The first term represents the discrete part of the wave function $|\psi>$, while the second term is the continuous part of $|\psi>$.

$|\psi_i>$ and $|\psi(\xi)>$ are basis vectors. In quantum physics theory, in addition to an ordinary space, there is a dual space which is conjugate to it. This dual space is written as $<\psi|$. So $|\psi>$ and $<\psi|$ are a unique one-to-one correspondence; hence, one can write,

$$<\psi| = \sum_i <\psi_i| C_i^* + \int_{\xi_1}^{\xi_2} <\psi(\xi)| la^*(\xi) d\xi \text{----------}(2).$$

It is clear that equation 2 is a complex conjugate to equation 1.

Now as mentioned in the previous chapters, the scalar (inner) product is defined as $<\psi_a|\psi_b> = <a|b>$ using Dirac notation.

The following are some important properties of scalar product:

1. $<\psi_a|\psi_b> = <\psi_b|\psi_a>^*$
2. $<\psi_a|\lambda\psi_b> \equiv \lambda<\psi_a|\psi_b>$
3. $<\lambda\psi_a|\psi_b> \equiv \lambda^*<\psi_a|\psi_b>$
4. $<\psi_a+\psi_b|\psi_c> \equiv <\psi_a|\psi_c>+<\psi_b|\psi_c>$

$\left. \right\}$ ----------- (3).

Now multiply (2) from the right by $|\psi_a>$. The following is obtained:

$$<\psi|\psi_a> = \sum_i <\psi_i|\psi_a> C_i^* + \int <\psi(\xi)|\psi_a>a(\xi)^* d\xi \text{--------}(4).$$

As mentioned earlier, the first term is the discrete part, and the second term is the continuous part. Also, in Schrödinger wave mechanics, the scalar product is represented integrally such as $(\psi_a(r),\psi_b(r)) = \int \psi_a^*(r)\psi_b(r)d^3r \geq 0$. This is the integral representation of the scalar product.

Another important physical concept is the norm; it is defined as $<\psi_a|\psi_b> = <a|b> \geq 0$. If it is 0, the vector is called a null vector.

But if it is 1, then the vector is a unit vector. Now using Schwartz inequality in mathematics, it is written as

$$<\psi_a|\psi_a><\psi_b|\psi_b> \geq |<\psi_a|\psi_b>|^2 \text{-------------} (5).$$

For example, given the vectors A and B as in the figure below:

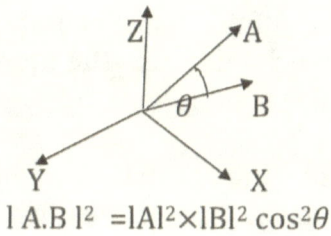

$|A.B|^2 = |A|^2 \times |B|^2 \cos^2\theta$

Since Cos θ is between 0 and 1; therefore, $|A|^2 \times |B|^2 Cos^2 \leq 1$. Hence, $|A.B|^2 \leq |A|^2|B|^2$----------------(6).

3.1.4. Hilbert Space:

It is already stated that this space is an infinite dimension space; it is a linear complex vector space. It can be shown graphically in the symbol of state vector and basis vectors, in Hilbert space, as in the figure bellow: $\left|\psi(r,t)\right>$

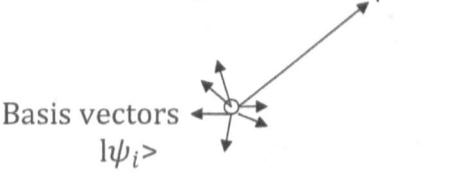

Basis vectors $|\psi_i>$

State vector in ket representation (Dirac notation)

The state vector $|\psi(r,t>$ can be written in terms of the basis vectors as in equation (2), i.e.,

$$|\psi> = \sum_i a_i \ |\psi_i> + \int_{\xi_1}^{\xi_2} a(\xi)|\psi(\xi)> \ d\xi.$$

Where $|\psi_i>$ and $|\psi(\xi)>$, as stated before, are basis vectors, and they are linearly independent. Also, the first term is the discrete part, and the second term is the continuous part. It is important to notice that the vector space usually obeys usual algebra. The complexities that indicate the coefficients are complex. The Hilbert infinite space is quite an important space where quantum mechanics is the tool of tackling the microscopic physical system.

On the other hand, the probability and the uncertainty principles play the basic role in dealing with such systems. Also, there is a dual Hilbert space given by,

$$\langle \psi \mid = \sum_i \langle \psi_i \mid a_i^* + \int_{\xi_1}^{\xi_2} \langle \psi(\xi) \mid a^*(\xi) \, d\xi.$$

As stated before, the scalar product (inner product) is given by, $\langle \psi_a \mid \psi_b \rangle = \langle \psi_b \mid \psi_a^* \rangle$; if the vectors are the same, i.e., $\mid \psi_a \rangle = \mid \psi_b \rangle$, so, $\langle a \mid a \rangle$=real quantity ≥ 0. It means $\langle a \mid a \rangle$ is a positive definite. Also, members of the basis vectors can be normalized as such:

1. $\langle \psi_i \mid \psi_j \rangle = \langle i \mid j \rangle = \delta_{ij} \left. \begin{cases} =1 & \text{if i=j} \\ 0= & \text{if i} \pm \text{j} \end{cases} \right\}$ Orthonormalization.

2. $\langle \psi_i \mid \psi(\xi) \rangle = \langle i \mid \xi \rangle = 0$, orthogonal

3. $\langle \psi(\xi) \mid \psi \xi^- \rangle = \langle \xi \mid \xi^- \rangle = \delta \ (\xi \text{-} \xi^-)$,Orthonormalization for the continues spectrum.

To prove 3, the following has to be proved:
(a.) Linearly independent vectors can be made orthogonal.

Take $\lambda \mid \psi_1 \rangle + \mu \mid \psi_2 \rangle \neq 0$, then construct,
$\mid \Psi_1 \rangle = a_{11} \mid \psi_1 \rangle + a_{12} \mid \psi_2 \rangle$ and $\mid \Psi_2 \rangle = a_{21} \mid \psi_1 \rangle + a_{22} \mid \psi_2 \rangle$.

These can be made orthogonal, such as
$\langle \Psi_1 \mid \Psi_2 \rangle = 0$, and normalized to unity, such as
$\langle \Psi_1 \mid \Psi_1 \rangle = \langle \Psi_2 \mid \Psi_2 \rangle = 1$

[Using Schmidt method (Messiah, Vol. 1. p. 173)]

In Hilbert space, the norms are finite which implies that the spectrum is discrete, so it can be written that

$\langle \psi(\xi) \mid \psi(\xi^-) \rangle = \delta \ (\xi - \xi^-) = 0$ if $\xi \neq \xi^-$, orthogonal,
as $\xi \rightarrow \xi^-$.

But Dirac assumed under such situation that
$\langle \psi(\xi) \mid \psi(\xi^- \rangle = \delta(\xi \text{-} \xi^-)$

where δ is Dirac (delta) function. It is a mathematical function with a very well-defined properties. (See Appendix A)

Now, take the following argument:

$<\psi| = \sum_i < \psi_i| \, a_i^* + \int < \psi(\xi^-) a(\xi^-)^* \, d\xi^-$, find its scalar product

$<\psi|\psi> = 1 = \sum_i |a_i|^2 + \int [|a(\xi)|^2 \underline{\int_{-\epsilon}^{\xi+\epsilon} < \psi(\xi^-)|\psi(\xi)> d\xi^-}] d\xi$

≥ 0, finite.

$\uparrow$

Therefore, $\int <\psi(\xi) |\psi(\xi^->d\xi^- = C$, where, $<\psi(\xi^-|\psi(\xi) > = c(\xi\text{-}\xi^-)$, indicates the norm cannot be finite; otherwise, the integral will be zero.

C can be normalized to 1. This implies that $|\psi(\xi)> = \frac{1}{\sqrt{c(\xi)}} |\psi(\xi)>$ and $<\psi(\xi^-)| = \frac{1}{\sqrt{c(\xi^-)}} <\psi(\xi^-|.$

Hence, $<\psi(\xi^-)|\psi(\xi) > = \delta(\xi\text{-}\xi^-)$ which is required for the continuous spectrum as shown in 3. It can be concluded that:

1. $|\psi> = \sum_i a_i \, \psi_i> + \int a(\xi) \, \psi(\xi)>d\xi$
2. The scalar product can be found to be

$<\psi_i|\psi> \equiv <\psi_i|[1]> \equiv \sum_j a_j <\psi_i|\psi_j> + \int a(\xi)<\psi_i|\psi(\xi)>d\xi$
$= \sum_j a_j \, \delta_{ij}) + 0 = \sum_j a_j \, \delta_{ij} \equiv a_i$
because $<\psi_i|\psi(\xi)> = 0$, orthogonal.

3. $<\psi(\xi)|\psi(\xi^- > = \int a(\xi^-))<\psi(\xi)|\psi(\xi^-)>d\xi^- =$
$= \int a(\xi^-)\delta(\xi^- - \xi)d\xi = a(\xi)$, using δ- function properties (Appendix A).

[As an exercise: Show that $<\psi|\psi> = \sum |a_i|^2 + \int d\xi |a(\xi)|^2 = 1$].

Here, a_i is the probability amplitude to find a configuration of ψ_i, or to find $|\psi_i>$ in $|\psi>$. The term $\int d\xi |a(\xi)|^2$ is the probability to find $|\psi(\xi)>$ in $|\psi>$, as $\xi \rightarrow \xi + d\xi$.

Since these operators are intrinsically associated with physical dynamical variables, and they operate on the eigenfunctions describing these dynamical variables producing their eigenvalues, which represent their measured values, so it is important to be acquainted with their intrinsic properties. Some of these properties are the following:

1. $A\psi = |\psi^->$, transformation property
2. $A[\lambda|\psi_1> + \mu |\psi_2>] = \lambda|\psi_1^-> + \mu|\psi_2^->$
3. $<\psi_a |A|\psi_b> = <\psi_a|(A|\psi_b>) = (<\psi_a|A)|\psi_b>$
4. $<\psi_a |A|\psi_b>^* = <\psi_b|A^*|\psi_a>$
5. $<A^* \psi_a |\psi_b> = <\psi_a|A|\psi_b>$
6. $<\psi^-|<\psi|A^*$. This means A^* acts on $<\psi|$ to the left.

Now, as it was shown before, if any operator H is Hermitian, then H = H*. Also, it is important to be noted that all physically observables are Hermitian operators. Their measured values are real. If UU* = 1, then U = U^{-1} is called a unitary operator as has been stated early.

It can easily be seen that if the operator is Hermitian, the eigenvalue is real. Let A_0 be the operator associated with the physical dynamical variable A described by the eigenfunction $|\psi_a>$, then the eigenvalue equation is written as,

$A_0|\psi_a> = a|\psi_a>$ ----------- (1)
$<\psi_a^-|A_0 = <\psi_a^-|a^-$ -----------(2).

Now multiply 1 by $<\psi_a^-|$; from left it is obtained
$<\psi_a^-|A_0|\psi_a> = <\psi_a^-|a|\psi_a>$ ---------(3).

Multiply 2 by $|\psi_a\rangle$ from right, then

$\langle\psi_a^-|A_0|\psi_a\rangle = a^*\langle\psi_a^-|\psi_a\rangle$ --------- (4).

Subtracting 4 from 3 leads to $(\langle a\,|\,a\rangle)(a-a^*)=0$
$\langle a|a\rangle = 1$, so $a-a^*=0$, then $a=a^*$, so a is real.

As an application, the solution of the time-dependent Schrödinger equation (Kernel solution) will be considered.

This equation is written as $H_0\psi = i\hbar\dfrac{\partial\psi}{\partial t}$ -------------------- (5)

$H_0 = P_0^2/2m + V(r) = [-\dfrac{\hbar^2}{2m}\nabla^2 + V(r)]$; therefore,

$(-\dfrac{\hbar^2}{2m}\nabla^2 + V(r))\,\psi = i\hbar\dfrac{\partial\psi}{\partial t}$ ----------- (6)

The solution of equation(6) is known to be

$\psi_k(r,t) = \psi_k(r)\, e^{-i\frac{E_k t}{\hbar}}$ ----------- (7).

Therefore,

$$\psi(r,t) = \sum_k \psi(r,t) = \sum_k C_K\,\psi_k(r)\, e^{\frac{-iE_k t}{\hbar}}$$

Where $e^{\frac{-iE_k t}{\hbar}}$ is the time-dependent solution; it is a phase factor. The space solution is $\psi(r) = \sum C_k\,\psi_k(r)$

Multiply both sides by $\psi_k(r)$. Then it is obtained that
$\sum_k(\psi_k(r),\psi_k(r))C_k = \int \psi_k^*(r)\psi_k(d^3r) \longrightarrow$

$\sum_k C_k = \int \psi_k^*(r,0)\psi_k(r,0)\ d^3r.$ ----------------------(8).

This is the coefficient at t=0.

Now $\psi(r,t)$ can be written as

$\psi(r,t) = \sum_k \psi(r^-)\,\psi_k^*(r)\psi_k(r)e^{\frac{-iE_k t}{\hbar}}$ ------------(9).

This can be written in the integral form also as

$$\psi(r, t) = \int d^3 r^- \underbrace{[\psi(r^-)\psi_k(r) \exp{-(iE_k t/\hbar)}]}_{I_{op}} \psi(r) \text{--(10)}.$$

The integrand (I_{op}) can be considered as an operator and acts on $\psi(r)$ at t=0, which gives $\psi(r, t)$ after time t. The integral operator is called the Kernel or the Green function. (This will be treated in the next chapters.) Therefore, equation 10 might be written as

$$\psi(r, t) = \int d^3 r^- K(r^-, r, t)\psi(r) \text{-----------------------------(11)}.$$

$K(r^-, r, t)$ is the Kernel function. So as far as Green function is known for Hamiltonian H, it implies everything about the system is open to be known. Hence, Schrödinger equation in terms of the Kernel (or Green function) is written as

$$H_o \int K(r^-, r, t)\psi(r^-)d^3r^- = i\hbar \frac{\partial}{\partial t} \int K\psi(r^-)d^3r^- \text{------(12)}.$$

It shows the Kernel function K is a particular solution of Schrödinger equation.

Dynamical Equation of Motion

The dynamical equation of a physical system is representing its state. It is well known in classical mechanics (see classical mechanics, Goldstein). It deals with measured values (almost expected average values). This expected value is given by, <A(t)>= $\int \psi^*(r, t)A_o\psi(r, t) d^3r$. A_o is the operator associated with A. Now take the derivative of the expected value with respect to time t:

$$\frac{d<A>}{dt} = \int \frac{\partial \psi^*}{\partial t}A_o\psi d^3r + \int \psi^* A_o\frac{\partial \psi}{\partial t} d^3r + \int \psi^* \frac{\partial A_o}{\partial t}\psi d^3r =$$

$$(\frac{\partial \psi}{\partial t} A_o, \psi) + (\psi A_o, \frac{\partial \psi}{\partial t}) + (\psi \frac{\partial A_o}{\partial t} \psi).$$

Now by substituting for $\frac{\partial \psi}{\partial t}$ from Schrödinger equation $H_0\psi = i\hbar\frac{\partial\psi}{\partial t}$, it is obtained that

$$\frac{d<A(t)>}{dt}=\frac{1}{i\hbar}(\psi[H,A_o]\psi)+(\psi\frac{\partial A_o}{\partial t}\psi) \text{ ----------- (13).}$$

Equation 13 is the quantum mechanical equation of motion of the dynamical variable A.

Theorem: If A is not explicitly time dependent, then $\frac{\partial A_o}{\partial t}=0$, which leads to $\frac{d<A>}{dt}=\frac{1}{i\hbar}(\psi[H_0,A_0]\psi)=0$. This implies $[H_0, A_0]=0$.

Which it means A_o is commuting with H_o, so A is constant of motion.

In classical mechanics, the corresponding dynamical equations are $q_p=\frac{\partial H}{\partial p_r}$, $p_r=\frac{\partial H}{\partial q_r}$, H is the Hamiltonian; q_p is a derivative of the generalized coordinate q with respect to the momentum p, and p_r is the derivative of the momentum p with respect to q or r.

More Formal Aspects of Quantum Theory

In general, one might meet that some physical aspects are repeated as interference of some ideas. These interference of some physical ideas and their mathematical formalisms are quite useful in learning the theory of quantum mechanics and through teaching them in another branch of science. So it is necessary to clarify that by some examples.

Suppose ψ can be written as $\psi=\sum_i \lambda_i \psi_i$, where ψ_i, as stated before, are basis vectors, and λ_i is the probability amplitude. The eigenvalue equation again is given by $A_0\psi_i = a_i\psi_i$; the scalar (inner) product is given as $\int \psi^*\psi d^3r=(\psi,\psi)=1$. It is normalized. This leads to $\sum_i \lambda_i^2 = 1$. Again, the normalization condition is

$$\int \psi_i^* \psi_j d^3r = (\psi_i, \psi_j) = \delta_{ij} = 1, \text{ if } i=j$$
$$= 0, \text{ if } I \neq j.$$

Now take $\psi = \sum_i \lambda_i \psi_i$, then $\psi^* = \sum_i \lambda_i^* \psi_i$; therefore, $<A> = <\psi| A_o|\psi>$, substituting for ψ and ψ^*. It is obtained $<A> = \sum_i a_i |\lambda_i|^2$ ---------------- (14).

Equation 14 indicates the state vectors (or the eigenfunctions) might be represented by numbers such as $\lambda_1, \lambda_2, \lambda_3, \ldots \lambda_n$. Now if the physical system might be represented by two dependent wave functions, ψ and ϕ, the total wave function (Ψ) describing the system is given by the well-known superposition principle as:

$$\Psi = a\psi + b\phi \text{ -------- (15).}$$

If $\phi = \sum_i \mu_i \psi_i$, and $\psi = \sum_i \lambda_i \psi_i$, where μ_i is represented by the elements,

$$\begin{pmatrix} \mu_1 \\ \mu_2 \\ \mu_3 \\ - \\ \mu_n \end{pmatrix} \quad \text{and } \lambda_i \text{ as} \quad \begin{pmatrix} \lambda_1 \\ \lambda_2 \\ \lambda_3 \\ - \\ \lambda_n \end{pmatrix}$$

λ_i is the ith component of ψ, and μ_i is the ith component of ϕ.

Therefore, $\frac{1}{\sqrt{2}}(\phi + \psi) = \frac{1}{\sqrt{2}}\sum(\lambda_i + \mu_i)\psi_i$, so there will be components of a matrix, such as

$$\begin{pmatrix} \lambda_1 + \mu_1 \\ \lambda_2 + \mu_2 \\ \lambda_3 + \mu_3 \\ ------ \\ \lambda_n + \mu_n \end{pmatrix}$$

Since $\sum |\mu_i|^2 = 1$ and $\sum |\lambda_i|^2 = 1$, this implies
$(1/2)\sum |\lambda_i + \mu_i|^2 = \text{finite} = 1/2\int |\psi + \phi|^2 d^3r = \text{finite quantity}$
$= \int |\psi|^2 d^3r + \int |\phi|^2 d^3r + \int \psi^*\phi d^3r + \int \phi^*\psi d^3r$-------------- (16).

From equation 16, it is to be noticed that:

1. The first and the second terms are finite;
2. The other two terms are interference terms.

By Schwartz inequality, it is $(|f^*g d^3r|^2) \le |f|^2 d^3r + |g|^2 d^3r = \text{finite}$.
Therefore, if ϕ and ψ are square integrable function, then $(\psi + \phi)$ is also square integrable function. This is a very important theorem in mathematics.

The conclusion out of these facts is if there is in a space two vectors dependent on each other, then it is possible to combine them to give a vector state in the space. This is considered an important conclusion in quantum mechanics. As an example of a vector space is the linear vector space represented by Hilbert space of infinite dimension.

This vector space is characterize by (as mentioned before):

1. $\psi_a + \psi_b = \psi_b + \psi_a$ ------addition property.
2. $\psi_a + (\psi_b + \psi_c) = (\psi_a + \psi_b) + \psi_c$ ---associative property.
3. $\mu(\lambda\psi_a) = (\mu\lambda(\psi_a) = (\mu\lambda)\psi_a, \lambda(\psi_a + \psi_b) = \lambda\psi_a + \lambda\psi_b$

This is a multiplication by complex number property.

4. Null vector ψ_0 indicates that $\psi_a + \psi_0 = \psi_a$
5. Linearly independency of a vector.

If $\psi_1, \psi_2, ----\psi_n$ are linearly independent, there exists a set of number (λ_i) such that $\lambda_1\psi_1 + \lambda_2\psi_2 + ------\lambda_n\psi_n = \psi_0 (=0$ null vector). And $\lambda_1 = \lambda_2 = ----\lambda_n = 0$. From this information, it is possible to define the dimension of space.

The dimensionality of space can be clarified by the following illustration of the concept:

Let N be the maximum number of linearly independent vectors say,

$$\begin{pmatrix} a_1 \\ a_2 \\ a_3 \end{pmatrix} = a_1 \begin{pmatrix} 1 \\ 0 \\ 0 \end{pmatrix} + a_2 \begin{pmatrix} 0 \\ 1 \\ 0 \end{pmatrix} + a_3 \begin{pmatrix} 0 \\ 0 \\ 1 \end{pmatrix}$$

Therefore, these three linearly independent vectors form a set of basis vectors of N=3!

Other important physical mathematical practical concept is the scalar (inner) product symbolically as (,). Although, it was mentioned earlier, it is important to summarize its main properties in the following:

1. $(\psi_a, \psi_b) = (\psi_b, \psi_a)^*$.
2. $(\lambda \psi, \phi) = \lambda (\psi, \phi)$---$\lambda$ is a complex number.
3. $(\psi_a, \lambda\psi_b) = \lambda(\psi_a, \psi_b)$
4. $(\lambda\psi_a, \psi_b) = (\psi_b, \lambda\psi_a)^* = \lambda^*(\psi_a, \psi_b)$
5. $(\psi\lambda, \phi) = \lambda^*(\psi, \phi)$
6. $(\psi_a, \psi_b + \psi_c) = (\psi_a, \psi_b) + (\psi_a, \psi_c)$, distributive property.
7. $[\psi(\phi_1+\phi_2)] = (\psi, \phi_1) + (\psi, \phi_2)$, distributive property.
8. $(\psi_a, \psi_a) \geq 0$; if $(\psi_a, \psi_a) = 0$, then it is a null vector. But if $(\psi_a, \psi_a) > 0$, it is a positive definite.
9. The eigenvector ψ_a has a norm given by (ψ_a, ψ_a). Its length is given by $\sqrt{(\psi_a, \psi_a)}$.
10. The orthonormalization property is given by $(\psi_i, \psi_j)=\delta_{ij}$

Now using the column vector idea, $\sum a_i^* b_i$ written as,

$$[a_1^*, a_2^*, a_3^* \text{-----} a_n^*] \begin{pmatrix} b_1 \\ b_2 \\ b_3 \\ - \\ - \\ b_n \end{pmatrix} = \sum a_i^* b_i$$

Note that,

$$\begin{pmatrix} a_1 \\ a_2 \\ a_3 \\ - \\ - \\ a_n \end{pmatrix}^* = [a_1, a_2, a_3 \text{--------} a_n]$$

Chapter Four

Schrödinger and Heisenberg Pictures (Representations) with Some Applications

As it was pointed out previously, the quantum theory was developed after the discovery of Planck; the energy is quantized in 1900 followed by de- Broglie's proposal in 1924 that any moving system (macro or micro) is associated with a wave of length $\lambda = \frac{h}{p}$, where h is the Planck constant and p is the momentum of the moving system.

In 1926, the wave mechanics was introduced by Schrödinger, based on sound wave equation in classical physics, to describe the physical microscopic systems corresponding to classical mechanics, according to the corresponding principle proposed by Bohr. At the same time, Heisenberg introduced different picture of mechanics to describe these microscopic systems based on the matrix in mathematical theory. Both approaches reach the same result. Then quantum mechanics was developed by other scientists such as Max Born and Dirac.

These developments were the bases of the present quantum theory. In the following, the Schrödinger and Heisenberg pictures will be introduced. It started as nonrelativistic quantum mechanics, and then it was developed to take into consideration the relativistic principle by Dirac in 1932. Also, it was the so-called intermediate picture developed. Using the previous information, let $|\mu>$ be the eigenstate of a system of Hamiltonian H, then the Schrödinger equation is

$$H_0|\mu> = E_0|\mu> = i\hbar \frac{\partial}{\partial t}|\mu> \text{------------- (1)}$$

H_0 is the operator of H. E_0 is the energy operator.
At $t = t_0$ (initial time) equation (1) takes the form.

$H_0|\mu(t_0)>=i\hbar\frac{\partial}{\partial t}|\mu(t_0)>$, $|\mu(t_0)>$ is the eigenstate at t_0.

The time dependence of the eigenstate means $|\mu(t)>=$ $U(t,t_0)|\mu(t_0)>$. The transformation of the eigenstate from t_0 to t occurs through the evolution operator $U(t,t_0)$.

The main important properties of the evolution operator are:-

1. $U(t_2, t_1)=U(t_2, t_1)\,U(t_1, t_0)$
2. $U(t, t_0)$ is a unitary operator, i.e., $U^+U=1, U^{-1}\,U=1$.

This can be shown as follows:

Take $<\mu(t)|\mu(t)>=$
$<\mu(t_0)|U^+(t, t_0)U(t, t_0)|\mu(t_0)>=<\mu(t_0)|\mu(t_0)>$, since $U^+U=1$.

3. When $t_2=t_0$, this implies $U^+(t, t_0)=U(t_0, t_1)U(t_1, t_0)=1$
$U(t_0, t_1)=U^{-1}(t_1, t_0)$. Hence, $U^+(t_2, t_0)$ can be written as:
$U^+(t_2, t_0)=U(t_2, t_1)U(t_1, t_2)----U(t_{n-1}, t_n)----U(t_n, t_0)----$ (1).

Now Schrödinger equation can be written as
$H_0U(t, t_0)|\mu(t_0)>=i\hbar\frac{\partial}{\partial t}U(t,)\,|\mu(t_0)>$, since $|\mu(t_0)>$ is the initial state can be canceled from both sides; therefore, Schrodinger equation is written in terms of the evolution operator U as
$H_0U(t, t_0)=i\hbar\frac{\partial}{\partial t}U(t, t_0)$ ------------------- (2).

It is time dependent; its solution is well known mathematically as $U(t, t_0)=\exp[-\frac{i}{\hbar}H_0(t-t_0)]$ ---------(3).

Remember that $e^k e^g=e^{k+g}$, only if k and g are numbers, but if they are not commuting operators, this is not valid. If they commute, then it is valid. Now,

$U(t_2, t_0)=U(t_2, t_1)U(t_1, t_0)$-----------------------------(4).

(Prove 4 as an exercise.)

Then $\exp[(-i/\hbar \, H_0(t_2-t_0)]=\exp[-(i/\hbar)H_0(t_2-t_1)]$. $\exp[-(i/\hbar H_0(t_1-t_0)]$. If $(t_2-t_1)=\delta t$ (very small), then $U(t, t_0)=1-i/\hbar \, (H_0 \delta t)$, which implies H_0, is infinitesimal generator of transformation in time.

Now let A be a dynamical variable. It is not explicit time dependent; its expected value under the effect of time development is usually given by:

$<A>_t=<\mu(t)|A_0|\mu(t)>=<\mu(t_0) | U^*(t, t_0)A_0U(t, t_0)|\mu(t_0)>$.

Taking the time derivative of the expected value of A, the following is obtained:

$$\frac{d<A>}{dt}=\frac{i}{\hbar}<\mu(t_0)|[U^+A_0U, H]|\mu(t_0)> \text{--------}(5)$$

Therefore, all information about time is contained in A. Accordingly, it is possible to imagine two pictures:

1. Time dependency, included in the eigenstate $|\mu(t)>$, and in the dynamical variable (A). This is the Schrödinger picture or representation.
2. Time dependency included in A, but $|\mu>$ is not time dependent but rather a mathematical matrix picture. It is the Heisenberg picture or representation.

It is important to find the relation between the operators in the two pictures. Let A_s the operator in Schrödinger picture, and H_h the operator in Heisenberg picture. Then

$A_h=U^+(t,t_0)A_s \, U(t_)=\exp[i/\hbar. \, (H \, t)]A_s\exp[-i/\hbar. \, (H \, t)]$. It is known too that $U^+|\mu(t)>_s=|\mu>_h \longrightarrow |\mu>_h=|\mu \, (t_0)>_s$. $A_h(t)$, in general, is not stationary even if A_s is not explicitly time dependent. Accordingly, $<_h|A_h|_h>=<_s|UU^+A_sUU^+|_s>=<_s|A_s|_s>$.

The equation of motion in Heisenberg picture is

$$\frac{dA_h}{dt} = (-i/\hbar)\,[A, H] = (i/\hbar)[H, A].$$

Remember that $U(t_0, t_0) = 1$, (see equation 3), the initial condition. This is Heisenberg equation, where A and $|\mu\rangle$ are time dependent.

3. There is a third picture; it is called the interaction or the intermediate picture. The Hamiltonian of the system is given by $H = H_0 + H_I$. Where H_0 is for free system $[V(r) = 0]$, and H_I is due to the interaction. Schrödinger equation is written as:

$$i\hbar \frac{\partial}{\partial t}|\mu(t)\rangle = H|\mu(t)\rangle = (H_0 + H_I)|\mu(t)\rangle = H_0|\mu(t)\rangle + H_I|\mu(t)\rangle.$$

Any unitary transformation of vectors and observables of Schrödinger and Heisenberg representation define a new picture. All these pictures furnish strictly equivalent description of quantum phenomena. Therefore, in the interaction picture, both state vectors and operators are time dependent. For an illustration, take an example an atom is subjected to a Maxwellian field, electromagnetic field, as shown in the diagram bellow. The Hamiltonian of the system is $H = H_0 + H_I$. H_0 is the free atom (no external force), and H_I is the interaction term due to the electromagnetic field.

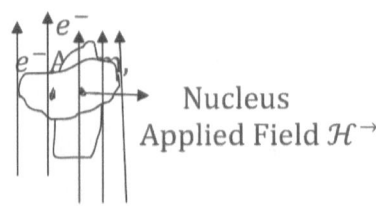

Nucleus
Applied Field $\mathcal{H}^{\rightarrow}$

The solution of Schrödinger equation for H_0 is well known and exact, but for the interaction term, the solution is found by using approximation methods to be treated in the following chapters. From previous section, the time evolution $U(t, t_0)$ is given as $U(t, t_0)=\exp[-(i/\hbar)H(t-t_0]$.

For time transformation, the eigenstate, the general coordinates, and the momentum take the forms:

$|\psi^-(t)>=U_0(t, t_0)|\psi(t)>$

$q_r^-(t) = U(t, t_0)q_r U(t, t_0)$ All these are time dependent.

$p_r^- =U(t, t_0)p_r U(t, t_0)$

Schrödinger equation for this picture is given as

$i\hbar\frac{d}{dt}|\psi^-(t)>=H_I^-(t)|\psi^-(t)>$. $H_I^-(t)$ is   a   kind   of   small perturbation.

As an exercise, show the following:

$i\frac{\hbar d}{dt}|q_{r(t)}^->=[q_{r(t)}^-, H_0(q_r^-, p_r^-)]$, free equation of motion,

$i\hbar\frac{d}{dt}<p_r>=[p_r, H(q, p)]$, independent of time.

$[q_r(t), q_r(t^-)]\neq0$, if $t\neq t^-$

but $[q_r(t),q_s(t)]=0$, $t=t^-$

and $H_0^-(q, p)=H_0(q_r^-, p_r^-)=H_0(q, p)$

Correspondingly, the following conclusion might be deduced. Any physical observable, dynamical observable, say A, its expected value is independent of time, i.e., $i\hbar\frac{d<d>}{dt}=0$, it is a constant of motion. In Schrödinger picture, this can be written as $i\hbar\frac{d<A>}{dt}=<[A, H]>+i\hbar<\frac{\partial<A>}{\partial t}>$, since A is a constant of motion, so $[A, H]=0$, which implies that A commutes with H. Therefore, any dynamical variable commutes with Hamiltonian is a constant of motion. In Heisenberg picture, the same conclusion is obtained. So from Schrödinger and Heisenberg pictures, many arbitrary pictures can be constructed.

4.2. *Some Applications of Schrödinger Equation*

The harmonic oscillator is a very excellent physical application to Schrödinger equation. It is a very representing system to a real physical phenomenon in nature. It is describing the vibration motion of the physical system in the macroscopic and the microscopic systems. Figure 1 represents the physical motion of a vibrating system.

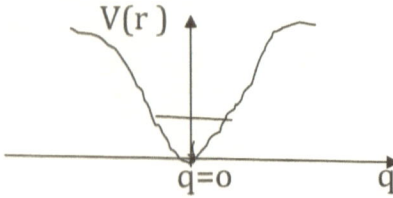

From the classical mechanics, it is known that the motion of the harmonic oscillator, under the potential effect, is given by

$V(r)=(1/2)m\omega^2 q^2$--------------- (6).

Where m is the mass of the oscillator, ω is its angular frequency, and q is the generalized coordinates. The Hamiltonian of the system is given by $H=p^2/2m+V(r)=$kinetic energy + potential energy. Also, it is known from the commutation relation that

$[q, p]=i\hbar$. It is the quantization rule in quantum mechanics.

The generalized coordinate classical representation is given by $q(t)=A \sin(\omega t + \delta)$, where δ is a phase shift. Therefore, Hamiltonian is given by:

$H = \omega\hbar[\frac{p^2}{2m\hbar\,\omega} + \frac{m\omega}{2\hbar}\,q^2] = \omega\hbar[P^2/2m + \frac{1}{2}Q]$ -----------(7).

Where, $P = p/(m\omega\hbar)^{1/2}$ and $Q = (\frac{m\omega}{\hbar})^{1/2}\,q$ ------------(8).

Then [Q, P] takes the form $[Q, P] = 1/\hbar\,[q, p] = i$ -------(9)

Now let a and a^+ be introduced as two operators defined as $a = 1/\sqrt{2}\,(Q+iP)$ and $a^+ = 1/\sqrt{2}\,(Q-iP)$.

Consequently, $Q = \frac{1}{\sqrt{2}}[a^+ + a]$ and $P = \frac{i}{\sqrt{2}}[a^+ - a]$ ---------(10), where $[a, a^+] = 1$. (Prove it.)

By substituting for Q, P from 10 into 7, H becomes
$H = \hbar\omega[a^+a + 1]$ ------------------------------------- (11).

(Prove 11 as an exercise.)
Introduce the number operator N given by
$N = a^+a$ --(12).

Hence, harmonic oscillator problem is mathematically described by $[a^+, a] = 1$ and $N = a^+a$.

Take $a = \lambda Q + \mu P$
$a^+ = \lambda^+ Q + \mu^+ P$, and $[a^+, a] = 1$

As an exercise, show that the only combinations show this relations are $a = \frac{1}{\sqrt{2}}\,(Q+iP)e^{i\phi}$ and $a^+ = \frac{1}{\sqrt{2}}\,(Q-iP)e^{i\phi}$, where $e^{i\phi}$ is a phase factor.

Now to find the eigenvalue of $N = a^+a$, where $[a, a^+] = 1$, let the eigenstate to be ln>. Hence, Nln>=nln> is the eigenvalue equation, where $n \geq 0$, and it is real.

To show this, take <lNln>=n<nln>=n. Remember <nln>=1, normalized. Also, <nla^+aln>=<lNln>=n or (<nla^+)(aln>)=n=<nln≥0; it is norm. This implies that $n \geq 0$. So it is proved that n is real.

Now if n=0, then ln>=l0> which represents the vacuum state. But if n>0, then aln>≠0∝ ln-1>. To prove this, take [a, a⁺]=1 [aa⁺-a⁺a]=1, aa⁺-N=1, N=aa⁺-1, multiply by a; it will be found that N a=(aa⁺a - a)= (aN-a)=a(N-1).

Now Naln>=a(N-1)ln> → N(aln>)=(N-1)(aln>) ⟹
Nln⁻>= (n-1)ln⁻> --------------(13). (aln>=ln⁻>)

As an exercise, following the same way, find that a⁺ln>∝ ln+1> if n>0, then n-1 must be one of the permissible value; hence, n≥1.

Now if n>1, then a⁺²ln>≠0 but∝ $n - 2$ where n-2≥ 0, so n≥ 2. Hence, [n=0, 1, 2, 3, ----] --------------- -------- (14).

Now it can be written that
aln>=0, which implies n=0, also <nla⁺a>=Nln>=n
aln>=c(n-1)   =lcl²<n-1ln-1>=lcl²
a⁺ln>=c*(n-1)   =1

Hence, c=$\sqrt{2}$, then, aln>=ln-1>$e^{i\vartheta_n}$,. By the same way, it can be found that lc*l²=n+1, so a⁺ln>=(n+1)(n+1) which implies that lc*l²=n+1; therefore, a⁺ln>=$\sqrt{n+1}$ln+1>$e^{i\vartheta_n}$, then a⁺aln>=$\sqrt{n}$ a⁺ln-1>$e^{i\vartheta_n}$=nln>.

Since H=$\hbar\omega$(a⁺a+1/2)=(N+1/2) $\hbar$ ω then
Hln>=(N+1/2)$\hbar\omega$ln>=(Nln>+1/2ln>)=(n+1/2)ω$\hbar$ ln>; hence,
Hln>=$\hbar\omega$(n+1/2) ln>-----------------------(15).

Note that equation 15 is the eigenvalue equation of the harmonic oscillator motion. And $\hbar\omega$(n+1/2) is the eigenvalue of the Hamiltonian operator H, which is the energy of the harmonic oscillato; ln> is its eigenstate. The energy is denoted by E_n =$\hbar\omega$ (n+1/2), so the ground state is l0>; its energy is $E_0=\frac{1}{2}\hbar\omega$. The other states take the values as n=1, 2, 3, 4,----n.

The following figure illustrates the energy spectra of the HO:.

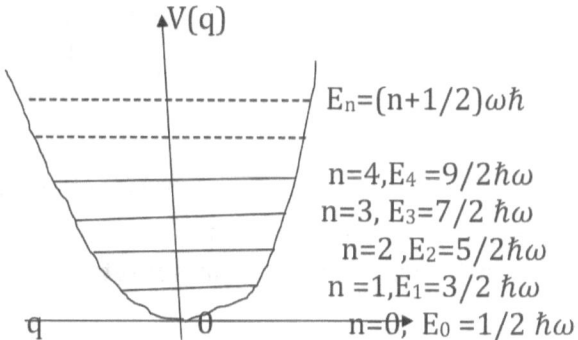

$E_n = (n+1/2)\omega\hbar$

$n=4, E_4 = 9/2\hbar\omega$
$n=3, E_3 = 7/2\hbar\omega$
$n=2, E_2 = 5/2\hbar\omega$
$n=1, E_1 = 3/2\hbar\omega$
$n=0, E_0 = 1/2\hbar\omega$

Note that the equal space between the states is $\hbar\omega$.
Now the matrix elements of the operators (a) and (a⁺) can be found by the following sequences operations:

Take $a^+|0> = |1>$, remember that $a^+|n> = (n+1)|n>$. It is a creating operator as shown before. So $a^{+2}|0> = a^+a^+|0> = a^+|1> = (|n+1>.)$
Remembering that $a^+|n+1> = (n+1)^{1/2}|n+1>$.

Also, it can be shown that $1/\sqrt{2}\ (a^+)^n|0> = |n>$, show that.

Now $<n|N|m> = n\delta_{nm}$,

$$N = \begin{pmatrix} 0 & 0 & 0 \\ 0 & 1 & 0 \\ 0 & 2 & 0 \\ 0 & 3 & 0 \\ 0 & 4 & 0 \\ - & - & - \end{pmatrix} \qquad \text{--------(16).}$$

$<m|a|n> = (n)^{1/2}\delta_{m,n-1}$, so:

$$a = \begin{pmatrix} 0 & \sqrt{1} & 0 & 0 & - \\ 0 & 0 & \sqrt{20} & 0 & - \\ 0 & 0 & \sqrt{3} & 0 & 0 & - \\ 0 & 0 & \sqrt{4} & 0 & 0 - \\ - & - & - & - & - & - \end{pmatrix}, \quad a^+ = \begin{pmatrix} 0 & 0 & 0 & 0 \\ \sqrt{1} & 0 & 0 & 0 \\ 0 & \sqrt{2} & 0 & 0 \\ 0 & 0 & \sqrt{3} & 0 \\ 0 & 0 & 0 & \sqrt{4} \\ - & - & - & - \end{pmatrix} \quad ---- (17)$$

[This might be shown as an exercise.]
Equations 16 and 17 are called N representation.

Therefore, the conclusions are

1. $a^+a=N$, where N is the number operator which takes the values 0, 1, 2, 3, 4,---------n.
2. $a|n> = \sqrt{n}\ |n-1>$, so a is annihilation operator
3. $a^+|n> = \sqrt{(n+1)}\ |n+1>$, so a^+ is a creation operator
4. $a|0> = 0|0>$, 0 the eigenvalue associated with the vacuum state
5. $a^+|o> = \sqrt{0+1}\ |0+1> = |1>$

4.3. *Similarity between HO and Quantum Field Theory (Brief Demonstration)*

In physics, the field of a force is defined as the domain where the effect of this force takes place on a particle. On this basis, the interactions instead between particles are interactions between the fields. Also in the case of the interactions between so many particles (say 1,000 or more), there will be this numerous number of equations to be solved which are mathematically fairly complex. Therefore, the use of the quantum field as state representing such many particles is the right way to deal with such problem.

As an example, take the Maxwell electromagnetic fields in vacuum. These fields are represented by $\mathbb{E}(x, t)$ as the electric field and $\mathbb{H}$ is the magnetic field. The average radiated energy is given by

$$H=\int 1/8[\mathbb{E}^2 (x, t)+\mathbb{H}^2 (x, t)]dx=\sum_\omega \hbar\omega [a_\omega^* a_\omega+\text{constant}]$$

Where, $[a_\omega, a_\omega-]=\delta_{\omega\omega}-$

From the solution of Maxwell equations (see: the classical electrodynamics by Jackson) $E(x)$ and $\mathbb{H}$ are given by

$$\mathbb{E}(x)=\sum_\omega a_\omega f(\omega, x)+a_\omega^* f^+(\omega, x)] \text{----------------}(18).$$
$$\mathbb{H}(x)=\sum_\omega a_\omega g(\omega, x)+a_\omega^* g^+(\omega, x)]=$$
$$i\sum_\omega[a_\omega f(\omega, x)-a_\omega^* f^+(\omega, x)] \text{----------------}(19).$$

This field is equivalent to an infinite harmonic oscillator, which means that the Hamiltonian H is

$$H=\sum_\omega \hbar\omega(n_\omega +\text{constant}) \text{----------------------}(20).$$

H represents system of particles, each particle of energy $(\hbar\omega)$. In fact, the field here is quantized which represents an arbitrary number of identical particles. Also, it is possible that certain particles either created or annihilated; n_ω in equation 20 stands for the number of particles. The (a) and (a^*) are the annihilating operator and the creating operator respectively.

4. Q-Representation

This is a generalization of the coordinates (x, y, z). Therefore, the equation of Schrödinger takes the form

$$[-\frac{\hbar^2}{2m}\nabla^2 +(1/2)m\omega^2 q^2](q)=E_n(q) \text{-----------}(21).$$
$$n=0, 1, 2, 3, 4,\text{-----}$$
$$\psi_n(\pm q)=\text{B.C. So as } q\longrightarrow \infty, \psi \longrightarrow 0$$

$E_n = \hbar\omega(n+1/2)$, n=0, 1, 2, 3, ----the energy.

$\langle q|n\rangle \equiv \psi_n(q)$;this is q representation of state $|n\rangle$.

$\psi_n(q) = N_n \exp[-\frac{m\omega\, q^2}{2\hbar}]\mathcal{H}_n$ (q) is the eigenfunction which is the solution of equation 21. $\mathcal{H}_n$ (q) is the well-known Hermit polynomial (See quantum mechanics by I. Schiff).

Matrix Elements

Quantum mechanics by Heisenberg in 1926, where the measured values are in fact are mathematical numbers, represented by the matrix elements. These values are measured as average values, which are usually found as

$\langle n|$ observable (q, p) $|n\rangle$.

Therefore, the matrix elements of the kinetic energy $p^2/2m$ is

$\langle n|p^2/2m\,|n\rangle = \int_{-\infty}^{+\infty} \psi_n(q)[-\frac{\hbar^2}{2m}\frac{d^2}{dq^2}]\psi_n(q)dq$

$= \langle n|1/2m(-m\omega\hbar/2)[a^+a^+-a^+a-aa^++aa]|n\rangle = -\frac{\hbar\omega}{4}\langle n|a^+a^+|n\rangle$

$= \frac{\hbar\omega}{2}(n+1/2)$

Problems: Show the following:

1. $\langle n|1/2m\omega^2\, q^2\,|n\rangle = E_n/2$
2. $\langle q_n\rangle = \langle n|q|n\rangle = 0$
3. $\langle p_n\rangle = \langle n|p|n\rangle = 0$
4. $\Delta_n q = \sqrt{\langle (q-\langle q \rangle_n)^2 \rangle} \geq \sqrt{\frac{\hbar\omega}{m\omega}}(n+1/2)^{1/2}$
5. $\Delta_n\, p = \sqrt{\langle (p-\langle p \rangle_n)^2 \rangle} \geq \sqrt{m\hbar\omega}(n+1/2)^{1/2}$

$\Rightarrow \Delta_n q_n \Delta_n p_n = \hbar(n+1/2)$, $\Delta_0 q_0 \Delta_0 p_0 = \hbar/2$; it is the minimum uncertainty (n=0, ground state).

4.5. *Two Dimensions Isotropic Oscillator**

The Hamiltonian of two dimensions isotropic oscillator is given by

$$H = 1/2m[p_1^2 + m^2\omega^2 q_1^2] + 1/2m[p_2^2 + m^2\omega^2 q_2^2] \text{----------(22)}.$$

Following the same steps in treating the one dimension harmonic oscillator, it is easy to find the eigenvalue and the eigenvectors which are common to N_1 of dimension q_1 and N_2 of dimension q_2. Remember that $N = a^+a$. The related data are in the table below:

E_n	$N+1$	(Eigenstate)	Remarks
$\hbar\omega$	1	l0, 0>	$n_1=n_2=0$, N=0.G.S.
$2\omega\hbar$	2	l10>, l01>	
		[N=n_1+n_2=2, if n_1=0, n_2=1, N=1, or n_1=1, n_2=0, N=1]	
$3\omega\hbar$	3	l20>, l11>, l02>	
(n+1) $\hbar\omega$	n+1	ln 0>ln-1, 1>------l0n>	

Remember:

1. $E_n = \hbar\omega \, (n+1/2)$,
2. $\psi_n = \psi_1\psi_2$
3. $[q_i, q_j] = [p_i, p_j] = 0$
4. $[q_i, p_j] = i\hbar\delta_{ij}$

Also,

5. $q_i = (\sqrt{\dfrac{\hbar}{2m\omega}})[a_i^* + a_i]$, $p_i = \sqrt{\dfrac{m\hbar\omega}{2}}(a_i^* - a_i)$
6. $[a_i, a_j] = 0 = [a_i^*, a_j^*]$
7. $= E_n = \sum_n \hbar\omega \, (n_i + 1/2) = \hbar\omega(n+1)$
8. $[a_i, a_j^*] = \delta_{ij}$
9. $H = \sum_i \hbar\omega(a_i^*a_i + 1/2)$

[*Based on the Quantum Mechanics, Vol. 1 p. 434 by Messiah.]

From the above data, it can be concluded that the ground state is n=0, where the zero point energy is $E_0=\omega\hbar$. So where $n_1=n_2=0$, the state is nondegenerate. Now let |⓪> be the total eigenstate of the two dimension isotropic harmonic oscillator. Let |0> be the individual eigenstate (one dimension) for the first and the second dimensions. Therefore, |⓪>=|0>|0> is the eigenstate of the two-dimension isotropic HO. This means the individual states are independent of each other.

The first excited state is n=1, $E_1=2\hbar\omega$,

if $n_1=1$, $n_2=0$, |1>=|1>|0>,
or $n_1=0$, $n_2=1$, |1>=|0>|1>. $\left.\right\}$ degenerate state of $E_1=2\hbar\omega$

The second excited state is $n=2=(n_1+n_2)$, so $E_2=3\hbar\omega$

If $n_1=2$, $n_2=0$, |2>=|2>|0>
if $n_1=1$, $n_2=1$, |1>=|1>|1>   degenerate state of $E_2=3\ \hbar\omega$
if $n_1=0$, $n_2=2$, |0>=|0>|2>.

Note: The degeneracy rank increases as the energy increases (for the highest excited states).

4.6. The Angular Momentum Operator

The angular momentum operator L_O is defined by the relation $L_3 =1/2(q_1p_2-q_2p_2)=i(a_1a_2^* - a_1^*a_2)$, where [H, L_3]=0, which indicates that L_3 is a constant of motion, since it commutes with Hamiltonian H.

Now let $i[a_1a_2^* -a_1^*a_2]|0><0|=0, \rightarrow |00>=|0>|0>$, which implies that only the linear combination of |0> and |0> is representing the eigenstate, so $i[a_1a_2^*-a_1^*a_2][\alpha|1>|0>+\beta|0>|1>]=\ell[\alpha|1>|0>+ \beta|0>|1>]=[\ell\alpha|1>|0>+\ell\beta|0>|1>] \equiv [-i\beta|1>|0>+i\alpha|0>|1>]$. (Show this as an exercise).

By comparison, the following relations are obtained:

$\ell\alpha=-i\beta,\ \ell\beta=i\alpha \longrightarrow \ell^2\alpha\beta=\alpha\beta \rightarrow \ell^2=1 \rightarrow \ell=\pm1$

For $\ell=1$, $\alpha\,|1>|0>+\beta|0>|1>=i\beta[|1>|0>+|0>|1>]$
(As an exercise, work it out for $\ell=-1$.)

Now it can be shown that N and L form another complete set of commuting observables. This can be done by introducing the following operators:

$$\left.\begin{array}{l} A\pm=\frac{1}{\sqrt{2}}\,(a_1\mp ia_2) \\[2mm] A^+_{\pm}=\frac{1}{\sqrt{2}}\,(a_1^+\pm ia_2^+)= \end{array}\right\} \quad \text{-------------------- (23)}$$

Where these operators satisfy the following commutation relations:

$$\left.\begin{array}{ll} [A_r,\ A_s^+]=\delta_{rs}, & r=+ \text{ or } r=- \\[2mm] [A_r,\ A_s]=[A_r^+,\ A_s^+]=0, & s=+ \text{ or } s=- \end{array}\right\} \quad \text{--------- (24)}$$

A_+ and $A^+_{\dotplus}$ are interpreted as destruction and creation operators of quanta type (+) respectively. Where A_- and $A^{\pm}$ are destruction and creation operators respectively of quanta type (-). On this bases. the operators $N_+=A^+_{\dotplus}A_+$ and $N_-=A^{\pm}A_-$ can be constructed to represent the numbers of positive quanta and negative quanta respectively. Since the commutation relations of these operators are identical to that of a and a^+, it indicates that it is possible to form common eigenvectors to N_+ and N_-, analogous to the case of N_1, N_2 where $N=$(number operator). Therefore, N_+ and N_-, each has its spectrum from the sequence of the non-negative integers, $n_+=0, 1, 2, 3,$---- and $n_-=0.1, 2, 3$-----

Also, these two observables form a complete set of commuting observables to each pair of quantum numbers (n_+, n_-). It can be written that $A_+|00>=A_-|00>=0$, thus the vector $|00>$ mentioned in the previous table is an eigenvector of the ground state $(n_+=n_-=0)$, which leads to

$|n_+ n_-> = (n_+!n_-!)^{-1/2} A_+^{+n_+} A_-^{\pm n_-} |00>$, which forms a complete orthonormal eigenstate set common to N_+ and N_-, where $N_+|n_+n_->=n_+|n_+n_->, N_-|n_+n_->=n_-|n_+n_->$ ---------------- (25).

Let N and L be the function of $N+$ and $N-$, such as $N=N_++N-$, and $L=N_+-N-$ ----------------------------------- (26).

Since N_+ and N_- form a complete set of commuting observables, their sum is N, and their difference is L; they have the same properties. Also, it is known that $n=n_1+n_2$, where $n(N)=n_++n_-$, but it is not necessary that n_+ and n_- are the same as n_1 and n_2, but their sum is the same. Hence, it can be shown that the Hamiltonian H can be written as:

$H=\hbar\omega[A^+_+A_++A_-^+A_-+1]$ ---(Work it out as an exercise.)

The set of vectors $|n_+n_->$ which are a complete orthonormal set are eigenvectors of H satisfy the eigenvalue equation.

$H|n_+n_->=\hbar\omega(n_++n_-+1)|n_+n_->=\hbar\omega(n+1)|n_+n->, n=n_++n-$
$E_n=\hbar\omega(n+1)$.

Also, $L|n_+n->=(n_+-n-)|n_+ n->$ (show that).
As an exercise, show that $[L, A^+_\pm]=\pm A^+_\pm, [L, A_\pm]=\mp A_\pm$.

Consequently, if $A_+{}^+$ and A_- act on an eigenvector of L, L will be increased by a unit; but if $A_-{}^+$ and A_+ act on it, it will be decreased by a unit. The interpretation of this is as follows:

In quantum theory of charged fields, where the field appears as a set of two dimension isotropic oscillators, N_+ is the number of positive-charged particles, and N_- is the number of negative-charged particles. But L is the total angular momentum. Hence,

A_+^+ creates a positive charge, and A₋ destroys a negative charge; but as it is known as destroying a negative-charged particle, it means creating a positive-charged one and vice versa; so both way, the charge increases by one unit.

Similarly, A_-^+ and A₊ decrease the charge by one unit. It is very important to notice that in the theory of the crystalline vibrations, the motion of the lattice is likewise represented by a set of isotropic two dimension oscillators. The quanta of the oscillation are called phonon. The representation in terms of phonons of type 1 and 2 corresponds to standing waves. But the representation in terms of phonons of type (+) and (-) corresponds to traveling waves (propagating) in one dimension or the other.

4.7. Three Dimensions of Isotropic Harmonic Oscillator (TDIHO)

The problem of this isotropic (equally oscillates in the three dimensions) oscillation is a problem of a particle located in a central potential $V(r) \propto r^2$, where r is the vector distance from the center. The Hamiltonian of this system is given by

$H=P^2/2m + 1/2 .m\omega^2 r^2 = H_X + H_Y + H_Z$ (H-components).
This might be written as $H_i = 1/2m[P_i^2 + m\omega^2 r_i^2]$ ----------(27).
(i=x, y, z)

The solution of this problem is quite similar to the solution of the one and two dimensions of the harmonic oscillator problems. Therefore, the eigenvalue of the Hamiltonian H is:
$E_n = \hbar \omega (n+1/2)$, where n=0, 1, 2, 3,-------,
it is $(1/2)(n+1)(n+2)$-fold degenerate.

The observables N_x, N_y, N_z form a complete set of constant of motion. The eigenvectors (n_x, n_y, n_z) of their basis labeled by three corresponding eigenvalues: n_x, n_y, n_z. The vectors are deduced by $|n_x\ n_y\ n_z> (n_x!n_y!n_z!)^{-1/2}a_x^{+n_x}a_y^{+n_y}a_z^{+n_z}|000>$

Where $|000>$ is the ground state vector and $a_x|000>=a_y|000>=a_z|000>=0$.
$L=\vec{r} \times \vec{p}$, the angular momentum.

The Hamiltonian H can be written as
$H=N+1/2$, where $N=a^+a$, in unit of $\hbar\omega$.

Actually, the eigenvectors $|nm>$ are common to H and are labeled by the usual quantum numbers (n, m). The eigenvalue of H, L, and L_Z are $(n+3/2)\ \hbar\omega$, $\ell(\ell+1)\ \hbar\omega$, and $m\hbar$, respectively.

Therefore, the vectors $|n>$ form a complete orthonormal set of eigenvectors of H. They are derived from the vectors $|n_xn_yn_x>$ by a unitary transformation. Remember that in $\{r\}$ representation, $|n_xn_yn_z>$ is represented by the wave functions $\psi_{n_x}(x)$, $\psi_{n_y}(y)$, and $\psi_{n_z}(z)$, which implies that $\psi_{n\ell m}(r)$ is represented by the ket $|n\ell\ m>$. This can be represented in terms of the radial solution (and the angular solution), i.e.,
$\psi_{n\ell m}^m(r)=(\frac{\mathcal{Y}_{n\ell(r)}}{r})\ Y_\ell^m\ (\theta,\varphi)$------------ (28)

where, $\mathcal{Y}_{n\ell}$ (r) is the solution of the radial equation,

$[-\frac{\hbar^2}{2m}\frac{d^2}{dr^2}+\frac{\ell(\ell+1)}{2mr^2}+(1/2)m\omega^2r^2]\mathcal{Y}_{n\ell}(r)$
$= [(n+1/2)\ \omega\hbar]\mathcal{Y}_{n\ell}(r\ )$ -----------------------(29).

This is the radial Schrödinger equation.
The value of ℓ and m can be determined for affixed n (principal quantum number), so it is possible to find different possible states of L for each energy level. Now consider (as it is known) that the operator $A_m(m=-1,0,1)$ is defined as

$A_1 = 1/(2)^{1/2}(a_x - ia_y)$, $A_0 = a_z$, $A_{-1} = 1/(2)^{1/2}(a_x + ia_y)$. The complex conjugate of A_m is A_m^+ which is given by

$A_m^+ = 1/(2)^{1/2}(a_x + ia_y)$, $A_0^+ = a_z$, $A_{-1}^+ = 1/(2)^{1/2}(a_x - ia_y)$.

Also A_m^+ and A_m satisfy the following commutation relations:
$[A_r, A_s] = [A_r^+, A_s^+] = 0$, $[A_r, A_s^+] = \delta_{rs}$.

The operators A_m and A_m^+ are interpreted as a destruction and creation operators respectively of quanta type (m).

Now the operator $N_m = A_m^+ A$ is the number of the quanta type (m). This implies that N_1, N_0, N_{-1} form a complete set of commuting observables. The Hamiltonian H and the number operator N are related through the relation

$H = (N_1 + N_0 + N_{-1} + 3/2) = [N + 3/2]\,\hbar\,\omega$
$[N = N_1 + N_0 + N_{-1}]$ ------------ (30).

To each triplet of eigenvalues (n_1, n_0, n_{-1}), there correspond an eigenvector common to these three observables, namely the vector $[(n_1 n_0 n_{-1})^{-1/2}]A_1^{+n_1} A_0^{+n_0} A_{-1}^{+n_{-1}}|000>$

These vectors form a complete set of eigenvectors of the Hamiltonian H; therefore

$H|n_1 n_0 n_{-1}> = (n+3/2)\hbar\omega\,|n_1 n_0 n_{-1}>$ --------------------(31).

In general, these eigenvectors are not eigenvectors for L^2, but they are eigenvectors for its z-component L_z because $L_z = (N_1 - N_{-1})\hbar$. (Show this as an exercise.) This means $m = (n_1 - n_{-1})$.

Now consider the subspace of the eigenvectors of H corresponding to the eigenvalue $(n+3/2)\hbar\omega$, so it can be concluded that the quantum number (m) can take the values $m = -n, -----, +n$.

Let C_m be the number of linearly independent vectors corresponding to each value of m. So it is easy to show that:

$$\left.\begin{array}{l} |m|=n, n-1, n-2, n-2s, n-(2s+1), n-(2s+2), \\ C_m=1, 2,-----s+1, s+2 \end{array}\right\}--- -------(32)$$

As it is known physically, each value of ℓ corresponds to a certain number of set of $(2\ell +1)$ vectors of well-defined angular momentum $(m\ell)$, where m in each set takes the $2\ell+1$ integral values contained between $-\ell$ to $+ \ell$).

Let $d\ell$ represents such number, then
$C_m=\sum_{\ell\geq m} d\ell \Rightarrow d\ell = C_\ell - C_{\ell+1}.$

By investigating equation 32, it is possible to find
$d\ell =1$ for $\ell=n, n-2,-----,n-2s$.

The conclusion is for all integral values of parity $(-)^n$ are contained between 0 and n, and that $d\ell =0$ for other values of ℓ. Also, it can be concluded that for each eigenvalue $(n+3/2)\hbar\omega$ of the energy corresponds $(1/2)(n+1)(n+2)$ states of well-defined angular momentum (ℓ_m). Hence, for the possible values of ℓ, there exist $(2\ell +1)$ eigenstate corresponding, respectively, to the $(2\ell +1)$ values of m ranging from $-\ell$ to $+ \ell$ (as shown before). Also, it is known that the value of ℓ takes the value n, n-1, n-2, 0 if $(-)^n=1$, $[(1/2)(n+2)]$ distinct values], and n, n-2,---, 1 if $(-)^n=-1$, $[1/2(n+1)$ distinct values]. The following table represents the spectrum of the three-dimension harmonic oscillator.

Table 1. The spectrum of the three-dimension HO.

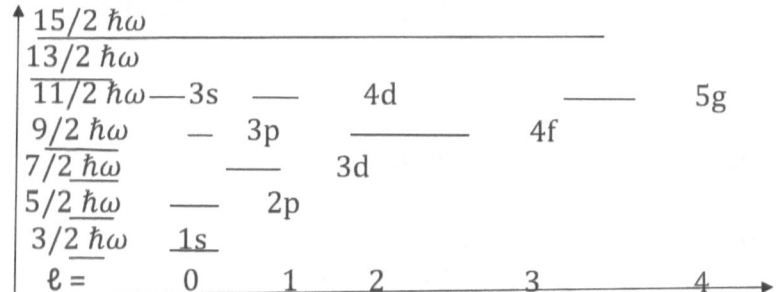

15/2 ℏω						
13/2 ℏω						
11/2 ℏω—3s	——	4d		——	5g	
9/2 ℏω —	3p	——	4f			
7/2 ℏω	——	3d				
5/2 ℏω —	2p					
3/2 ℏω 1s						
ℓ = 0	1	2	3	4		

Chapter Five

Angular Momentum in Quantum Mechanics
(Properties and Applications)

5.1 *Definition of Angular Momentum*

The angular momentum is defined as
$\vec{L} = \vec{r} \times \vec{p}$, where $\vec{r}$ represents the position and
$\vec{p}$ *is the momentum*, i.e.,
$\vec{r} \equiv (\vec{x}, \vec{y}, \vec{z})$ and $\vec{p} \equiv (\vec{p_x}, \vec{p_y}, \vec{p_z})$.

In vector operator representation, the angular momentum is written as, $\vec{L} = -i\vec{r} \times \vec{\nabla}$, where $\vec{\nabla} = \frac{\partial}{\partial r}$, in unit of $\hbar$.

$\vec{L}$ has three components which are differential operators satisfy the following commutation relations:

$[L_x, L_y] = iL_z$, $[L_Y, L_z] = iL_x$, $[L_z, L_x] = i\, L_y$ ----------------(1)
And $L^2 = L_x^2 + l_y^2 + L_z^2$, $[\vec{L}, L^2] = 0$.

[As an exercise, prove that:
For N-particles, $[N = n_1, n_2, n_3, ----]$

$L^n \equiv r^n \times p^n$ is the angular momentum for n^{th} particle. Therefore, the total angular momentum of the system of N-particles is the vector sum of the angular momenta of these:

$\vec{L} = \sum_{n=1}^{N} L^{(n)}$ ------------------------------- (2).

The commutation relations (equation 1) are

$$[L_x, L_y] = \sum_n \sum_{n^-} [L_x^n, L_y^{n^-}] = \sum_n [L_x^n, L_y^n] = i \sum_n L_z^n i$$
$$[L_y, L_z] = \sum_n \sum_{n^-} [L_y^n, L_z^{n^-}] = \sum_n L_y^n, L_z^n] = i \sum_n L_x^n \quad\quad \Bigg\}$$
$$[L_z, L_x] = \sum_n \sum_{n^-} [L_z^n, L_x^{n^-}] = \sum_n [L_z^n, L_x^n] = i \sum_n L_y^n.$$

$$\text{-----(4).}$$

Accordingly, it can be concluded that the angular momentum is the vector which its components are observables satisfying the commutation relations in equation 3.

5.1.1 *The Characteristic of Algebraic Relations*

The angular momentum is a physical dynamical observable; it is of important use in physically describing a microscopic systems which are usually in moving states. So it plays main roles in quantum mechanics; therefore, it is quite necessary to understand their physical and mechanical properties. The total angular momentum $J^{\rightarrow}$, is written as $J^2 = J_x^2 + J_y^2 + J_z^2$. It commutes with its components: $[J^2, J_x] = [J^2, J_y] = [J^2, J_z] = 0$, and $[J^2, J]. = 0$. (Prove it.)

Also, J^2 commutes with the functions of its components,
i.e., $[J^2, F(J_x, J_y, J_z)] = 0$.

Now from this, the total angular momentum can construct very important operators; they are the so-called the raising and the lowering operators. They play basic roles in building the status of the described physical systems. They are defined as

$$J_+ = J_x + i J_y, \quad J_- = J_x - i J_y \text{ -------------------- (4),}$$

where J_+ is the raising operator and J_- is the lowering operator.

The use of these operators will be used in details later.

The operators J_+, J_- and the z-component J_z define very well the total angular momentum vector J in a complete way. From equation 4, the following commutation relations can be deduced such as

$$[J_z, J_+]=J_+, \ [J_z, J_-]=-J_-, \ [J_+, J_-]=2J_z \ \text{------------(5)}$$
$$[J^2, J_+]=[J^2, J_-]=[J^2, J_z]=0 \ \text{--------------------(6)}$$
$$J^2=1/2[J_+J_-+J_-J_+]+J_z^2 \ \text{--------------------(7)}$$
$$\left.\begin{array}{l} J_-J_+=J^2-J_z(J_z+1) \\ J_+J_-=J^2-J_z(J_z-1) \end{array}\right\} \ \text{----------------------(8)}$$

(As an exercise, show these relations.)

5.1.2. The Spectrum of J^2 and J_z

Since J^2 commutes with its components, as shown above. It is quite possible to build up a complete set of eigenfunctions common to all, as it is a well-known conclusion to the commutation relations. Also, from 1, J_x, J_y, and J_z are Hermitian operators put hard limitation on the eigenvalue of the spectrum. Hence, it is so important to be noticed that since J_x, J_y, and J_z are Hermitian operators, so they are positive definite operators, where for any eigenstate la>, $<a|J_i^2|a> \geq 0$. This represents the norm of the vector ($J_i$la>≥ 0, (i=x, y, z). These leads to eigenvalues ≥ 0.

This implies that the eigenvalues can be written as
$\varepsilon_v=j(j+1)$, j>0, and it is real quantity. Why?

Hence, for any eigenvector state ljm>, which is common to J^2 and J_z (why?), the eigenvalue equation is given by

J^2ljm>=j(j+1)ljm> in $\hbar$ unit.
$J_z$ljm>=mljm>   =====

It is very important to note that if J^2 and J_z do not commute, there exist several linearly independent state ljm>, implying ljm> is a particular ket vector chosen once and for all in the subspace of angular momentum(jm). So for any vector thus chosen, the following is valid with subject to the condition imposed on ljm>, that J^2ljm>=j(j+1), $J_z$ljm>=mljm>, as shown above.

Now take J.J$_+$ and J$_+$J$_-$, operate on ljm> using equation 8. The following results are obtained (show that)

J.J$_+$ljm>=[j(j+1)-m(m+1)]J.ljm>=(j-m)(j+m+1)J.ljm>--------(9)
J$_+$J.ljm>=[j(j+1)-m(m-1)]J$_+$ljm>=(j+m)(j-m+1)J$_+$ljm>----- (10)

Taking the norms of the vectors J$_+$ljm> and J.ljm>, one gets that
<jmlJ$_+$J.ljm>=(j+m)(j-m+1)<jmljm>
=(j+m)(j-m+1)≥0.

Similarly, <jmlJ.J$_+$ljm>=(j-m)(j+m+1)≥0.

It complies with the well-known Axiom of the Hilbert space; the norm is positive definite. From this, it can be concluded that -j≤m≤j, (prove it). Now in the case, the norm is =0; the vector is null, and then J$_+$ljm>=0 if and only if (j-m)(j+m+1)=0, which implies that j=m, i.e., *m* is maximum. By the same way, if J.ljm>=0 if and only if (j+ m)(j-m+1)=0, which implies that j=-m, i.e., *m* is minimum. These results ensure the value of *m* is between (-m) as minimum and (+m) as maximum, i.e., -j≤ m ≤ j. This can be illustrated as

J$_+$ljm>=0 if m=j, J. ljm>=0, if m=-j -------- ------------- (11).

If j≠m, J$_+$ljm> is not null vector, but it is a vector of angular momentum (j, m+1), so (J^2J_+)ljm>=J$_+$$J^2$ljm>=j(j+1)J$_+$ljm> in $\hbar$ unit.

Using the fact that $[J_z, J_+]=J_+ \longrightarrow J_zJ_+=J_+(J_z+1)$, then
$J_zJ_+ljm>=J_+(J_z+1)ljm>=(m+1)J_+ljm>$---------------------- (12).
Using $[J_z, J_-]=-J_-$, following the same way, one gets
$J_zJ_-ljm>=(m-1)J_-ljm>$ --------------------------- (13).

From 12 and 13, the following can be concluded:

1. If ljm> is a vector of angular momentum (jm) with norm N, it is necessary that the range value of m is $-j$ to $+j$, $-j\leq m\leq +j$.
2. If $m\neq j$, then $(J_+)ljm>$ is a vector of angular momentum (j, m+1) and its norm $[j(j+1)-m(m+1)]$.
3. If $m=-j$, $(J_-)ljm>=0$, but if $m\neq-j$, then $J_-ljm>$ is a vector of angular momentum (j, m-1), and its norm $N=[j(j+1)-m(m-1)]$.

By repeating the application of J_+ acting on ljm>, it is possible to get a vector $J_+^p ljm>$, where p>0. It is an integer.

Therefore, it can be shown that

$J_+ljm>, j_+^2 ljm>, j_+^3 ljm>,$---------$j_+^p lm>$ --------------(14).

It is already pointed out that in equation 11, $J_+ljm>=0$ if $m=j$; but if $m<j$, then $J_+ljm>\neq0$, i.e., it is not a null vector, with angular momentum (j, m+1). This is very quite indicating that $J_+ljm>$ has the properties (1–3), which are characteristics of any common eigenvector of J^2 and J_z. This implies the necessary requirement that $m+1<j$. If $m+1=j$, then ljm>=0, but if $m+1<j$, it means ljm>$\neq0$. It is not null vector with angular momentum (j, m+2), which also has the properties (1–3) as mentioned before.

Now by following the continuous analysis of the properties of the vectors (14), then it is restricted by the fact that $-j\geq m\leq +j$; therefore, this sequence has to be terminated somewhere. Otherwise, it is possible to construct eigenvectors of J_z with eigenvalues larger than any given number. This is in a contradiction with the fact that m≯j. This indicates that there exists an integer p(≥0) such that J_+^p ljm>≠0 with angular momentum (j, m+p).

Also, lj, m+p> leads to j=m+p or m=j-p, p ≥0,
J-m=p≥0; hence, j-m is an integer >0, a positive one.

Accordingly, the p –vectors J_+ljm>, J_+^2ljm>, J_+^3 ljm>--J_+^pljm>
are states of a well-defined J with the same eigenvalue j(j+1) of J^2 and that m+1, m+2, m+3, m+4,-----m+p=j of J_z.

Following the same steps with the same concepts, it can be shown that J_-ljm>, J_-^2ljm>,-----J_-^q ljm> represent states of a well-defined J with the same eigenvalue j(j+1) of J^2 and that m-1, m-2, m-3,-------, m-q=-j of J_z. Also, j+m=q ≥0.

Since p and q are positive numbers, then (p+q) ≥0 =2j
[m-q-m-p=-j-j=-2j, - (p+q)=-2j ⟶ p+q=2j]

If these results are taken as facts, the following theorems can be deduced:

1. The only possible eigenvalues of J^2 are given by j(j+1), where j is a positive integer or half integer numbers such as J=0, 1/2/, 1, 3/2, 2, 5/2-------∞, in ℏ unit.
2. The only possible eigenvalues of J_z are the integer and half integer numbers such as m=0, ±1, ±3/2, ±2, --∞ in ℏ unit.
3. If j(j+1) and m (in ℏ unit) are the eigenvalues of J^2 and J_z respectively, which correspond to a common eigenstate of J^2 and J_z, with angular momentum (jm), where the possible values of m are given by (2j+1), takes the range –j, -j+1,--,+j.

5.1.3. *Eigenvectors of J² and Jz, Construction of Invariant Subspaces*

If ljm> is a given vector with a well-defined angular momentum (AM), then all (2j+1) vectors can be constructed with all their well-defined angular momentum by repeating application of J₊ and J₋. In general, J₊,J₋ are not normalized to unity. But they can be normalized as follows:

Let <jmljm> be a norm of the vector ljm>; it is one, as it is known, i.e., <jmljm>=1, if m<j then J₊ljm> is a vector with angular momentum (j, m+1), denoted by lj, m+1>, as shown previously; its norm is 1. It can be defined as:

J₊lj m>=Cmlj m+1>. Taking its norm, shown before, Cm can be found as:

$$lC_ml^2=[j(j+1)-m(m+1)]----------------------(15)$$

where Cm is real >0. By fixing the phase of ljm+1>, then

J₊ljm>=$\sqrt{j(j+1)-m(m+1)}$ ljm+1>. Multiplying both sides by J₋, using equation 9, then it will be found that:

J₋lj m+1>= $\sqrt{j(j+1)-m(m+1}$ lj m>.

Now if m+1=j, J₊lj m+1>=0; but if m+1≠ j, the vector ljm+2> can be formed with angular momentum (j, m+2). As usual, its norm is one; by continuing building state vectors, the vector is ljj>, i.e., m=j (maximum). Similarly, we can get that for J₋, ljm-1>, lj m-2>, ljm-3>,----lj,-j>, with norm equals to one and angular momentum (jm-1),-----(j,-j).

Hence, starting from lj, m>, it is possible to form a series of (2j+1) orthonormal vectors, lj j>, lj j-1>----lj m>---lj,-j>, which satisfy the eigenvalue equation:

$J^2|j, \mu>=j(j+1)|j \mu>$, in $\hbar$ unit ------------- (16).

$J_z|j, \mu>=\mu |j \mu>$ --------------------------(17).

The relative phases are chosen so that one can obtain these vectors, each one from other, by the relations:

$J_+|j \mu>=\sqrt{j(j+1)-\mu(\mu+1)}|j \mu+1>$ ---------(18).

$J_-|j \mu>=\sqrt{j(j+1)-\mu(\mu-1)} |j \mu-1>$-----------(19).

These (2j+1) vectors span certain subspace. As it will be seen, these operators J_+, J_-, and J_z transform a vector from *the subspace* $\mathcal{E}^j$ into another vector of the subspace itsself. This kind of transformation is called invariant transformation. It follows that any function F(J) of the components j, being a function only of J_+, J_-, and J_z, also leaves $\mathcal{E}^{(j)}$ invariant under all rotations of quantum mechanical system corresponds to the application of an operator of the F(j) to the state vectors.

This implies that any operation of rotation of the whole system leaves (subspace) invariant. In the case that J^2 and J_z do not commute, i.e., $[J^2, J_z]\neq0$, which implies that many systems of basis vectors are common to these operators; even if they form a complete set, the phase of each basis vector may be arbitrary fixed. In this case, there are certain manipulations of the angular momentum which is preferably called standard representations $\{J^2, J_z\}$. This representation of the basis vectors is corresponding to the specified quantum number j; it can be grouped into one or several series of (2j+1) vectors, which are connected through the relations in equations 17, 18.

This leads to the conclusion that to each series, there corresponds a subspace $\mathcal{E}^j$. So the Hilbert space is a direct sum of such subspaces. If those eigenvectors of the eigenvalues j(j+1) with $j_z=j$ are considered, then a certain subspace X^j of Hilbert space can be formed with a dimension (1, 2, 3,---- ∞).

In this case, it is always possible to choose a complete set of orthonormal vectors |τjj> in $\mathcal{X}^j$, where τ distinguishes vectors of the angular momentum (jj) from each other. It takes the values 1, 2, 3,-----∞, (discrete or continous). But here, it is to be considered discrete. Hence,

<τjj|τ¯jj>=$\delta_{\tau\tau^-}$, orthonormalization condition.

Therefore, with each of the vectors |τjj>, 2j vectors can be associated, which is obtained by repeating the application of J-, as it was shown previously to form a (2j+1) dimensional subspace $\mathcal{E}^j$. To distinguish between these subspaces, it is possible to denote each one with $\mathcal{E}(\tau, j$). Hence, the (2j+1) basis vectors of this subspace are:

|τjj>, |τj j-1>,--------, |τ j, -j> ------------(20).

These basis vectors are orthonormal, satisfying the following fundamental relations:

J^2|$\tau j\mu$>=j(j+1)|$\tau j\mu$>, in ℏ unit -------- (21).
J_z|$\tau j\mu$>=μ|$\tau j\mu$>.   -------------------- (22). in ℏ unit
J_+|$\tau j\mu$>=[j(j+1)-$\mu(\mu + 1)$]$^{1/2}$|$\tau j\mu$>------ (23).
J_-|$\tau j\mu$ > =[j(j+1)-$\mu(\mu - 1)$]$^{1/2}$|$\tau j\mu$ >   (24).

By using equations 23 and 24), the following important relations can be obtained:

1- |$\tau, j, \pm\mu$>=$[\frac{(j+\mu)!}{(2j)!(j-\mu)!}]^{1/2}J_{\mp}^{j-\mu}$ |$\tau, j, \pm j$>--------(25).

2- |$\tau, j, \pm j$>=$[\frac{(j+\mu)!}{(2j)!(j-\mu)!}]^{1/2}J_{\pm}^{j-\mu}$ |$\tau, j \pm \mu$>------(26).

(As an exercise, prove 25 and 26).

5.1.4. *Angular Momentum and Rotations*

Naturally, any physical system is, in general, subjected to external and internal forces, which might affect the status of the system. This effect disturbs the stability of the system; it might rotate its position coordinates in the space. Therefore, it is very important to know the status of the system in its new frame of coordinates. So the motion of any physical system is intrinsic character.

For this physical mechanical phenomenon, a mathematical formalism was developed on the two physical natures: the macroscopic and the microscopic system. Here, it is dealing with the microscopic systems, such as nuclear particles, nuclei, atoms and molecules.

The angular momentum plays main role in the description of the rotations of the rotated system under the mathematical formalism of the quantum mechanics. Also, the angles of rotation, the so-called Euler angles, are very important too.

Definition of Rotation: Euler Angles (α, β, γ)

The rotation about a given point (O) means an overall displacement of the points of the space in which the point (O) stays fixed. So each point (P), due to rotation, takes up a new position (P^-). Hence, P $\longrightarrow$ (P^-). Where (R_o) is the rotation operator. It is a one-to-one correspondence transformation. Also, the rotation about a point (O) is as one-to-one correspondence between points of the space, in which point (O) corresponds to itself is conserving both distances (and therefore the angles) and the sense of the coordinates axes. This can be illustrated by figure 1 below.

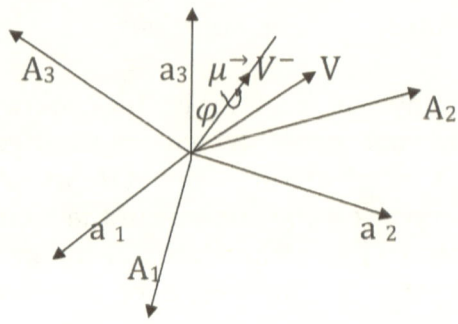

Figure 1 Rotation Illustration.

Let R be the rotation function, which is a function of a unit vector $(\vec{\mu})$ and an angle (φ) (see Figure 1). If φ is an infinitesimally small, then an infinitesimal rotation occurs, where $\sin\varphi = \varphi = \epsilon$; it is an infinitesimal displacement.

Then $V \longrightarrow V^- \equiv V + \epsilon(\mu \times V)$, $(\epsilon \ll 1)$---------------(27).

Notice, there are infinite number of ways specifying the rotation (a_1). $R_\mu(\varphi) = R_{\mu_1}$, if $\mu = \mu_1$

If $\varphi_1 = \varphi + 2\pi n$ or $\varphi_1 = -\varphi + 2\pi n$, (n is any integer).
Also, there is another method based on Euler angles to describe the rotation. These angles are defined in figure 2.

The system 0xyz) is obtained by the rotation of the system (OXYZ), where ou is one of the two directed axes perpendicular to the plane (OZ). By noticing the figure 2, the Euler angles are $\alpha = (Y, 0, ш)$, $\beta = (Z, 0, z)$, and $\gamma = (ш, o, y)$.

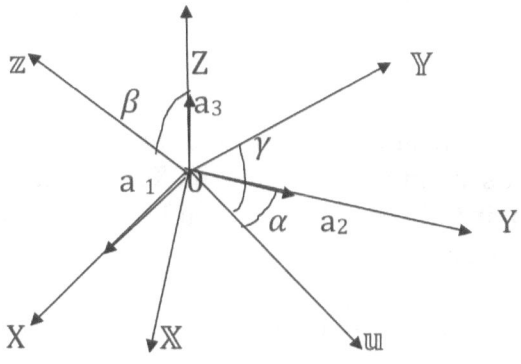

Figure 2. Euler Angles

The sequences of the three rotations are

1. A rotation of angle (α) about (0Z), $R_z(\alpha)$ (0Y→0ɯ).
2. A rotation of angle(β) about (0 ɯ)$R_ɯ(\beta)$ (0Z→ 0ℤ).
3. A rotation of angle (γ) about (0Y) $R_z(\gamma)$(0ɯ →0 y).

Therefore, R(α, β, γ) performs the rotation as follows:

$$R(\alpha,\beta,\gamma)=R_z(\gamma)R_ɯ(\beta)R_z(\alpha) \quad \text{------------ (28)}.$$

It is to be noticed that α β γ are algebraic quantities; they might carry (±) signs. It depends on the rotation about the axes 0z, 0ɯ, and 0z, positive or negative direction. Such rotation can be defined by several different sets of Euler angles. The necessary and sufficient condition for R(α, β, γ)=R(α_1, β_1, γ_1) is

$\alpha_1=\alpha+2\pi n_\alpha$, n is any integer

$\beta_1=\beta +2\pi n_\beta$, n is any integer

$\gamma_1=\gamma +2\pi n_\gamma$, = = = =

With each rotation, R can be associated with a certain 3×3 matrix. Consider figure 2, take the system (0xyz) with unit vectors a_1, a_2, and a_3 in the direction of 0X, 0Y, and 0Z, respectively.

Under rotation, these unit vectors transform to A_1, A_2, and A_3, (figure 2). This means the system (0XYZ) goes to the system (0X̃ỸZ̃). Also, it indicates that A_j is a linear combination of a_1, a_2, and a_3, such as

$a_iR_{ij}=a_1R_{1j}+a_2R_{2j}+a_3R_{3j}$ where A_j can be written as:
$A_j\equiv R[a_j]=a_iR_{ij}\Rightarrow R_{ij}=(a_iA_j)$--------------------(29).

Equation 29 represents the matrix elements of R.

The coefficients R_{ij} of these three linear combinations are the elements of the 3×3 matrix, denoted by the same letter R, which is denoting the rotation itself. Therefore, the rotation is completely defined by this matrix. So if $V=a_iV_j$ is any vector in space, specified by its coordinates (V_1, V_2, V_3) along the axes (0XYZ). It is transformed by the rotation into the vector,
$V^-=R[V]=A_jV_j=a_iR_{ij}V_j$ ---------------(30).
So the components are $V_i^-=R_{ij}V_j$------------- (31).

These are obtained from the components of V by applying the matrix (R), since A_j forms a Cartesian system as shown in figure 1. It shows that the real matrix R is orthogonal and unimodular. It implies $R=R^*$, $R^\sim=R^{-1}$, det. $R=1$--------- (32).

Therefore, R_{ij} is real, i.e., $R=R^*$
It is orthogonal, i.e., $R^\sim=R^{-1}$
It is unimodular, i.e., det. $R=1$

The 0XYZ system is chosen once and for all. It implies that the matrix associated with the rotation is uniquely defined.

Conversely, to each real orthogonal, unimodular matrix, there is a corresponding one and only one rotation.

Equation 28 clearly shows the matrix elements associated with $R(\alpha, \beta, \gamma)$ are explicitly a function of Euler angles.

As an example, take the rotation of V; it demonstrates that

$$\left.\begin{array}{l} V_1^- = V_1\cos\alpha - V_2\sin\alpha \\ V_2^- = V_1\sin\alpha + V_2\cos\alpha \\ V_3^- = V_3 \end{array}\right\} \qquad \text{-------------- (33).}$$

So $V_i^- = R_{ij}V_j$ ------------------ (34).

Also, it is important to note that if $R=R_2R_1$ operationally indicates R_1 operates first, then R_2 acts. Now go back to figure 1. It can be shown that

$$A_i = \sum_j R_{ij}a_i = \sum_j C_{ij}a_j, \text{ where } C_{ij} = (A_i, a_j) = R_{ji} = \tilde{R_{ij}}$$

Where $A^\sim$ is the transpose to A.
Hence, V can be written as

$$V \equiv \sum_i V_i a_i \rightarrow V^- = \sum_i V_i A_i = \sum_{ij} V_i\, R_{ij}a_j \text{ ---------- (35).}$$

5.1.5. Rotation Operator in Hilbert Space

Hilbert space, as defined earlier, is a linear space of infinite (∞) dimension. The rotation of a physical system (like the hydrogen atom H) about its center of mass is equivalent to a rotation of the system (0XYZ) axes, without any perturbation. Thus, the transformation in Hilbert space is the wave function which describes the system is represented by $|\psi\rangle \longrightarrow |\psi^-\rangle$, and the transformation of an observable Q is $Q \longrightarrow Q^-$.
However, this type of transformation does not change the quantity of the observables, such as the expectationvalues,.i.e., $\langle\psi\,|Q|\psi\rangle = \langle\psi^-\,|Q^-|\psi^-\rangle$. The simplest way to make this occurs is to use a unitary operator (R), such that $RR^+=1$.

So $|\psi^-\rangle = R|\psi\rangle$; therefore, $\langle\psi|Q|\psi\rangle = \langle\psi|R^+Q^-R|\psi\rangle =$

$\langle\psi^-|Q^-|\psi^-\rangle \longrightarrow Q = R^+Q^- R$. Hence,

$[Q^- = RQR^+]$--------- (36).

Therefore, it can be concluded that if the system is rotated without any perturbation, there is a unitary operator in Hilbert space.

Important note: In quantum mechanics, it is needed to be more careful in defining the rotation of a physical system. This is due to the fact that the relation between the dynamical variables and the dynamical states is being less direct. As an example, suppose a is a dynamical state of a single particle, and $\psi(r)$ is the corresponding wave function. If this system is subjected to a rotation(R), where

$a \longrightarrow a^-$ and $|\psi(r)\rangle \longrightarrow |\psi^-\rangle$, $a^- \longrightarrow = R[a]$, $\psi^- = R[\psi(r)]$.

Now consider a position measurement: let $|\psi|^2 = P(r)$ and $|\psi^-(r)|^2 = P^-(r)$ be the probability distribution for the states a and a^- respectively. Now $|\psi^-(r)|^2 = |\psi(r_1)|^2$, $r_1 = R^{-1}r$. Also, for momentum space, $|\varphi^-(p)|^2 = |\varphi(p_1)|^2$, $p_1 = R^{-1}p$. The value of $\psi^-(r)$ at r equals to the value of ψ at r_1. This leads to $\psi(r)$ $= R[\psi(r_1)] = \psi(R^{-1}r)$--------------------------(37).

Equation 37 sets up one-to-one correspondence between ψ and $\psi(r_1)$, which is linear. So there exists an operator R such that it is a unitary operator such that $[\psi^- = R\psi]$.

In general, $\psi(r^1, r^2, ---r^n)$ is transforming by the rotation operator ($\mathbb{R}$) into:

$\mathbb{R}[\psi(r^1, r^2, r^3, -----r^n)] = \psi[\mathbb{R}^{-1}r^1, ------\mathbb{R}^{-1}r^n]$

$= \mathbb{R}\psi(r^1, r^2, ----r^n)$--(38).

Where $\mathbb{R}$, is a rotation operator, it is linear and unitary.
In general, this type of operator is usually associated with the rotation (R) of a given physical system. Therefore,

$$\mathbb{R}\mathbb{R}^+ = \mathbb{R}^+\mathbb{R} = 1, \; la^->=\mathbb{R} \, la> \text{------------------------ (39).}$$

From equation 39, the law of the density operator transformation, referring directly to its definition, can be deduced. Let ρ be the density operator of a certain (pure or mixed) state of the system, and ρ^- is the density corresponds to the state resulting out of the rotation.

Hence, $\rho^- = R(\rho) = \mathbb{R} \, \rho \mathbb{R}^+$ ----------------------------(40).

5.1.6. *Rotation of Observables*

In the preceding section, the law of transformation, which transforms the state vectors la>, are clearly defined. It is found necessary to find the transformation law of the observables, which are representing various measuring operations that can be performed on the physical systems.

Let Q be an observable, and Q^- is the transformed Q given by $Q^- = R[Q]$. So the average value of Q in state la> is [<Q>] and the average value of Q· in the state la·> is $<Q^->$ are [<Q>]$\equiv$ $<Q^->$, where $la^-> \equiv \mathbb{R}[la >]\}$.

Therefore, $[<a \; lQla>] = <a^- \; lQ^- \; la^->$, since $la^-> \equiv \mathbb{R}la >$
$<a^- \; l = <al\mathbb{R}^+$. This must hold for every state la>.
$<a \; lQla> = <a \; l\mathbb{R}^+Q^-l\mathbb{R}la>$. Remember, $<a^- \; l = <al\overleftarrow{\mathbb{R}^+}$.

Hence, $Q = \mathbb{R}^+ Q^- \mathbb{R} = Q^-$ ---------------------------- (41).

Conclusion: From the previous sections, the following remarks might be noted:

1. In the rotation R, the observables are transformed as the state vectors transform under the same unitary transformation.
2. If an observable S is scalar quantity, it is invariant under rotation, i.e., $S^- = \mathbb{R}\, S\, \mathbb{R}^+ = S$ ------------ (42).

Hence, equation 42 implies that the quantities which are invariant under rotation (R) are scalar or pseudoscalar. But the scalar quantity does not change sign under the reflection through the origin, while the pseudoscalar does. Now since $\mathbb{R}$ is a unitary operator, then $[\mathbb{R}, S] = 0$ ------------------ (43).

Equation 43 states that any observable which is invariant under rotation (R) commutes with the rotation operators.

Suppose $\mathbb{K}$ is given as a vector operator; its components are $k_i(\mathrm{k}.\, a_i)$. Apply the rotation R to the operator K_1, the component of $\mathbb{K}$ along OX. In general, $R\,(\mathrm{k}.\, a_i) = \mathrm{k}.\, a_i$, where $a^- \equiv R[a\,]$, thus $k_i^- \equiv R[\mathrm{k}_i] = \mathbb{K}.\ \mathrm{A_i} = \mathrm{K_j R_{ji}} \longrightarrow$
$k_i^- = \mathbb{R}\mathrm{k_i}\mathbb{R}^+ = R_{ij}^{\sim}\mathrm{k_i}$ --- (44).

Where $R^{\sim} = R^{-1}$ [transpose].
Note that the component k transforms in the rotation R, just like the transformation of a vector in the rotation.

It is quite different from the transformation represented by equation 34.

For more elaborations, consider the expectation value of $\mathbb{X}$ and how it is related to the expectation value of $\mathbb{X}^-$.

Given $<\psi|\mathbb{X}a_i|\,\psi> \,=\, <\psi^-|\mathbb{X}\ \mathrm{A_i}|\psi^-{>}$, $\longrightarrow$ a_i is a projection operator. This can be written as

$< \psi|R^+(\mathbb{X}Ai)R|\psi >$, *remember* $< \psi^-| =< \psi|R^{+\leftarrow}$,
$$\psi >= R\, l\, \psi >.$$

$\Rightarrow \mathbb{X}A_i = \sum_j \mathbb{X}R_{\tilde{ij}}\, a_j = \sum_j R_{\tilde{ij}}\, (\mathbb{X}a_j)$

As it is known, the transformation law is $X_i = \sum_j R_{\tilde{ij}}\, R^+\, X_j R$, but in classical mechanics, it is given by $X_i^- = \sum_j R_{ij} X_i$.

By inversion of the matrix, it can be found that $R^+ X_i R = \sum R_{ij} X_j = X_i^-$. Now the inversing with respect to R leads to $X_i = \sum R_{\tilde{ij}}\, R^+ X_j R$. Hence, $R X_i R^+ = \sum_{ij} R_{\tilde{ij}} R R^+ X_j X_i^- \equiv \sum R_{\tilde{ij}} X_j$; this is a vector operator transformation under the rotation R.

Now to show that $\psi^-(\chi) = \psi(R^{-1}\chi)$, consider $<\chi|\psi^-> = \psi^-(\chi) = <\chi|R|\psi> = <R^{-1}\chi|\psi>$. Hence,
$$\psi^-(\chi) = \psi(R^{-1}\chi), \text{ as required} \text{----------------- (45).}$$

For more illustration, consider the rotation in three dimensions (figure 3).

Using equation (34) i.e.,
$V_i^- = \sum_i R_{ij} V_j \quad ,\{R_{ij}\} = R.$
   Do it as excersice.

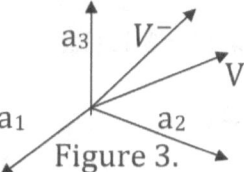

Figure 3.

5.1.7. Generators of Infinitesimal Rotation Operators

As it is physically known, the angular momentum plays a main role in the rotational phenomenon. Therefore, it is quite necessary to establish its basic relation with the infinitesimal rotation operators of the rotated system. To do that, consider a single particle referring to $\psi^-(r) \equiv R[\psi(r)]$ as transformed wave function under the effect of the rotation R. So the rotation $R_z(\alpha)$ of an angle (α) about $0Z$ (figure 2) transforms the wave function $\psi(x, y, z)$ into $R_z(\alpha)[(\psi x, y, z)]$ which (referring to equation 33) is given by

$R_z(\alpha)[\psi(x, y, z)] = \psi$ (x cosα +y sinα-x sinα+y cosα, z)---- (46).

Consider the angle (α) is very small, such that sinα $=\alpha=\epsilon$. Where $\epsilon << 1$, it is an infinitesimal quantity, so the rotation
$\qquad R_z(\alpha) \longrightarrow R_z(\epsilon)$.

Under this limitation, the Tylor expansion can be used to the right hand side of the equation 46 about the point (x, y, z), neglecting the higher orders in ε); and then after mathematical manipulation, R_z is found to be

$R_z(\epsilon) = (1-i\epsilon L_z)\psi(x, y, z)$, $L_z = y\frac{\partial}{\partial x} - x\frac{\partial}{\partial y}$, in $\hbar$ unit. (Show this as an exercise.) This represents the infinitesimal rotation operator.

In general, the infinitesimal rotation about(μ) gives the operator of rotation as

$\qquad \mathbb{R}_\mu(\epsilon) = 1-i\mu(L.\mu)$ ---------------(47).

where L is the total angular momentum of the system. It is too important to be recognized that, when the system has no classical analogy, this fundamental relation serves as a definition of the total angular momentum. If the total angular momentum of the system is J, its component along any axis (μ) is related to the operator of the infinitesimal rotation about this axis by the relation

$\qquad \mathbb{R}(\epsilon) \approx 1-i\epsilon(J.\mu)$ --------------------- (48).

Since $(J.\mu)$ represents the component of $(J^{\rightarrow})$ along μ, it leads to the conclusion that to each infinitesimal rotation $\mathbb{R}(\epsilon)$, there corresponds one and only one infinitesimal rotation given by $\mathbb{R}(\epsilon)$. This implies that the operation of rotation given by

$(R_\mu(\epsilon)=R_x(\epsilon\mu_x)R_y(\epsilon\mu_y)R_z(\epsilon\mu_z)\approx 1-i\epsilon(\mu_x J_x+\mu_y J_y+\mu_z J_z))$. Also, it was shown that (equation 43) any scalar operator commutes with the component of $J^\rightarrow$, so $[(\mu.J^\rightarrow), S] = 0$------------------(49).

The rotation shown in equation 48 also provides the commutation relation rules for the components of J with those of a general vector operator $\mathbb{K}$. Let $K_a=\mathbb{K}.a=$
a component along a given unit vector a, so k_a^- can be written as $[k_a^-]=R_\mu(\epsilon)k_a R_\mu^+(\epsilon) \approx k_a-i\epsilon[J_\mu\,.k_a]$.

It is shown before that $a^- \cong a + \epsilon(\mu \times a)$, $\epsilon<<1$, then
$k_a^-=k.a^- \cong k.[a + \epsilon(\mu \times a)]$; it leads to
$[(\mu.J), (a.k)]=i\{(\mu \times a)\,.K\}$ -------------------(50).

Let $K=J\rightarrow[(\mu\,.J), (\,a\,.J)]=i[(\mu\times a\,).J]$, which gives the commutation relation characteristic of the angular momentum represented by $[a\,.J, b\,.J]=i[(a\times b).J]$.

Hence, the rotation operator $\mathbb{R}$ is given by
$\mathbb{R} =1-i\epsilon\,/\hbar(J.\,\mu)=1-i\epsilon\,(J.\mu)$, $\hbar=1$, $\epsilon<<1$; μ is a unit vector.
$\mathbb{R}^+ =1+i\epsilon\,(J^+.\mu)$.

Now is J a Hermitian operator? The answer is yes.
The proof: $\mathbb{R}\mathbb{R}^+ =1=(1-i\epsilon((\,J.\mu))(1+i\epsilon(J.\mu))=$
$1+i[(J^+-\,J).\mu]+$(high terms in $\epsilon=0$).$]=1+i\epsilon[(\,J^+-J.\mu\,]$. Hence, $1=1+i\epsilon[(\,J^+_-\,J).\mu]\Rightarrow i\epsilon[(J^+-J).\mu]=0$, since $i\epsilon \neq 0$ and $\mu \neq 0$; therefore, $(J^+-J)=0$, so J is a Hermitian operator. Its eigenvalue is real given by $j(j+1)\hbar\geq 0$. Also, the total angular momentum is the generator of the infinitesimal rotation. As shown before, the commutation relation between its components is given by $[J_1, J_2]=iJ_3$, $[J_2, J_3]=iJ_1$, $[J_3, J_1]=iJ_2$.

Take an example the hydrogen atom H, where its rotation is symmetric, (the energy does not change), i.e., the Hamiltonian is the same. Therefore, $<\psi^-|H|\psi^->=<\psi| H |\psi>$, $\psi \rightarrow \psi^-$, under rotation. This indicates that $H=\mathbb{R}^+H\mathbb{R}$ which implies that

$\mathbb{R}H = H\mathbb{R}$, remember, $\mathbb{R}^+ = \mathbb{R}^{-1}$. Then $[\mathbb{R},H] = 0$. So the components of J commute with H, i.e.,

$[J^\rightarrow, H] = 0$, $\Rightarrow J$ is a constant of the motion. Hence, the only physical meaning of J is it is the angular momentum of the system.

5.1.8. *Construction of the Operator Transformation* $\mathbb{R}(\alpha, \beta, \gamma)$

From what so far has been said about the infinitesimal rotation $\mathbb{R}$, it can be restated that any finite rotation might be looked up as a succession of infinitesimal rotation. The rotation operator $(\mathbb{R})$ is the product of the corresponding infinitesimal rotation operators. From equation 48,

$(\mathbb{R}_\mu) = 1 - i\epsilon \ (J.\mu)$, it is well recognized that the infinitesimal rotation operator is a function of J. From figure 2, the rotation $(R_\mu(\varphi))$ is a succession of infinitesimal rotations about μ–axis, particularly $R(\varphi + d\varphi) = R_\mu \ (d\varphi) R_\mu(\varphi)$.

Using equation 48 and let $J_\mu = (J.\mu)$, a component of J, where $\mu = $ i, j, k, x, y, z. Accordingly, for $d\varphi << 1$, $\mathbb{R}_\mu \ (\varphi + d\varphi) = [1 - iJ_\mu$ $d\varphi \ ]\mathbb{R}_\mu \ (\varphi) \Rightarrow \frac{d}{d\varphi}\mathbb{R}_\mu(\varphi) = -i(J_\mu \mathbb{R}_\mu \ (\varphi))$.

The solution is $\lim(\mathbb{R}(\varphi) = -iJ_\mu \varphi + c$, as $\varphi \to 0$, $c \longrightarrow 0$, then $[\mathbb{R}_\mu(\varphi) = e^{-i\varphi J_\mu}]$ -------------------------------- (51).

The rotation $R(\alpha, \beta, \gamma)$ found by Euler angles about $0z$, 0μ, and $0Z$ axes respectively gives the rotation operator $\mathbb{R}(\alpha, \beta, \gamma) = (\mathbb{R}_z(\gamma) \ \mathbb{R}_\mu(\beta)\mathbb{R}_z(\alpha)$. Using equation 51, the rotation operator takes the form

$\mathbb{R}(\alpha, \beta, \gamma) = e^{-i\gamma Jz}e^{-i\beta J_\mu}e^{-i\alpha Jz}$ ------------------- (52).

It should be noted that the operation of the rotation operator (52) takes place in order, right to left. Let 51 be formally generalized, where only the components of the angular momentum along the coordinate axes appear. Using the law of transformation of the operators, i.e., $Q=\mathbb{R}^+Q^-\mathbb{R}$ or

$$Q^-=\mathbb{R}Q\mathbb{R}^+ \Rightarrow J_\mu = \mathbb{R}_z(\alpha) J_y \mathbb{R}^+(\alpha)=e^{-i\alpha J_z}J_y e^{+i\alpha J_z}.$$

Thus, $e^{-i\beta J_\mu}=e^{-i\alpha J_z}e^{-i\beta J_y}e^{+i\alpha J_z}$.

By substituting this into 51, the operator $\mathbb{R}(\alpha,\beta,\gamma)$ takes the form $\mathbb{R}(\alpha,\beta,\gamma)=e^{-i\gamma J_z}e^{-i\alpha J_z}e^{-i\beta J_y}$.

By successive applications of the rotation, $R_z(\alpha)$ and $R_\mu(\beta)$ can be obtained from J_z by eliminating it just like what has been done with J_μ above. Therefore, $(\mathbb{R})$ will take the form

$$\mathbb{R}(\alpha\beta\gamma)=e^{-i\alpha J_z}e^{-i\beta J_y}e^{-i\gamma J_z}=e^{-i[\alpha+\gamma]J_z}e^{-i\beta J_y} \text{----------- (53).}$$

Consider a rotation through an angle of 2π and a half-integer angular momentum, using equation 51, $\mathbb{R}(2\pi)=e^{-i2\pi J_\mu}$. It might be noticed that the rotation through the angle (2π) about the axis brings the system back to the starting point. But $\mathbb{R}$ is not necessary $=1$. It is a matrix diagonal in a matrix representation in which its diagonal elements are +1 or -1. It depends on whether the eigenvalue of J_μ is integer or half integer.

Consider D as a function of J^2 with eigenvalue +1 for j integer and -1 for j half-integer, remembering that the eigenvalue of J^2 is $j(j+1)\hbar$. Where D is an observable with the following features:

1. $1/2(1+D)$ is the projection onto the subspace of integer j.
2. $1/2(1-D)$ is the projection onto the subspace of half-integer j.
3. $D^2=1$.

4. D commutes with all the rotation operator, i.e.,

$$\left.\begin{array}{l} [D, \mathbb{R}]=0 \\ \mathbb{R}(2\pi)=D \end{array}\right\} \qquad \text{----------------------------- (54)}.$$

So to have $2\pi=1$, J increases to take the integral value only. $\mathbb{R}(4\pi)=D^2=1 \Rightarrow D=\pm1$.

5.1. *Spin, Hypothesis, and Applications*

Quantum mechanics was invented and developed in 1926–1927 by Schrödinger based on the duality property of moving matter and proposed by de- Broglie in 1924 and by Heisenberg based on the mathematical matrix theory. Also, it is based on the correspondence principle proposed by Niles Bohr in 1915–1916, which states that the quantum physics is the general theory of physics where the classical physics is the limit when $h \longrightarrow 0$ and $n \rightarrow \infty$.

Schrödinger's wave quantum mechanics tackled the wave motion for many simple natural problems. But for the complex systems, like the complex atoms, it failed to describe them well, even for the nonrelativistic cases. Therefore, there are two important modifications to be taken in consideration.

These modifications are intrinsic characters in quantum physics and have no analogue in classical physics. They are

1. Pauli Exclusion Principle (PEP). It is retaining only those solutions of Schroödinger equation that have certain well-defined symmetrical properties in permutation of the electrons coordinates. This restricts the distribution of the electrons in the atomic orbits, which then is generalized to all particles with half-integer angular momentum. These particles are called fermions after the name of the well-known scientist Fermi.

114

2. Hypothesis of the electron spin. This hypothesis was proposed by Stern-Gerlach because of their noticing the complex behaviors of the atoms in the magnetic field (Zeeman effect), where the splitting of the energy levels occurred.

As it is pointed before, Schrödinger equation is developed for the microscopic systems (atomic and subatomic) at the time where there was no spin concept yet (1927).

So Schrödinger equation is written as $H_0\psi_n = E_n\psi_n$.

The Hamiltonian here is considered for the nucleus of the atom because most of the mass (99 percent) in fact is concentrated into the nucleus. According to this fact, the position will coincide with the center of the mass. Hence, the Hamiltonian H is given by

$$H_0 = \sum_{i=1}^{z} \frac{P_i^2}{2m} - \frac{ze^2}{r_i} + \sum_{i<j} \frac{e^2}{|r_i - r_j|} \text{------} \text{---------}(55).$$

From the electromagnetic field, if a magnetic field (static) described by the potential A(r) is applied on a certain atom, the momentum P_i will take the form $P_i \rightarrow p_i - \frac{eA}{c}$.

For a constant magnetic field, A is given by the form $A(r) = 1/2(\mathcal{H} \times r)$ and $(p-eA/c)^2 = p^2 - e/c[(A.p)+(p.A)] + \frac{e^2}{c^2}A^2 = p^2 - \frac{e}{c}(\mathcal{H}.L) + \frac{e^2}{4c^2}\mathcal{H}^2 r_\perp^2$, where $r_\perp$ is the projection of r on the plane to the field.

So H_0 can be written as

$$H_0 = \frac{-e}{2mc}\mathcal{H}.L + \frac{e^2}{8mc^2}\sum_i^Z r_i^2 \perp j + H \text{---------} (56)$$

115

Or $H = H_0 - \dfrac{e}{2mc}(\mathcal{H}.L) +$ neglected high order terms(very small quantities). $H = H_0 - \dfrac{e}{2mc}\mathcal{H}.L)$ ------------- (57)

$L = \sum i(r_i \times p_i) =$ total angular momentum of Z-electrons.

It is quite well known that any charged particle circulating an orbit will induce a magnetic moment (m).

It is related to its orbital angular momentum as
$\overrightarrow{\mathrm{m}} = \dfrac{e}{2mc}\overrightarrow{\ell}$, where e is the charge, and m is the mass of the circulating charged particle. Also, $\overrightarrow{\mathrm{m}}$ can be written as $\mathrm{m} \propto \ell$ $= g_m\ell$, g_m is the gyromagnetic ratio. This is for one electron. For an atom of Z-electron, it is given by
$\overrightarrow{\mathrm{m}} = \dfrac{e}{2mc}\sum_i^Z \overrightarrow{\ell_i} = \dfrac{e}{2mc}L$. L is the total angular momentum of the atom. By substituting $\overrightarrow{\mathrm{m}}$ in 57, it is obtained that $H = H_0 - \mathrm{m}.\mathcal{H}$; hence, Schrödinger equation is

$H\psi = [H_0 - \mathrm{m}.\mathcal{H}]\psi$ -------------------------- (58).

If the magnetic field is directed along Z-axis, then the wave function is a common eigenfunction to H, L^2 and L_Z. This eigenfunction can be represented by lnLM>.

Hence, equation 58 takes the form

H lnLM> $= H_0$ lnLM> $- \dfrac{e}{2m_e c} L_Z.\mathcal{H}$ lnLM> $\Rightarrow$

$E^{nLM} = E_{nLM} = E_0^{nL} - M\mu_B\mathcal{H}$ ---------------- (59).

$\mu_B = \dfrac{e\hbar}{2m_e c}$, Bohr magneton.

E_0^{nL} is the ground state energy (free).

The term $(-M\mu_B\mathcal{H})$ is due to the perturbation by the applied magnetic field ($\mathcal{H}$), where M is the eigenvalue of L_Z, which takes all integral values from $-L$ to $+L$, i.e., it has $2L+1$ values.

Accordingly, each level is split under the effect of the magnetic field ($\mathcal{H}$) into 2L+1 level, each level of the atomic spectrum is split by the applied magnetic field into a multiplet of 2L+1 equidistant levels.

These levels are distributed on either side of [E_0^{nL} level].
In such a way, their distances from 0 level average to zero. The distance between two neighboring levels is a quantity represented by ($\mu_B \mathcal{H}$). It is independent of the atom which is considered. It is proportional to the applied magnetic field ($\mathcal{H}$).

These predictions are only partially and experimentally confirmed. The two important discrepancies are

In odd Z atom, all multiples are even. It indicates that the total angular momentum (L) is half integer.

The distance between neighbors in the same multiplet is found to be ($g\mu_B \mathcal{H}$), where (g) is the Lande factor. It is varying from one multiplet to another, within the large limits.

As an example, take L=3 (f-state) where the values of M given by 2L+1=7 represented by -3, -2, -1, 0, +3, +2, +1.
Diagramically, it can be represented by the figure 4.

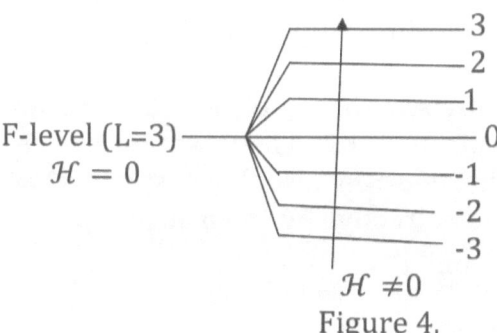

Figure 4,

Hence, the existence of the half-integer angular momentum is experimentally confirmed (Stern-Gerlach experiment).

As the atoms are building up the beam, all of them are almost in the ground states. The numbers of the spots on the screen are actually representing the multiplicity of the ground states. For example, the silver (Au) atoms are observed in both two spots, which means the ground state of the silver atom has angular momentum of 1/2 (in $\hbar$ unit). In general, the atoms of odd Z give even number of spots. This is due to the characteristic of the half-integer angular momentum. These points are observed simultaneously in the study of Zeeman effect (anomalous). Therefore, the spectrum results, in general, permit simultaneously the determination of the multiplicity of the states, where between them, the optical transitions are affected and also their respective Lande factor.

To face these difficulties, it is necessary to introduce half-integer angular momentum and gyromagnetic ratios differ from $e/2m_e$. This was faced by the finding experimentally by Uhlenbeck and Goudsmit in 1925 that the electron has an intrinsic spin denoted by S or angular momentum $L_e=S=(1/2)\hbar$. With this spin, S is associated a magnetic moment $\vec{\mu_s}$ which is given by $\vec{\mu_s}=g_s\dfrac{e}{2mc}\vec{S}$ ------------- (60).

The g_s is an adjustable constant. To match the theory with the experiment, the proposed value 2 gives an excellent matching. In 1932, Dirac used the relativistic quantum mechanics to study the electron spin hypotheses.

This Dirac relativistic theory of the electron spin gives a very good explanation to $g_s=2$. Then experiment showed that the proton and the neutron also have spin $S=1/2$. This can be easily revealed directly by measuring μ_p and μ_n, where $\mu_p=g_p\dfrac{e}{2m_p c}S_p$, and $\mu_n=g_n\dfrac{e}{2m_n c}S_n$.

The experiment gives $\mu_p=5.59$ and $\mu_n=-3.83$; in nuclear magneton ($e/2m_p c$), m_p is the proton mass.

5.3. Nonrelativistic Theory of Particles with Spin ½ (Pauli theory)

5.3.1. Spin 1/2 and Pauli Matrices

S is an intrinsic angular momentum (spin vector) of a particle with a spin $1/2$ in $\hbar$ unit. Analogue to the orbital angular momentum L, S^2 has the eigenvalue $s(s+1)\hbar=1/2(1/2+1)=3/4$ in $\hbar$ unit. The Z-component of S is S_z has the eigenvalue $\pm 1/2$, spinning up or down respectively. It is supposed to be nondegenerate. This implies that the components of S are operators acting in a two-dimensional space, in which a possible basis is contributed by the two eigenvectors $l+>=l1/2,1/2>$, and $l->=l1/2, -1/2> \equiv lS, S_z>$. It is quite clear that $l+>$ and $l->$ are eigenvectors of S^2, S_z. Also, the components of S are S_X, S_Y, and S_Z, where $S^2=S_x^2+S_y^2+S_z^2=(3/4)\hbar$.

It is possible, as in J's case, to construct the operators S_+ and S_- such as $S_+=S_X+iS_Y$, $S_-=S_X-iS_Y$. It is possible to show that $S_+^2 +S_-^2=0$, since $S_+^2 =S_x^2 -S_y^2 +i(S_XS_Y+S_YS_X)=i/4$, then $(S_XS_Y+S_YS_X)\neq 0$; hence, these operators are anticommute.

Now introduce the Pauli matrices $\sigma \equiv (\sigma_x, \sigma_y, \sigma_z)$. The physical relation between the spin and the Pauli matrices is
$S=(1/2)\sigma$ --------------------------------(61).

Note: In all formulation, $\hbar$ is considered one.
The Pauli matrix components are given by

$$\sigma_x = \begin{bmatrix} 0 & 1 \\ 1 & 0 \end{bmatrix}, \sigma_y= \begin{bmatrix} 0 & -i \\ i & 0 \end{bmatrix}, \sigma_z = \begin{bmatrix} 1 & 0 \\ 0 & -1 \end{bmatrix} \quad ----(62).$$

As exercise show that

$[\sigma_x^2 =\sigma_y^2 = \sigma_z^2 =1, \sigma_x\sigma_y - \sigma_y\sigma_x =i\sigma_z,$
$\sigma_y\sigma_z - \sigma_z\sigma_y =i\,\sigma_x, \sigma_z\sigma_x-\sigma_x\sigma_z=i\sigma_y.$

And $\sigma_x \sigma_y \sigma_z = i$, $T_r\sigma_x = T_r\sigma_y = T_r\sigma_z = 0$

Also, det. σ_x =det. σ_y =det. σ_z =1

From these, it can be deduced a general identity for the vectors A and B, where they are any vectors.

$$(\sigma . A)(\sigma . B) = (A.B) + i\sigma.(A \times B) \text{--------------} (63).$$

However, for two vector operators, their components commute with σ components, the order of A and B in both sides of equation 63 to be respected. As an example, take the position r and the momentum p using equation 63. It will be
$(\sigma . r)(\sigma . p) \equiv (r.p) + i\sigma.(r \times p)$ -------------------- (64).

Since S is considered as angular momentum, so the rotation operator $\mathbb{R}_\mu^s(\varphi)$ is affecting the transformation of the vectors of the $(S^\rightarrow)$ space in the rotation $R_\mu(\vartheta)$, such as

$$\mathbb{R}_\mu^s(\varphi) = e^{-is\varphi} = e^{-i\frac{\varphi\sigma_\mu}{2}} \text{--------------------} (65),$$

where $\sigma_\mu = (\mu . \sigma)$—components, $\mu = 1, 2, 3 \equiv$ x,y,z.

By expanding the exponential and separately summing the terms as even and odd in σ_μ, using the relations $\sigma_\mu^{2p} = 1$ and
$\mathbb{R}^s(\varphi) = \cos\frac{\varphi}{2} - i \sigma_\mu \sin\frac{\varphi}{2}$ ------------------(66).

Therefore, $\mathbb{R}^s(2\pi) = -1$.
Now the operator representing the rotation $R(\alpha, \beta, \gamma)$ is

$$\mathbb{R}^s(\alpha, \beta, \gamma) = e^{-\frac{i\alpha\sigma_z}{2}} e^{\frac{-i\beta\sigma_y}{2}} e^{\frac{-i\gamma\sigma_z}{2}} \text{----------} (67).$$

The spin space is of two dimensions; it is called spinor.
The orbital motion and the spin motion independently occur, so their orbital dynamical variables [position distance (r) and

the momentum (p)] and the intrinsic (spin) dynamical variables are commuting for r and p, $[r_i, p_j]=i\delta_{ij}$, in $\hbar$ unit.

Also, as for r and p, it is known that $[S_i, S_j]=i\varepsilon_{ij}S_k$, $S^2=3/4$.

Therefore, the space vectors ($\mathcal{E}$) of the particle is given by the tensor product of the orbital space ($\mathcal{E}^0$) and the spinor (because they are independent of each other).

So $\mathcal{E}=\mathcal{E}^0\otimes\mathcal{E}^s$. To build up the vectors ($\mathcal{E}$), it is important to choose the representation with r and S_z diagonal. Hence, *each vector l ψ > is represented by the wave function.*

$\psi(r,\mu)$ =<rμl ψ> -- (68).

It is a function of the continuous variables r (x, y, z) and the discrete variables (μ) (related to the eigenvalues of S_z which takes the values $\pm1/2$, accordingly. The total angular momentum of the particle is given by $\vec{J}=\vec{\ell}+\vec{s}$ ----- (69).

The fundamental variables of this physical systems (particles), are: x, y, z, p_x, p_y, p_z, S_x, S_y, S_z, and J as given in equation 68.

Now since L and S commute, it can be deduced from equation 53 that the rotation operator is given by

$\mathbb{R}(\alpha,\beta,\gamma)=\mathbb{R}^s(\alpha,\beta,\gamma)\mathbb{R}^o(\alpha,\beta,\gamma)$ ----- ----------- (70).

(S for spin, o for orbit). It can be shown that the rotation through 2π gives $\mathbb{R}^o=1$ and $\mathbb{R}^s=-1$, all kets l> change sign in such rotation. However, all the fundamental variables are invariant under a rotation through angle 2π. Therefore, it is convenient to write $[\psi(r, \pm1/2)]=\psi_{\pm}(r)$, and $\psi(r,\mu)$ can be written as two components form, such as $\psi \equiv \begin{bmatrix} \psi_+(r) \\ \psi_-(r) \end{bmatrix}$

This leads to the statement that for each value of r, ψ represents a ket vector of the space $\mathcal{E}^s$, namely

$<rl\psi> \equiv \psi_+(r)l+>+\psi_-(r)l->$ ------------------------(71).

The wave function $<rl\psi>$ may be regarded as spinor field, where in the rotation $R(\alpha,\beta,\gamma)$ the spinor field ψ transforms into $R[\psi]=\mathbb{R}\psi=\mathbb{R}_i^{1/2}\begin{bmatrix}\psi_+(R^{-1}r)\\\psi_-(R^{-1}r)\end{bmatrix}$

 It is a direct result of equation 65.

$\mathbb{R}^{\frac{1}{2}}$ is the rotation matrix corresponding to J=1/2. This transformation might be compared with the law of transformation of scalar field, i.e., $\psi^-(r)$ =R[$\psi(r)$]=$\psi(R^{-1}r)$.

This will be clarified in the next section. So far, this was for a single particle (system). It can be extended for a system of Z-particle of spin 1/2. This can be tackled straight forward such as, $\mathcal{E}^z=\mathcal{E}^1\otimes\mathcal{E}^2\otimes\mathcal{E}^3----\mathcal{E}^{z-1}\otimes\mathcal{E}^z$------------------ (72).

Equation 72 represents the space of the Z-system. It has dimensions which are a tensor product of the z individual spin spaces. Accordingly, the total spin is given by

$S^\rightarrow=1/2\sum_{i=1}^z\sigma_i$ -- (73)

The overall rotation through (2π) of the spins is represented by the operator$(-)^z$.

5.3.3. Comparison between Particle of Spin 1/2 and Vector Fields

Let A(r) be a vector field associated with a physical system, like a magnetic, electric field, or a wave function of particle of spin 1/2. For a comparison between these vector fields and a particle of spin 1/2, first is to see how A(r) field transforms under a rotation such as R[A], where R represents the rotation as a physical phenomenon.

So $A^-(r_1)$ is a field at the point (r_1). It can be obtained by applying the rotation R to the vector A (r_1), which represents the field at the point $r_1 \equiv R^{-1}r$.

Therefore, $A_i^-(r)=R_{ij}A_j(R^{-1}r)$, (i=x, y, z). Going back to equation 66, it can be shown that

1. The rotation through angle (α) about 0z leads to:

$A^- =\mathbb{R}_z(\alpha)[A]$, r₁=(x cosα +y sinα −x sinα+y cosα, z) ⟶

$$\left.\begin{array}{l} A_x^-(r)=A_x(r_1)\cos\alpha-A_y(r_1)\sin\alpha \\ A_y^-(r)=A_x(r_1)\sin\alpha+(A_y(r_1)\cos\alpha \\ A_z^-(r)=A_z(r) \end{array}\right\} \text{-------(74)}.$$

2. In particular, the infinitesimal rotation about 0z gives:

$$\mathbb{R}_z(\epsilon\,)[A]=1\text{-i}\,\epsilon(\,L_z+S_z)]\,A \quad \text{----------------(75)}.$$

Where L_z is the Z-component of L, it is a differential operator, i.e., $L_z=i\frac{\partial}{\partial\varphi}$, where φ is the azimuthal angle, L_z in $\hbar$ unit. S_z is an operator defined as

$$S_z \begin{bmatrix} A_x(r) \\ A_y(r) \\ A_z(r) \end{bmatrix} = \begin{bmatrix} -iA_y \\ iA_x \\ 0 \end{bmatrix}$$

This indicates clearly that S_z transforms each component of the field A(r) at a given point into a particular linear combination of the three components of the field at the same point. The vector field A(r) is defined by its Cartesian components, A_X, A_Y, A_Z, and S is represented by the following matrices:

$$S_X = \begin{pmatrix} 0 & 0 & 0 \\ 0 & 0 & -i \\ 0 & i & 0 \end{pmatrix}, \; S_y = \begin{pmatrix} 0 & 0 & i \\ 0 & 0 & 0 \\ -i & 0 & 0 \end{pmatrix}, \; S_z = \begin{pmatrix} 0 & -i & 0 \\ i & 0 & 0 \\ 0 & 0 & 0 \end{pmatrix} \text{--- (76)}$$

(As an exercise, show that S_x, S_y, and S_z verify the commutation relations, which are characteristic of the S components. Also, show that $S^2=2$ corresponds to s=1.)

As it is stated earlier, by definition, S is an intrinsic angular momentum or the spin of the vector field A(r). This vector field describes a particle of spin one, just like the photon.

$A_i(r)$ can be represented such as A(r, i), i=x, y, z. Therefore, A(r, i) is a wave function depending not only on the position variables but also depending on an index (i) which might take three values, which constitute internal variable that are describing the orientation of the particle. The scalar product of the two wave functions of this type is given by

<A, B>=$\sum_i \int B^+(r, i)A(r, i)dr = \int (B^+ . A) \, dr$ ------------(77).

A and B are considered as those two wave functions. Since the orbital motion and the intrinsic motion are independent of each other, so [L, S]=0, they commute, hence they have a common wave function. Also, the infinitesimal rotation operator is $\mathbb{R}_z(\epsilon)) = [1-i\epsilon(L_z+S_z)]$ as shown before. Now $A(r,\mu)=A_\mu(r)$, (μ=+1,0,-1). From the definition given by equation 66, it can be written as:

<r|A>=$A_+(r)$|+>+$A_0(r)$|0>+$A_-(r)$|->, which leads to:

$A_+ = \frac{1}{\sqrt{2}} (A_x - iA_y)$

$A_0 = A_z$ ---------------------------------- (78)

$A_- = \frac{1}{\sqrt{2}} (A_x + iA_y)$

5.3.4. Inclusion of the Spin in the Nuclear Reactions (As Example)

The simplest example of a nuclear reaction is the nucleon-nucleon reaction, i.e., p.p., n.n., and n.p. collision or interaction. This type of interaction is quite important in nuclear physics in general, where a lot of information about the nuclear force features is expected to be obtained. Let figure 5 illustrates the coordinate's structure of the nucleus.

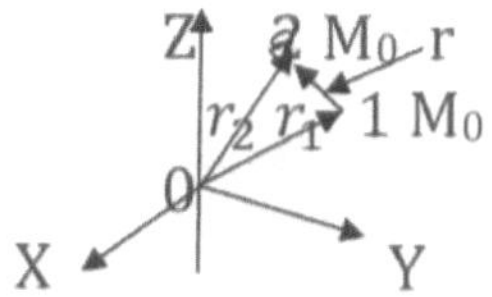

Figure 5

Let M_0 be the mass of the nucleon (p or n) and $\vec{r} = \vec{r_1} - \vec{r_2}$ is the relative distance between the nucleons 1 and 2 (figure 5).

$P = 1/2 (p_1 - p_2)$ is the relative momentum.
$S_1 = (1/2)\sigma_1$ is the spin of the nucleon 1.
$S_2 = (1/2)\sigma_2$ is the spin of the nucleon 2.

As it is well known, there are, for such system, two kinds of motion: the center of the mass motion and the relative motion.

Physically, it is possible to separate between them easily. The total orbital angular momentum $\vec{L}$ is $\vec{L} = \vec{r} \times \vec{p}$, and the total spin S is $S = (1/2)(\sigma_1 + \sigma_2)$, where σ_i is Pauli matrices. So the total angular momentum of the system (two nucleons) is given by $\vec{J} = \vec{L} + \vec{S}$; the Hamiltonian H is given by $H = P^2/2M_u + V(r)$, where M_u is the reduced mass $= M_0/2$. All these are for the relative motion.

Still a complete information about nuclear force is not available. It is now not the force combining the nucleons inside the nucleus only, but there is the strongest one combining the quarks and the gluons inside the nucleons, which is treated by new type of quantum mechanics, the so-called quantum chromo dynamic (QCD). So in fact, the nuclear force combining the nucleons is the remaining of this very strong force. Therefore, the potential due to this force is not well defined but different proposed potentials are considered such as

$V_1(r)$ --- central potential ------------------ (79a).

$V_2(r)(\sigma_1.\sigma_2)$ ------- including --------------- (79b).

$V_3(r)(L\,S)$ ---------- spin ----------- (79c).

$V_4(r)[3\frac{(\sigma_1.r)(\sigma_2.r)}{r^2} -\sigma_1.\sigma_2]$ effect ----- ----------- (80d).

It is quite good to notice that equations 79b and 80d are written in their traditional forms. In fact, they can be expressed in different way as follows:

Take $S=1/2(\sigma_1 +\sigma_2)$, $S=S_1+S_2$, $S_i=(1/2) \sigma_i$,
$S^2=1/4(\sigma_1^2 +\sigma_2^2 +2\sigma_1 .\sigma_2)$, also $\sigma_1^2 =\sigma_2^2=3$.

Therefore, $S^2=1/4[3+3+2\sigma_1 .\sigma_2]=(1/2)[3+\sigma_1.\sigma_2]$,
or $\sigma_1. \sigma_2=2S^2-3$. Notice in equations 79b and 79a.

Now $\vec{J}=\vec{L} +\vec{S}$, so $J^2=L^2+S^2+2\vec{L} .\vec{S}$.
Hence, $\vec{L} .\vec{S}=1/2[J^2-L^2-S^2]$ -------------------(81).

Now go back to $S=(1/2) (\sigma_1 +\sigma_2)$; remember, $S=S_1+S_2$.
Now $(S)=(1/2)[(\sigma_1)+(\sigma_2)]$; therefore,
$(S.r)^2=1/4[(\sigma_1. r) +(\sigma_2, r)]^2=1/2[(\sigma_1.r)(\sigma_2.r)+r^2]$ $\longrightarrow$
$(\sigma_1 . r)(\sigma_2 .r)=2(S.r)^2 -r^2$

(Show this as an exercise.) Then it is obtained that
$S_{12}= 2 [3\frac{(S.r)^2}{r^2} -S^2]$ ----------------------------(82).

Where S_{12} is denoted as a tensor operator. The interaction (80) is of tensor potential type or a tensor force (F=-∇V). If the potential V(r) is a linear combination of all (79) parts, it indicates that the Hamiltonian of the system is invariant under rotation and reflection (by reflection r ➝ -r,p ➝ -p, but $S_{Op.}$ unchanged. Why?)

The following remarks are quite useful, with the notice that the interactions (79) are arranged in the order of decreasing symmetry, implying that

1. V(r) is independent of the spin; its potential shape might take the square well—Gaussian, Yukawa exponential, and the hard-core shape.

2. $V_2(r)(\sigma_1.\sigma_2)$ commutes with L and S separately; it implies that it is invariant under all rotations, orbital and spin.

3. If $V=V_1(r)+V_2(r)(\sigma_1.\sigma_2)$, the eigenfunction of the Hamiltonian (H) might be looked for among the common eigenfunctions of L^2, S^2, L_z and S_z. The corresponding eigenvalues have a rotational degeneracy of the order $(2L+1)(2S+1)$.

4. If $V=(1)+(2)+V_3(LS)$, H still does commute with L^2 and S^2 but not separately invariant under the rotation of the orbital and the spin. The eigenfunction of H is among the common eigenfunction of L^2, S^2, J^2, and J_z, but its eigenvalues have a rotational degeneracy of order $(2J+1)$ only.

5. If $V(r)=V_4[3(\frac{(\sigma_1.r)(\sigma_2.r)}{r^2}-\sigma_1.\sigma_2]$, it is the least symmetrical potential among the others four potentials. It commutes with S^2. From equation 79d, it can be shown that $[S^2, S_{12}]=0$. (Prove it.) But $[L^2, S_{12}]\neq0$. (Why?) For this potential, the eigenfunction of H may be sought among the common eigenfunction of the parity, S^2, J^2, and J_z. The parity operator operates (π_o) on the wave function as $\pi_o \psi=\pm\psi(-r)$, += even, -=odd, $[H, \pi_o]=0$.

As it was mentioned, the nuclear force is not very well defined, but its general features are deduced from the nuclear interactions experiments results. The higher energy used, the more accurate information are obtained, and more understanding of the nature of this force is to be acquainted with. This is why there are several proposed potentials. So the potentials (V_1, V_2, V_3, V_4) proposed to take in consideration all possible interactions (79a and 80d). The Hamiltonian H takes the form $H=V_1(r)+V_2(r)(\sigma_1.\sigma_2)+V_3(r)(L.S)+V_4(r)S_{12}$. Consider the system that is shown below:

σ_1 o———————— r ————————o σ_2, σ_i is Pauli matrices,

where $J=J_1+J_2$, $|JM>=|j_1m_1>|j_2m_2>$, and $J=j_1+j_2$, j_1+j_2-1---$|$ j_1 -$j_2|$. And $M=(+J)$, $+J-1$, $+J-2$,-- $(-J)$, M takes $(2J+1)$ value.

The dimension is given by $\sum_{J=j_1-j_2}^{j_1+j_2}(2J+1)=(2j_1+1)(2j_2+1)$.

The eigenfunction of the system is the linear combination of the eigenfunctions of particle 1 and particle 2. It takes the form $|JM>=\sum_{M=m_1+m_2} C_{m_1m_2}|j_1m_1>|j_2m_2>$, where $C_{m_1m_2}$ is the known Clebch-Gordon coefficients. It will be clarified in the next chapters. Take the case when J and M are maximum, i.e., $|J=j_1+j_2$, $M=J>=|j_1j_1>|j_2j_2>$. For more illustrations, some applications are to be dealt with.

$J_{-}|JM>=((J_1-iJ_2)|JM>=\sqrt{J(J+1)-M(M+1)}\hbar|JM-1>$.
For the spin, $S_1+S_2=\frac{\hbar}{2}(\sigma_1+\sigma_2)=S\hbar$.

Take $S=1/2+1/2$, $1/2+1/2-1$,-------$|1/2, -1/2>$, $S=1, 0$, $M=1, 0, -1$, triplet, $M=0$, singlet. Go back to the system of the two particles (1, 2) with $\vec{\ell_1}$ and $\vec{\ell_2}$ respectively. In general,

$L^2|\ell m>= \ell(\ell+1)|\ell m>$ in $\hbar$ unit.
$L_z|\ell m>=m|\ell m>$, $-\ell\leq m\leq \ell$,

where ℓ is integer or half integer.

Also, $L_+ = \ell_1 + i\ell_2$ ------ raising operator.
$L_- = \ell_1 - i\ell_2$, ----- lowering operator.

As it had been done before, it can be shown that
$L_+|\ell m> = C_m|\ell m>$, multiply by $|\ell m>$, then
$C_m = <\ell m + 1| L_+|\ell m> = \sqrt{\ell(\ell + 1) - m(m + 1)} = 0$ if $\ell = m$, $C_m = 0$,
(maximum m).

Therefore, $L_+|\ell\ell> = 0$. (As an exercise, show that $L_-|\ell, -\ell> = 0$.)

Now take two systems (1 and 2) with the eigenstate $|j_1 m_1>$ and $|j_2 m_2>$ respectively. The eigenstate of the new combined system of the two is given by $|\ell_1 m_1>|\ell_2 m_2> = |\ell_1 m_1, \ell_2 m_2>$.

This leads to $L_1|\ell_1 m_1, \ell_2 m_2> = \ell_1(\ell_1 + 1) |\ell_1 m_1>|\ell_2 m_2>$.
$L_2|\ell_1 m_1, \ell_2 m_2> = \ell_2(\ell_2 + 1)|\ell_1 m_1, \ell_2 m_2>$, $L^2 = L_1^2 + L_2^2$
$L_{1z}|\ell_1 m_1, \ell_2 m_2> = m_1|\ell_1 m_1, \ell_2 m_2>$, $L_z = L_{1z} + L_{2z}$
$L_{2z}|\ell_1 m_1, \ell_2 m_2> = m_2|\ell_1 m_1, \ell_2 m_2>$.

5.4. Addition of Angular Momentum

In many problems in quantum physics, the Hamiltonian is invariant under rotation. Therefore, the Hamiltonian (H) commutes with the components of the angular momentum, J_x, J_y, and J_z. Hence, the eigenfunction of H is the common eigenfunction of J^2 and the component J_z. On this basis, the eigenvector of the angular momentum J can be formed as $|JM>$, where $J = \sum_i j_i$ (for spinless particles in a central field). Remember that the eigenfunction of the orbital angular momentum L is given by $R(r)Y_\ell^m(\theta, \varphi)$, where $R(r)$ is the known radial function which is usually the spherical Bessel function, and Y_ℓ^m is the angular function represented by the spherical harmonic fucntion. Now taking the spin of the system

into consideration then for two particles, the total angular momentum is given by $J=\sum_i j_i=\sum_i \ell_i+s_i \Rightarrow$
$J=L+S=L+1/2(\sigma_1+\sigma_2)$.

For individual angular momentum, $\psi(r,\theta,\varphi)=$
$R(r)\ Y_\ell^m(\theta,\varphi)\ |\mu_1\rangle|\mu_2\rangle$, where $|\mu_1\rangle$ and $|\mu_2\rangle$ stand for the spin eigenket, so these may take (+1/2) ↑ or (-1/2) ↓ as whether the spinning is up or down respectively.

Now it is necessary to form an eigenfunction for a system of two particles with j_1 and j_2 to particle 1 and particle 2 respectively. Let the eigenvectors be represented in general by $|\alpha j_1 j_2 m_1 m_2\rangle$. They are eigenvectors to j_1^2, j_2^2, j_{1z} and j_{2z}. But α is an additional quantum number which might be necessary to specify the dynamical state completely or it is an eigenvalue of an observable A, it might form a complete set of commuting observables with, j_{1z} and j_{2z}. This eigenvector also forms a standard basis with respect to the total angular momentums (1 and 2). Therefore, to each set of quantum numbers $(\alpha\ j_1\ j_2)$, there corresponds as many vectors as there are distinct pairs $(m_1 m_2)$. These vectors can be obtained from one another by repeating the applications of $j_{1\pm}$ or $j_{2\pm}$ and span a subspace $\mathcal{E}(\alpha j_1 j_2)$ of $(2j_1+1)(2j_2+1)$ dimensions.

As it is already pointed out, A, j_1^2, j_2^2 commute with the total angular momentum J; therefore, their eigenfunction is common between them, so it is easy to find the eigenvectors of J and J_z. Each subspace has to be treated separately. Consider a particular subspace (ε), and for simplicity, let the vectors $|\alpha j_1 j_2 m_1 m_2\rangle$ for this subspace be $|m_1 m_2\rangle$, and $|JM\rangle$ is the eigenvector for the total angular momentum in it. The question now is what are the possible values of the pair $|JM\rangle$, and what are their respective orders of degeneracy?

Before constructing these eigenvectors, it is quite necessary to point out the following important principles (based on two observations):

1. Each vector $|m_1 m_2\rangle$ is an eigenvector of J_z with eigenvalue $M=m_{1+}m_2$. This is easy to be verified by using the fact that $J_z = J_{1z} + J_{2z}$. So
   $J_z|m_1 m_2\rangle = M|m_1 m_2\rangle = J_{1z}|m_1 m_2\rangle + J_{2z}|m_1 m_2\rangle = m_1|m_1 m_2\rangle + m_2|m1m2\rangle = (m_1 + m_2)|m_1 m_2\rangle$ ➤ $M = m_1 + m_2$.
2. To each value of J, there corresponds a certain number $N(J)$ of linearly independent series of $2J+1$ eigenvectors of J. Also, the vectors of a given series is actually obtained from one another by repeated applications of J_+ or J_-. Accordingly, the $2J+1$ possible values of M are obtained, where M takes the values $-J, -J+1, -J+2, ----- .+J$. Therefore, if $n(M)$ denotes the order of the degeneracy of the eigenvalue M, then
   $n(M) = \sum_{J \geq M} N(J)$. Consequently,
3. $N(J) = n(J) - n(J+1)$ ------------------------(83).

To obtain $N(J)$, $n(M)$ must be determined for each possible value of M. From 1, $n(M)$ is the number of the pair $(m_1 m_2)$ such that $M = m_1 + m_2$ (as was proven before). To find this, it can be drawn with a diagram between m_1 (as x-axis) and m_2 (as yaxis). Figure 6 shows this clearly. (Refer to Q.M. Volume 1 by Messiah).

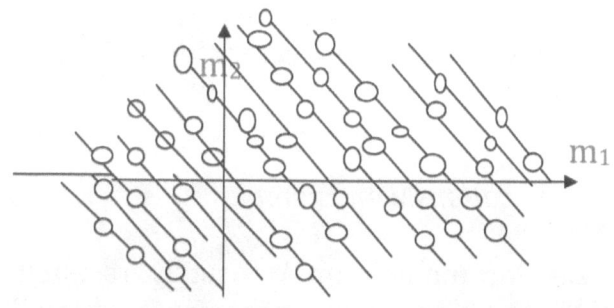

Figure 6 m_1 versus m_2 (schematic)

In this diagram, each pair $(m_1 + m_2)$ is represented by a point of abscissa m_1 and ordinate (m_2), i.e., $n(M)$ is the number of the points located on the diagonal x+y=M.

For the case $j_1 > j_2$, it is found that

$$n(M) = \begin{cases} 0 \text{ if } |M| > j_1 + j_2 \text{ Why?} \\ j_1 + j_2 + 1 - |M| \text{ If } j_1 + j_2 \le M \ge |j_1 - j_2| \\ 2j_2 + 1 \text{ If } |j_1 - j_2| \ge |M| \ge 0 \end{cases}$$

By substituting in 83, it is obtained that
$N(J) = 1$, for $J = j_1 + j_2,\ j_1 + j_2 - 1, \text{------} |j_1 - j_2|$.

From these, it can be concluded that the fundamental theorem of the angular momentum addition states the following: in the $(2j_1 + 1)(2j_2 + 1)$ dimensional space spanned by the vectors $|\alpha j_1 j_2 m_1 m_2>$ [where $(\alpha j_1 j_2)$ is fixed, but $m_1 m_2$ are variables]. The following are noticed:

1. The possible values of J are: $j_1 + j_2,\ j_1 + j_2 - 1, \text{-----} |j_1 - j_2|$. This indicates the total angular momentum J forms with j_1, j_2 triangle as

   Where
   $j_1 + j_2 + J$ is an integer.

2. To each of these values, there corresponds one and only one series of $2J + 1$ eigenvectors $|JM>$ of the total angular momentum J.

5.4.1. Some Applications with Examples

The total angular momentum is usually resulted from the addition of the individual angular momenta. So it will be either integer or half integer. It depends on the number of the added angular momenta whether it is even or odd. As an example, take the addition of two 1/2 spins. It leads to a space of four dimensions $[(2s+1)(2s+1)=4]$. The total spin is either 0 or 1. It depends on their spinning antiparallel or parallel respectively. Therefore, the total spin is S=0, 1.

The singlet state is S=0, and the triplet state is S=1. The possible state is always given by 2S+1, so it is singlet for S=0, given by l00>, and it is triplet for S=1given by l11>, l10>, and l1,-1>. Let P_0 be the projector operator that acts on the singlet state l00>, and P_1 is the projector operator that acts on the triplet state. These operators can be written in terms of S or $\sigma_1.\sigma_2$. Also, S^2li>=s(s+1)li>, where s(s+1) is the eigenvalue. It is 0 for the singlet state and 2 (in ℏ unit) for the triplet state. Using the fact that $\sigma_1.\sigma_2 =2S^2-3$, the projectors can be written as

$$P_0=1-(1/2) S^2 \quad =(1/4)(1-\sigma_1.\sigma_2) \left.\vphantom{\begin{matrix}a\\b\end{matrix}}\right\} \text{------------- (84).}$$
$$P_1=(1/2)S^2 \quad =(1/4)(3+\sigma_1.\sigma_2)$$

It might be noted that $P_1+P_0=1$ and $\sigma_1.\sigma_2=P_1-3P_0$.

For more illustration, consider a particle of spin 1/2 and orbital angular momentum L. The total angular momentum of the particle will be J=L+S (coupling of L and S). J takes the values L+S or L-S, i.e., J=L+1/2 or J=L-1/2. If L=0 (S-state) where J=1/2. This implies that J can take all the half integer values from 1/2 to ∞, and to each of them, there corresponds two terms [and two series of (2J+1) vectors] of opposite parity.

Another example, take a system of two nucleons, each nucleon with orbital angular momentum L and spin S. Suppose l_1 and s1 are the orbital angular momentum and the spin of the first nucleon respectively, the l_2 and s2 are the same for the second nucleon. The total orbital angular momentum of the system is L=L1+L2. Its total spin is S=s1+s2,; therefore, its total angular momentum is J=L+S.

Again, S=0 is the single state, and S=1 is the triplet state. The coupling of L with S is called LS coupling or Russell coupling. It can take all positive or zero integral values. Hence, for each pair of values (LS), there corresponds $(2L+1)(2S+1)$ vectors, which can be linearly combined to obtain the eigenvector of the total angular momentum (J). For this case, triplet state S=1; hence, J=L-1, L, L+1 if L≠0, J=1, if L=0.

For more illustration, let us construct the spectroscopy notation for J=0, 1, 2, 3. The multiplicity is usually given by (2s+1). For example, represent the sample for these notations, J=0, s=0, L=0; therefore, the notation is

J=0, $3P_0$, S_0^1
J=1, $3S_1$, $3D_1$, $3P_1 1P_1$ spectroscopic notations
J=2, $3P_2$, $3F_2$, $3D_2$, $3D_1$
J=3, $3D_3$, $3G_3$, $3F_3$, $1F_3$

As an exercise, for J=1/2, 3/2, 5/2, 7/2, s=1/2. Show the spectroscopy notations are $s_{1/2}$ $p_{1/2}$ $p_{3/2}$ $d_{3/2}$ $d_{5/2}$ $f_{5/2}$ $f_{7/2}$ $g_{7/2}$.

5.4.2. *Eigenvectors of (TAM) J and Clebch-Gordon Coefficients (CGC)*

Given two systems of eigenvectors $|j_1 m_1>$ and $|j_2 m_2>$, what is the eigenvector of the total angular momentum (J) which is denoted by the ket $|JM>$? It is the coupling of the two eigenvectors of the two systems. The two types of the coupling are

1. -The weak coupling which gives,

$|JM> = \sum_{M=m_1+m_2} C_{m_1 m_2} |j_1 m_1> |j_2 m_2>$------------------(85).
$M_{max} = (m_1+m_2) = j_1 + j_2 = J_{max}$.
$J_{max} = j_1 + j_2$. The next possible quantum number is j_1+j_2-1.
$J = j_1+j_2, j_1+j_2-1, j_1+j_2-2, ------|j_1-j_2|$
$M = +J, j-1. ---,-J$, and $\sum(2J+1) = (2j_1+1)(2j_2+1)$.

2. The Strong Coupling: applying this coupling gives,

$|JM> = \sum_{M=m_1+m_2} C_{m_1 m_2} |j_1 j_2 m_1 m_2>$ -------------(86).

Where $C_{m_1 m_2}$ is Clebch-Gordon coefficients used in both types of coupling.

By multiplying equation 86 from the left by $|j_1 j_2 m_1 m_2>$, the (CGC) is obtained as $C_{m_1 m_2} = <j_1 j_2\, m_1 m_2 |JM>$ ------------- (89).

As it has been shown before, using the addition theorem of the angular momentum for each pair $|JM>$ of the total angular momentum (J), there is an eigenvector $|\alpha j_1 j_2 JM>$. Its norm is 1, i.e., $<\alpha j_1 j_2 JM |\alpha j_1 j_2 JM> = 1$ --------------------------(90).

This eigenvector, as is the case with $|\alpha j_1 j_2 m_1\, m_2>$, the $|\alpha j_1 j_2 JM>$ is an orthonormal basis of the space $\mathcal{E}(\alpha j_1 j_2)$.

The transformation from one basis to the other has to use the unitary transformation; therefore

$|\alpha j_1 j_2 JM> = \sum_{m_1 m_2} |\alpha\, j_1\, j_2 m_1 m_2> <\alpha j_1 j_2 m_1 m_2 |JM>$.

The coefficient $<\alpha j_1 j_2 m_1 m_2 |JM>$ has pure geometrical characters. It depends only on the angular momentum and its orientation. It does not depend on the physical nature of the dynamical variable from which the angular momenta are constructed. These coefficients are called CGC or the vector addition coefficients (VAC). This shows the importance of the Clebch-Gordon coefficients in the process of transformations. They are independent of(α), but they depend on j_1, j_2, m_1, m_2 and M. In the subspace $\mathcal{E}(\alpha j_1 j_2)$, the $|\alpha j_1 j_2 m_1 m_2>$ are the basis vectors of the standard representation in which the components of j_1 and j_2 are represented by matrices independent of α, such as:

$<\tau j\mu|J_z|\tau^- j^-\mu^->=\mu\delta_{\tau\tau^-}\delta_{jj^-}\delta_{\mu\mu^-}$

and

$<\tau j\mu\ |J_\pm|\tau^- j^-\mu^->=\sqrt{j(j+1)-\mu\mu^-}\ \delta_{\tau\tau^-}\ \delta_{jj^-}\delta_{\mu\mu^-_{\pm-1}}$

$\left.\right\}$ ---- (91).

Also, the matrix representations of J^2 and J_z are independent of α, which implies that the components $<\alpha j_1 j_2 m_1 m_2|\alpha j_1 j_2|JM>$ of their common eigenvectors have the same properties. Now using the (CGC) $|j_1 j_2\ m_1 m_2|JM>$, the following is obtained:

$|\alpha j_1 j_2 JM>=\sum_{m_1 m_2}|\alpha j_1 j_2 m_1 m_2><j_1 j_2 m_1 m_2|jm>$ --------------(92).

The conditions CGC$\neq 0$ are $M=m_1+m_2$, $|j_1-j_2|\leq J\leq j_1+j_2$.
Since the Clebch-Gordon coefficients are a type of unitary transformation, they obey the orthonormalization relations:

$\sum_{m_1 m_2} <j_1\ j_2 m_1 m_2|JM><j_1 j_2 m_1 m_2|J^- M^->=\delta_{JJ^-}\delta_{MM^-}$ -----93).
$\sum_{JM} <j_1 j_2 m_1 m_2|JM><j_1 j_2 m_1^- m_2^-|JM> = \delta_{m_1 m_1^-}\delta_{m_2 m_2^-}$, (94)

For simplest cases, it is possible to determine the linear combinations (83) directly. Take the case where, $J_{max.}=j_1+j_2$.

Then $|\alpha j_1 j_2, j_1+j_2 j_1+j_2> = |\alpha j_1 j_2 JM>=$ -------- (95).

By repeating the application of $J_-=j_{-1}+ij_{-2}$ to both sides of 95, it is possible to construct all the $|\alpha j_1 j_2 JM>$ correspond to $J=j_1+j_2$. So by following this procedure, all the vectors of the series $J=j_1+j_2-1$ can be constructed, starting with the one corresponding to the case $M=J$, which is clearly defined by the phase condition, $<j_1 j_2 j_1\ j_2|JJ\geq 0$ (it is real and positive definitive), where (m_{max}) and $M_{max}=J=j_1+j_2$. This phase has the orthogonal property to $|\alpha j_1 j_2,\ j_1+j_2,\ j_1+j_2-1>$. Take a simple example, addition of two 1/2 spins. The eigenvector of the total S, where $s_1=1/2$, $s_2=1/2$ or $s_1=-1/2$, $s_2=-1/2$.

This implies the eigenvectors l++>, l-->, l+->, l-+>. These four vectors are due to the fact that $(2s_1+1)(2s_2+1)=4$, represents the number of dimension.

As it is known, S takes the values S=1, triplet state, l11>, 1-1>, and l10>, S=0 singlet state l00>. It depends on the direction of the spinning parallel or antiparallel respectively. This can be summarized in table 1.

Table 1.

M	S=1	S=0
1	l11> = l++>	
0	l10> = (l+ ->+l- +>)/$\sqrt{2}$	l00>= (1+->+l -.+>/$\sqrt{2}$
-1	l1-1>= l+->	

Chapter Six

Identical Particles in Quantum Mechanics

6.1. *Introduction*

The identical particles have no problems in the classical mechanics. This is based on the fact that the physical concept about the laws of nature considers the describing of any physical system is quite determined. It is the base of the determinism philosophy. Also, accordingly there is no problem in distinguishing between identical particles in the most macroscopic systems through many known ways. But in quantum mechanics, where the described systems are microscopic systems, the identical particles, with similar properties, it is not possible to distinguish between them. This fact raises a real problem in quantum mechanics, where the probabilistic concept is to consider a collision between two identical particles (1 and 2) in classical mechanics. In this case, let the dynamical state of the system 1 defined as (r_1, p_1) at any time, where r_1 and p_1 are the position and the momentum.

Likewise, let (r_2, p_2) be the dynamical state of the system 2. The evolution of the system, in general, is determined by certain Hamiltonian function, depending on the two systems (particles) variables. Hence, H is written as:

$H(\xi^{(1)}, \xi^{(2)}) \equiv H(r_1 \times p_1, r_2 \times p_2)$, if the system is subjected to certain potential $V(r)$, which only depends on the relative distance (r) between the two particles, i.e., $r=|r_1-r_2|$. Then

$$H(\xi^{(1)},\xi^{(2)}) = \frac{p_1^2}{2m} + \frac{p_2^2}{m} + V(r) \ \text{----------------------- (1).}$$

Now if particles 1 and 2 are identical, it means the dynamical properties of the system $(1+2)$ are not modified when they are permuted, i.e., $\xi^{(1)}$ is ascribed to 2 and vice versa. These leads to $H(\xi^{(1)}, \xi^{(2)})=H(\xi^{(2)}, \xi^{(1)})$. This clearly means that H is invariant under permutation.

On the other hand, the state of the system at any time can be known only to within a permutation of the two particles. Observation can state that at a given time, one particle might be in state ζ^a, and the other might be in state ζ^b, *not allowed to determine the exact state for any particle.*

This seems to raise a problem, but it is an apparent one. Take the wave function $\psi(r_1, r_2)$ which describes the system.

This function naturally is consisted of two parts.
The symmetrical part is $\psi_s (r_1, r_2)$, and the anti-symmetrical part is $\psi_A(r_1, r_2)$. Therefore, the wave function is:

$$\psi(r_1, r_2)=\frac{1}{2}[\psi[r_1, r_2]+ \psi(r_2, r_1)]]+\frac{1}{2}[\psi[(r_1, r_2)-\psi(r_2, r_1)]]=$$
$$\psi_s(r_1, r_2)+\psi_A(r_1, r_2)\text{-------------------------------------(2)}.$$

If $\psi(r_1, r_2, t_0)=\psi_{AS}(r_1, r_2, t_0)$ then $\psi_s(r_1, r_2, t)=\psi_A(r_1, r_2, t)$.

So at time t_0, one of the particles is in state ζ_0^a, and the other is in state ζ_0^b. This implies that there are two possibilities: either particle 1 is in state ζ_0^a, or it is particle 2. These are corresponding to the same physical situation due to the fact that $H(\zeta_0^a, \zeta_0^b)=H(\zeta_0^b, \zeta_0^a)$, i.e., H is symmetrical, which means the laws of motion at t_0 for the particles in both states are the same.

For more precision, suppose particle 1 or particle 2 is in the state $\psi_0(r)$ and the other in the state $\phi_0(r)$. Consider the two particles are spinless. Hence, practically these wave functions are considered as wave packets localized at different region of space. It leads to

$$\psi_0(r_1, r_2) \equiv [\psi_0(r_1)\psi_0(r_2)]\Rightarrow \phi_0(r_1, r_2)=\phi_0(r_1)\phi_0(r_2).\text{--- (3)}.$$

The functions in 3 are linearly independent. It is to be noticed that the initial observation does not permit to decide whether the system is in ψ_0 or inϕ_0 states. The observation deals with simultaneously measuring certain compatible set of variables, where these eigenfunctions belong to the set of values given by this measurement. Therefore, any linear combination of these functions, say $(\lambda\psi_0+\mu\phi_0)$, has the same property.

Also, the initial observation is not allowed to decide which of these linear combinations represent the initial state of the system, which clearly indicates there is an exchange of degeneracy. The solution of Schrödinger equation, $i\hbar\frac{\partial}{\partial t}\psi(t)=H(t)\psi(t)$, is known to be $\psi_{sA}(t_0)=(\psi_{sA}(t_0)+(t-t_0)\frac{1}{i\hbar}H\psi_{sA}$. It is the time evolution of the system.

Let ψ (r_1,r_2) and ψ^- (r_1, r_2) be the solutions of the Schrödinger equation correspondence to ψ_0^a and ψ_0^b respectively, which are the initial state conditions. Since H is symmetrical operator, it implies that ψ and ψ^- can be obtained from one another by permuting r_1 and r_2. Now let ψ_s (symmetric) and ψ_A (antisymmetric), where $\psi_s=\frac{1}{\sqrt{2}}(\psi+\psi^-)$ and $\psi_A=\frac{1}{\sqrt{2}}(\psi-\psi^-)$ are the solutions of the Schrodinger equation. The initial conditions are $\psi_0^{(s)}=\frac{1}{\sqrt{2}}(\psi_0+\psi_0^-)$, $\psi_0^A=\frac{1}{\sqrt{2}}(\psi_0-\psi_0^-)$.

If ψ_0 is the state of the system at t_0 (initial), then $\psi_0=\alpha\psi_0^A+\beta\psi_0^s$, $[|\alpha|^2+|\beta|^2=1.]$

At later time (t), the system will be in the state:
$\psi(r_1, r_2)=\alpha\psi^A+\beta\psi^s$ -------------------------- (4).

Let $P(r_1, r_2)$ be the probability density of finding one of the particles at r_1 and the other at r_2, which is given by

$P(r_1,r_2)=|\psi(r_1, r_2)|^2+|\psi(r_2,r_1)|^2=$
$2[|\alpha|^2|\psi^A(r_1,r_2)|^2+|\beta|^2|\psi^s(r_1,r_2)|^2]$------------(5).

By the symmetry properties, it is known that $[\psi(r_2, r_1) = -\alpha\psi^A(r_1, r_2) + \beta \psi^s(r_1, r_2)]$. The condition makes (5) independent of α and β is that $[|\psi^A(r_1, r_2)|^2 = |\psi^s(r_1, r_2)|^2]$. This holds only for r_1 and r_2 when the two particles are not interacting, and the wave packets ψ^a and ψ^b do not overlap. In the case, there is no interaction between the particles. The potential is zero, i.e., $V(|r_1 - r_2|) = 0$. This means they move toward each other free. So the wave function of the system is given as

$[\psi(r) = \psi^a(r_1, t)\,\psi^b(r_2, t)]$. It implies that for certain period of time, the wave packets will overlap to exist a region of space where these functions are deferent from zero if r is a point in that region. Hence,

$$|\psi^s(r, r)| = \frac{1}{\sqrt{2}} |\psi^a(r) + \psi^b(r)| \neq 0 \text{ ----------(6)}.$$

Where $|\psi^A(r, r)| = 0$. Why?

The existence of an exchange of degeneracy raises a source of a real difficulty. Because it prevents making a precision of theoretical prediction for the statistical distribution of the results that is clue of the measurements performed after collision or the scattering problem in the vicinity of central potentials. This kind of difficulty can be overcome by the symmetrical postulation, which fixes once and for all the coefficients in the combination(3). It can be easily verified experimentally. Accordingly, it is possible to conclude that: *the dynamical states of two identical particle systems are necessary, either all symmetrical or antisymmetrical in the permutation of the two particles.* Any possibility of these to occur it depends on the nature of the particle itself. But in general, this postulate can easily be extended to systems of any numbers of identical particles. For example, take the helium atom (He) its wave function was found to be antisymmetric, i.e.,

$\psi \equiv \psi_A(r_1, r_2)$.
He

er$_1$ r$_2$e

He Nucleus

Therefore, in general, for the particles with half-integer spin (1/2, 3/2, 5/2, 7/2-----etc.), the wave function usually is antisymmetric; but for the particles with integer spin (1, 2, 3, 4,----n), the wave function is symmetric $\psi_s(r_1, r_2,----r_n)$. Statistically, the half-integer spin particles obey Fermi-Dirac statistics and are called fermions, but those particles with integer spin obey Bose-Einstein statistics and are called Bosons. These physical phenomenon are called the correlation of the spin with the statistics.

Therefore, a system of identical particles is equivalent to the relativistic quantum field. (It is important to keep it in mind.)

6.2. Symmetrization Postulate (SP)

To study a system of N-identical particles, all what so far has been stated has to be taken in consideration. The dynamical variables which are describing the ith particle are usually functions of its position r_i, its momentum p_i, and its spin s_i. Therefore, the dynamical variables ξ^i are written as $\xi^i(r_i, p_i, s_i)$. So given the magnitude of s_i, it is possible to construct the space if its dynamical states are ζ^i. This leads to the possibility of writing the dynamical states of the whole system ($\mathcal{E}$) as a tensor product of ζ^i, i.e., $\mathcal{E} = \zeta^1 \otimes \zeta^2 \otimes \zeta^3$ ------ $\otimes \zeta^n$ ------------ (7)

If two particles have the same spin, they are similar. (Note that similarity does not necessary imply identity.) Hence, r_i, p_i, s_i, and the state vectors of one are one-to-one correspondence with that of the others (observables and states). Therefore, it is possible to exchange two similar particles. Generally, if *n* particles are similar, then *n!* Permutations of these (n) particles are well-defined operations. This implies that to each of these operations, there corresponds a certain operator of the space $(\mathcal{E})$.

6.21. Basis Vectors in Hilbert Space

For two identical particles (1 and 2), their observables are (r_1, p_1, s_1) and (r_2, p_2, s_2) denoted as ξ^1, ξ^2 respectively.

The space of its set state vectors is ζ. Let q be a complete set of commuting observables of this space, and $lq_x>$ is the eigenvectors of basis (q_x). The corresponding eigenvalue is q_x. The index (x) or the set of the x indices are necessary to distinguish between the eigenvalues. Hence:

$<q_\alpha l q_u> = \delta_{au}$ ----------------------------(8).
q might, for example, be x, y, z, or r, taken together with s_z.

This implies that each particle a(a=1, 2, 3,--------N) of the system has its own set of commuting observables q^a.

Now let $Q(q^1, q^2,-----q^N)$ as a complete set of commuting observables for the space $(\mathcal{E})$ and the vectors

$lq_\alpha^1 q_\beta^2$ -----$q_v^n> = lq_\alpha^1 > lq_\beta^2 > -----lq_v^n >$-------(9)

are formed by taking the tensor product of the basis vectors of the spaces $\zeta^1, \zeta^2, \zeta^3,------\zeta^N$, forming the basis for a certain representation of the vectors and the operators of the space $(\mathcal{E})$. It is the Q-representation which is a type of a symmetrical representation. Therefore, $lr_1, s_z>, lr_2, s_z>-----lr_i, s_z>$ form complete sets of commuting observables such as $lr_l, s_z> \equiv lq_\alpha^1 >$, $lr_2, s_z> \equiv lq_\beta^2 >$, or in general,

$lq_\alpha^1, q_\beta^2, ---> \equiv lq_\alpha^1 > lq_\beta^2 >$ for two particles (1 and 2).

6.2.2. *Particle Exchange Operators Permutation Operators*

By the investigation of equation 9, it is quite clear that:

$|q_\alpha^1\rangle$ is the state of the particle (1).
$|q_\beta^2\rangle$ is the state of the particle (2).

$|q_\nu^\nu\rangle$ is the state of the particlen(ν).

If N particles are permutated, their distribution will be modified among $|q_\alpha\rangle$, $|q_\beta\rangle$,-------$|q_\nu\rangle$. It implies that equation 9 is replaced by another. In general, different vectors of the basis q-representation. Note that when the N-individual states $|q_\alpha\rangle$--------$|q_\nu\rangle$ are not all different, then certain permutation will leave equation 9 unchanged.

But if they are all identical, equation 9 will not change by any permutation. The permutation operation establishes one-to-one correspondence between the vectors of the orthonormal basis (9). This usually defines certain linear and unitary operation in state-vector space. It implies that with each permutation of the N particles is associated a permutation operator denoted by P, its property is defined as

$PP^+ = 1$, it is a unitary operator-------------------------- (10).

Consider P(12) as the permutation operator between the particle 1 and the particle 2. So by definition,

$P(12)|q_\alpha^1, q_\beta^2, q_\nu^\nu\rangle = |q_\beta^1, q_\alpha^2$,----->.

The vector (9) transforms in a vector representing state in which particle 1 replaces particle 2 and vice versa. But the remaining particles stay in their initial states. To clarify this physical phenomenon, consider a very simple example, a system of three particles, i.e., N=3, so by definition, their permutation operated is defined as P(123), where particle 1 takes place (3) and particle 2 occupies state 1, but particle 3 occupies state 2. This operation is represented by $P(123)|q_\alpha^1 q_\beta^2\ q_\gamma^3> = |q_\gamma^1 q_\alpha^2 q_\beta^3>$ If $|\psi>$ is a vector of$(\mathcal{E})$(in Hilbert space), then

$$P(123)|\psi> = \sum_{\alpha\beta\gamma} P(123)\ |q_\alpha^1 q_\beta^2 q_\gamma^3><q_\alpha^1 q_\beta^2 q_\beta^3\ |\psi>. \text{-----}(11)$$

Note: $<x, y, z|\psi> \equiv (x, y, z)$, as shown before.
Some of P properties:

1. $PP^+ = 1$
2. $P(123)q^{(1)}P^+(123) = q^{(2)}$
3. $P(123)q^{(2)}P^+(123) = q^{(3)}$ $\left.\vphantom{\begin{matrix}1\\2\\3\\4\end{matrix}}\right\}$ ----------------(12).
4. $P(123)q^{(3)}P^+(123) = q^{(1)}$

In general, $P(123)F(q^{(1)}q^{(2)}q^{(3)}) = F(q^{(2)}\ q^{(3)}\ q^{(1)})$.
Also, $\psi^-(q_\alpha^1, q_\beta^2, q_\gamma^3) \equiv <q_\alpha^1, q_\beta^2, q_\gamma^3|\psi^- >$
So $<q_\alpha^1, q_\beta^2, q_\gamma^3 \text{----}|P(123)|\psi> \equiv <q_\beta^1, q_\gamma^2, q_\alpha^3|\psi>$

Which implies that $P|\psi> = |\psi^->$ ------------------ (13).
$<\psi|p^+ = <\psi^-|$

Also, it is easily to show that

$$\left.\begin{matrix} P(12)q^{(1)}P^+(12) = q^{(2)}. \\ P(12)q^{(2)}P^+(12) = q^{(1)}. \end{matrix}\right\} \text{-------------------- (14)}.$$

Therefore, if there are *n* particle,. there will be *n!* (factorial) permissible permutation operations.

6.2.3. *Some Algebra of P*

1. $P(123)=P(12)P(23)$
2. $P(123)q^{(2)}P^+(123)=P(12)P(23)q^{(2)}P^+(23)P^+(12)=$
   $P(12)q^{(3)}P^+(12)=q^{(3)}$.

This means that $q^{(3)}$ is invariant under $P(12)$ permutation.

3. The parity of P is $(-)^p$, it is even if p is even and odd if p is odd. The evenness and the oddness of parity is unique. So it can be said even permutation or odd permutation.
4. $P^2(ij)=1$; hence, $P(ij)=\pm 1$. Show this.

The symmetrizer is defined as $S_{ij}=(1/2)[1+P(ij)]$.
Where $P(ij)S_{ij}|\mu>=S_{ij}|\mu>$, $P(ij)=+1$.
The antisymmetrizer $A(ij)$ is defined as
$A(ij)=(1/2)[1-P(ij)]$, where, $P(ij)A(ij)|\mu>=-A(ij)|\mu>$, $P(ij)=-1$.

From these simple algebra manipulations, it might be concluded that

A. Eigenvectors of the eigenvalue $+1$ are invariant in the transposition (ji).
B. It means they are symmetric in *i* and *j*.
C. This indicates the projector onto the space of all vectors that are symmetrical in *i* and *j* is the symmetrization operator, as defined above.

2. For those eigenvectors with eigenvalue (-1), change sign in the transposition (ij) indicates:

(a.) They are antisymmetric in *i* and *j*.
(b.) The projector onto the space of all such vectors are in fact antisymmetrization operator, as defined before, is given by:
$A(ij)=(1/2)[1-P(ij)]$.

As an exercise, verify the following mathematical relations between S(ij), A(ij), and P(ij):

(1.) S(ij)+A(ij)=1,
(2.) S(ij)-A(ij)=P(ij),
(3.) P(ij)S(ij)=S(ij)P(ij)=S(ij),
(4.) P(ij)A(ij)=A(ij)P(ij)=-A(ij).

It is important to notice here that any vector is the sum of antisymmetric vector in i and j and a symmetric vector in i and j. In general, it is necessary to extend the notion of symmetry and antisymmetry of dynamical states to take in consideration the more general cases of the permutation.

For more general case, take the vector $|\mu>$ of the space $(\mathcal{E})$. Now propose $\mathcal{E}_\mu$ as a subspace spanned by the vector $|\mu>$. Also by the permutation operation, all vectors can be obtained. The number of the dimensions of the subspace $(\mathcal{E}_\mu)$ is equal or inferior to N!, according to whether N! vectors $P|\mu>$ are linearly independent or not. The extreme case is when all $P|\mu>$ represent the same state, i.e.,

$P|\mu>=c_p |\mu>$, for any p ------------------------ (15).

The restriction conditions on c_p are:

1. If P is transposition, c_p takes the value ±1.
2. Since P(ij)=P(1i)P(2j)P(12) P(2j)P(i1)=P(12), then

$C(ij)=C (ij)^2C(2j)^2C(12)=C(12)$-------------(16).

This means C is constant; it is the same for all transportations. Hence, $C_{tr.}=+1$ or -1.

Therefore, the permutation P is product of transpositions; hence, c_p is a certain power of C_{tr}. This power is even if p is even, and it is odd if p is odd. So $P|\mu> = c_p 1\mu>$ is fulfilled only in the following two cases:

1. For any P, $c_p = 1$ ⟶ $P|\mu> = 1\mu>$------- (17).
2. For any P, $= c_p = (-)^p$ ⟶ $P|\mu> = (-)^p l> $ ----(18).

$1\mu>$ is asymmetric vector if the permutation of N particles behaves in 1. But it is antisymmetric if the permutation of N particles as in 2. Also, the symmetrical vectors form a subspace $\mathcal{E}^s$ of $\mathcal{E}$, while the antisymmetrical vectors form a second subspace which is orthogonal to $\mathcal{E}^s$. Accordingly, the corresponding projectors are given by

$$S = \frac{1}{N!}\sum_p P, \quad A = \frac{1}{N!}\sum(-)^p P $$------------------ (19)

Where S is the symmetrizer and A is the antisymmetrizer. The $\sum P$ extends over the N! Possible permutations. Multiplying it by P_1 from the left or the right, one can get

$$P_1 S = S P_1 = S, \quad P_1 A = A P_1 = (-)^p A $$--------- (20).

Replacing P by its inverse, i.e., $P^+ = P^{-1}$, (merely changing the order of elements) and since P^+ and P have the same parity, it leads to $S = S^+$, $A = A^+$ --------------------(21).

From 17 and 18, it is easy to find (do it as exercise) that $S^2 = S$, $A^2 = A$ and $SA = AS = 0$---------------(22).

Hence, S and A are orthogonal projectors.

If $1\mu>$ is in the subspace $\mathcal{E}^s$, then from 17
$$S|\mu> = \frac{1}{N!}\sum_p P(\mu) = (1/N! \sum_p p)|\mu> = 1\mu>.$$

Conversely, if $1\mu>$ is an arbitrary vector, then from 20, $PS|\mu> = S|\mu>$; hence, S is indeed the projector onto the subspace $\mathcal{E}^s$.

Also, $PA|\mu>=A|\mu>$; hence, A is indeed the projector onto the subspace. For example in the case $N=3$, $N!=6$, so S and A are written as

$S=(1/6)[1+P(123)+P(231)+P(312)+P(213)+P(132)+P(321)]$
$A=(1/6)[1-\{p(123)+P(231)+P(312)+P(213)+P(132)+P(321)\}]$.

Adding S and A gives $S+A=(1/6)(2)=1/3>0$, but it is<1, i.e., $0<1/3<1$. This indicates that $S+A$ projects onto the space of the states which are invariant under an even permutation of N particles, a subspace of $\mathcal{E}$ space, when $N>2$. Going back to equation 18, considering the space $\mathcal{E}_\mu$, it is possible to deduce that for any P, $SP|\mu>=S|\mu>$ and $AP|\mu>=(-)^P A|\mu>$. It implies that the vectors $P|\mu>$ which are spanning the space $\mathcal{E}_\mu$ have the same projections onto the symmetric space $\mathcal{E}^s$, within the same sign of the projection onto $\mathcal{E}^A$, the antisymmetric space. Therefore, according to whether $S|\mu>$ is different from zero or not, the space contains one and only one symmetrical vector or none what so ever. Also, it is likewise for $A|\mu>$.

This conclusion implies that if $N!$ vectors $P|\mu>$ are linearly independent, then particular combination of $S|\mu>$ and $A|\mu>$is$\neq 0$. Hence, the space contains $S|\mu>+A|\mu>$, i.e., by nature, it is so, symmetric vector plus antisymmetric vector.

6.3. Identical Particles and the Symmetrization

6.3.1. Postulate of the Symmetrization

The N-identical particles so far mentioned are not only similar but identical too. Therefore, the dynamical properties of this system are not modified due to the permutation of these particles. Hence, from these invariance characteristics, it can be concluded important remarks related to the law of motion and the observables of the system.

Now let $|\psi_{t_0}\rangle$ be the state of the system at the initial time t_0. Let $|\psi_t\rangle$ be the state of the system at later time t. Using the evolution operator $U(t, t_0)$, the transforming from t_0 to t can be shown as $|\psi_t\rangle = U(t, t_0)|\psi_{t_0}\rangle$.

If $|\psi_{int}\rangle = P|\psi_{t_0}\rangle$, by comparison, $U(t, t_0)$ is as P, except for P at time (t), it will be in the state $P|\psi_t\rangle$; therefore, $U(t, t_0)P|\psi_{t_0}\rangle = P|\psi_t\rangle = PU(t,t_0)|\psi_{t_0}\rangle$.

Since this must hold whatever $|\psi_{t_0}\rangle$ is, it implies that

$[P, U(t, t_0)] = 0$, i.e., P commutes with $U(t, t_0)$ -------------- (23).

$U(t, t_0)$ is evolution operator. It is known that this operator is also a solution of the Schrödinger equation,

$i\hbar \frac{dU(t,t_0)}{dt} = HU(t, t_0)$, where $U(t_0, t_0)=1$, initial condition. From this equation with the use of the relation $PU(t, t_0)=U(t, t_0) P$, it can be shown that $[P, H] = 0$ --------------------(24).

So P is a constant of motion. Conversely, if H commutes with P, it implies that $U(t, t_0)$ and its transformation under permutation (PUP^+) are equal because both satisfy the same Schrödinger equation with the same initial condition. Then U commutes with P. This implies the necessary and sufficient condition that the equation of motion be invariant under permutation is equation 24 satisfied for all P, i.e., always commutes with Hamiltonian of the system H.

Consider a physical observable ($\mathbb{O}$) of a given system. Let $|\mu\rangle$ be the eigenvector for this observable, where $\mathbb{O}|\mu\rangle = o|\mu\rangle$, where o is the corresponding eigenvalue of the observable $\mathbb{O}$

This means that the measurement of $\mathbb{O}$ will certainly produce o. Now if the system is in state $P|\mu\rangle$ which is obtained by applying the permutation P to $|\mu\rangle$, the measurement of $\mathbb{O}$ will be the same. In other words, each vector of the space generated by applying all possible permutations of N particles to $|\mu>$ must also be an eigenvector of $\mathbb{O}$ corresponding to the same eigenvalue o (exchange degeneracy). This indicates that the necessary and sufficient condition for this statement to be true for all eigenvalues o is $[\mathbb{O},P]=0$ for all P, i.e., $\mathbb{O}$ commutes with P. This means the measurement of P and $\mathbb{O}$ can be done at the same time.

From previous chapter, it was shown if two operators P and Q both has an inverse, then the product PQ has an inverse too. Hence, $[P,Q]^{-1}=Q^{-1}P^{-1} \rightarrow [P, Q]^{+}=Q^{+}P^{+}$. This fact leads to the statement that N particles are identical if Hamiltonian (H), and all possible physical observables of the system are symmetrical with respect to these particles. It leads to the symmetrization postulate :- Identical particles mean,

$[P, \mathbb{O}]=0$, $[P.H]=0$ and that P is constant of motion. Also,

$|\mu>_s$ is forever symmetric, and $|\mu>_A$ is forever antisymmetric. It is important to notice that it is not possible to make a relation between $|\mu>_s$ and $|\mu>_A$, but a mixture of them is possible.

Postulate: The physical state should be either symmetric or antisymmetric, i.e., $|\mu>_s$ or $|\mu>_A$, respectively. It depends on the particles themselves. If they are leptons (fermions) where the spin is half integer, their eigenkets are experimentally confirmed to be antisymmetric. This result is in good agreement with the quantum mechanics. But if the particles are Bosons with integer spins like alpha particle, the eigenket is confirmed to be symmetric. Therefore, the wave function $\psi_{s\,or\,A}(q_1, q_2, q_3, \text{----} q_n) \equiv <q_1, q_2 \text{---} |\psi>_{s\,or\,A}$, where q_i represent the coordinates and the spins, so

$$\psi_{s\,or\,A} \equiv <q_1, q_2, \text{----} q_i q_j |P(ij)P(ij)|\psi>_{s\,or\,A}$$
$$=\pm<q_1, q_2, \text{----} q_i q_j |\psi>_{s\,or\,A} =\pm\psi(q_1, q_2, \text{---} q_i\,q_j)$$

Here, q_1, q_2, q_3, -----q_N form a complete set of commuting observables in the space $\mathcal{E}$. But they are only symmetrical functions of these observables and can be physical observables that do not form a complete set in the space. Therefore, the symmetrization postulate will consist in limiting the state vector space to a certain subspace of the space where the physical observables in the question form a complete set. In this case, the exchange degeneracy is completely removed.

If there are N particles, then n_1 of N occupies state $|q_1\rangle$, n_2, occupies state $|q_2\rangle$,---n_x, occupies $|q_x\rangle$,$(n_1 + n_2 + --+ n_x = N)$.

The identity of the particles in each of these states will remain undetermined. Hence, there are $(\dfrac{N!}{n_1! \, n_2! -- n_x!})$ basis vectors of the Q-representation and have the desired property to build up the eigenstate. Consider $\mathcal{E}(n_1, n_2, n_3, ---- n_x)$ as the space spanned by these vectors (generated by the application of the N! permutation to any of them.) The state of the system is certainly represented by one of the vectors of this space. But as it should be, the observation which might be made does not allow to decide which particle is it. However, the predictions of the theory depend on which particle is observed. This ambiguity is, therefore, representing a source of real difficulty. By using the symmetrization postulate, this ambiguity can be solved. Hence, the symmetrization might be restated such that the states of a system with N identical particles are necessary, be it either all are symmetric or all are antisymmetric with respect to the permutation to N particles.

As it was pointed out earlier, the applications of the symmetrization postulates depend on the nature of the identical particles which are classified according to their spins, which are as follows:

1. The fermions, particles with spins of half-integer, such as the leptons (e⁻, e⁺,ˉ,). They obey Fermi-Dirac statistic.

2. The Bosons, particles with spins of integer such as the photon. They obey Bose-Einstein statistic with a symmetrical state $<q_1^1, q_2^2,\text{---}|\psi>_s=<q_2^1, q_1^2,\text{---}|\psi>_s$.

Even if $q_1=q_2$, the particle 1 and 2 occupy the same state because they are excluded from the effect of the Pauli exclusion principle. These leads to the statement: there is no limitation on the particle number which might occupy the same state. But in the case of the fermions, they are affected by the Pauli Exclusion Principle. Their states are antisymmetric, such as $<q_1^1, q_2^2, q_3^3, \text{------}|\psi>_A =-<q_2^1, q_1^2,\text{----}|\psi>_A$

If $q_1=q_2$, then $<q_2^1, q_1^2,\text{------}|\psi>=0$. This indicates that the probability of finding the two particles occupying the same state is zero. Also it implies that two identical particles with half integer spin cannot occupy the same quantum state. Hence, as a conclusion, the space $(n_1, n_2, n_3,\text{---}n_x)$ is either has one symmetrical vector (V_s) or one antisymmetrical (V_A). Therefore, the symmetrization postulate completely removes the exchange of degeneracy. The question that might be asked is, is this in conflict with the fundamental postulates of the quantum? The answer is it does not.

Consider the Bosons, as it was defined before, that the projector onto $|\psi>_s$, where it is a particular combination of the permutation operators $(S=\frac{1}{N!}\sum_p P)$. This implies that S commutes with the evolution operator $U(t, t_0)$, i.e., $[S, U(t,t_0)]=0$, ---------------------------------- (25).

and with the physical observables $\mathbb{O}$, i.e., $[S, \mathbb{O}]=0$. ------------------------------------ (26).

Equation 25 indicates that if the system is initially in a symmetrical state $|\psi>_s$, it remains in it as long as it is left undisturbed. Fermions can be treated in the same way.

Equation 26 shows that S and $\mathbb{O}$ have at least one common set of basis vectors. Why? If the state of the system is symmetrical, the expansion of the state vector in terms of this basis contains only symmetrical eigenvectors of the observable ($\mathbb{O}$), so $\mathbb{O}|\psi>_s = o|\psi>_s$. It indicates the symmetrization is left unchanged; consequently the system is left in a symmetrical state.

6.3.2. Bosons and Bose-Einstein Statistics(spins:0,1,2,3,---n)

Given a system of N bosons, where its states span the subspace $\mathcal{E}^s$ of the space $\mathcal{E}$. So a basis of this subspace can be formed from the vectors of the Q-representation as follows:

In each subspace (n_1, n_2,----n_x) only, one normalized symmetrical vector(defined to within a phase). It is the vector

$$[\frac{N!}{n_1! \, n_2! -- \, n_x!}]^{1/2} S |q_1^{n_1} q_2^{n_2} ----- q_x^{n_x} > ----------------(27),$$

where $|q_1^{n_1}, q_2^{n_2} -- q_x^{n_x}>$ is the basis vector of Q-representation. S is the symmetrization operator, as defined before, as $S = 1/N! \sum_p P$, and $[\frac{N!}{n_1! \, n_{2!--n_x!}}]^{1/2}$ is a normalization constant. Remember, $n! = n \, (n-1)(n-2)----(n-n)$, $0! = 1$

Note: Interchange of two particles occupying the same state leaves $|q_1^{n_1}, -- q_x^{n_x}>$ unchanged, but for two particles occupying different states give another state of the basis vector of Q-representation.

Now define $\Pi n_x! \equiv n_1! n_2! n_3! ---- n_x!$

Therefore, in general, this vector is invariant under any of $\Pi x!$ permutations, not changing the distribution of the N particles, among the individual dual states $|q_1>$, $|q_2>$, $--|q_x>$.

154

But any other permutation changes it into another basis vector of the Q-representation. Also, by applying each of the N! permutation to $lq_1^{n_1}, q_2^{n_2}, ----q_x^{n_x}>$, one can form $[\frac{N!}{\Pi n_x!}]^{1/2}$ basis vectors of ($n_1, n_2, n_3, ----n_x$). Each one vector of them is obtained times the vector in equation 27. It is given by,

$[\Pi n_x! \ N_i]^{1/2}$ S $lq_1^{n_1}, q_2^{n_2}, ----q_x^{n_x}>$. It is symmetrized and normalized to unity. Thus, to each sequence $n_1, n_2, n_3, --- n_x$ of non-negative integers such as $n_1 + n_2 + n_3 + --------+n_x=N$, there is a corresponding one and only one symmetrical state of the system, which is represented by equation 27. Hence, the formed vectors constitute an orthonormal basis in $\mathcal{E}^s$. The norm is given by

$$[\frac{N!}{n_1!n_2!--n_x!}]<q_1^{n_1}q_2^{n_2},---q_x^{n_x}l \ S^2 \ lq_1^{n_1}q_2^{n_2}----q_x^{n_x}>=1$$

And where, $S^2=S=(\frac{1}{N!}\sum_p P)=\frac{n_1! \ n_2!--n_x!}{N!}$.

Now if there are N particles of Bosons, where N is very large, it is possible to talk about Boson gas obeys Bose-Einstein statistics, in case the internal interaction is neglected (free gas), as a first approximation. Accordingly, the Hamiltonian is given by H=$h^1+h^2 ++h^i---+h^N$----------(28).

From the classical Boltzmann statistical theory in thermodynamic, it is known that the thermodynamically equilibrium is realized when the system is in the most probable (macroscopic state), where in fact the macroscopic state is a set of quantum state or microscopic states sufficiently close, such that it is impossible to differentiate between them at the macroscopic level. Also, it is well known that the microscopic states of the same energy are equally probable. It leads to the conclusion that the probability of a given macroscopic state is proportional to the number of distinct microscopic states composing macroscopic state. Therefore, this number plays an essential role in the determination of the thermodynamically equilibrium.

6.3.3. *Fermions and Fermi-Dirac Statistics, Exclusion Principle.*

The concept of the antisymmetrization plays main role in the wave function that describes the system of half-integer particles. So equation 27 for this case takes the form

$$[N!]^{1/2} \, A \, lq_1^1, q_2^2, ---- q_x^x > ------------(29).$$

Of course, these particles with half-integer spins usually obey Fermi-Dirac statistics, and as shown before in this case:

$< q_1^1 q_2^2, --- l\psi > = 0$, and the norm is given as:
$N! \, < q_1^1 q_2^2 ---- q_x^x \, l \, A^2 \, lq_1^1 q_2^2 ---- q_x^x > \rightarrow N! < --- lAl --->$

Since $A^2 = A$, as shown before, where $A = 1/N! \sum_p P$
Hence, $N!/N! < -- \sum l(-)^p Pl --> = 1$.

Therefore, $[N!]^{1/2} Alq_1^1, q_2^2, q_3^3, ---- q_x^x >$
$= 1/\sqrt{N!} \, [lq_1^1, q_2^2 ---- q_x^x > -lq_2^1 q_2^1, ---- >] = 0$ for $q_1 = q_2$.
due to the exclusion principle. Therefore,

$Alq_1^1, q_2^2 ---- q_x^x >$ is the sum of N! Mutually orthogonal vectors are not zero unless $q_1 = q_2$. Its norm is $1/N!$. But if $lq_\alpha q_\beta q_\nu >$ are the normal individual occupied states, so $\sqrt{N!} \, Al ---->$ is the corresponding antisymmetrical states, where its norm, as before, is unity.

This vector can be written in the form of N-determinant, (it is Slater-type determinant), as below:

$$\Psi(\mathbf{x}_1, \mathbf{x}_2, \ldots, \mathbf{x}_N) = \frac{1}{\sqrt{N!}} \begin{vmatrix} \chi_1(\mathbf{x}_1) & \chi_2(\mathbf{x}_1) & \cdots & \chi_N(\mathbf{x}_1) \\ \chi_1(\mathbf{x}_2) & \chi_2(\mathbf{x}_2) & \cdots & \chi_N(\mathbf{x}_2) \\ \vdots & \vdots & \ddots & \vdots \\ \chi_1(\mathbf{x}_N) & \chi_2(\mathbf{x}_N) & \cdots & \chi_N(\mathbf{x}_N) \end{vmatrix} \equiv |\chi_1 \ \chi_2 \ \cdots \ \chi_N|.$$

-- (30).

This represents antisymmetrical state which is the only one that corresponds to each set $|q_\alpha\rangle, |q_\beta\rangle, \text{------}, |q_\nu\rangle$ of N different states chosen from among the individual states: $|q_1\rangle, |q_2\rangle, |q_3\rangle$ ----$|q_\nu\rangle$. This set of vectors are so constructed, constituting an orthonormal basis in the subspace $\mathcal{E}^A$.

As an example, take three identical particles; their states are described by $\phi_a(x), \phi_b(x), \phi_c(x)$. The total function of the system of these three particles is $\Phi(x_1, x_2, x_3)$, which is given by Slater determinant as

$$\Phi(x_1, x_2, x_3) = \frac{1}{\sqrt{3!}} \begin{bmatrix} \phi_a(x_1) & \phi_b(x_1) & \phi_c(x_1) \\ \phi_a(x_2) & \phi_b(x_2) & \phi_c(x_2) \\ \phi_a(x_3) & \phi_b(x_3) & \phi_c(x_3) \end{bmatrix} \text{--------} (31)$$

The question that might be asked is, is it always necessary to symmetrize the wave function? To find the proper answer, take a system of n identical particles. If these particles are electrons (half-integer spins), their eigenstate which describes them is the vector $|\psi\rangle_A$ which is antisymmetric. However, these electrons usually are not the only electrons in the universe; there are others electrons in the universe that might affect the n electrons. But if their effect is neglected and the n electrons system is treated as a distinct entity from the rest of the universe electrons, that is to say the dynamical properties of the n electrons system are not affected by the presence of the others electrons in the universe.

But another question might arise; it is whether such a hypothesis is really well founded or whether the symmetrization postulate in establishing a certain correlation between these n electrons and the others renders it invalid.

As a practical example, let n electrons be inside a certain spatial domain D, where all dynamical properties of these n electrons are measured inside this domain. So other electrons outside D are ignored, i.e., no interaction with those n electrons inside D. This concept and its result is a general one. It is also applicable to the bosons as well as to the fermions. This can be proved for a special case of a system of two fermions as they are well known to us (e, p, n,).

For fermions (ignoring the existence of all particles), the dynamical state of two fermions is denoted as $|\varphi_{1,2}>_A$. 1 and 2 represent the coordinates and S_z of particles 1 and 2.

As pointed out before, in general, a given a state of the system, say X is $X^A(1, 2)$ where $<,(X^A(1,2), X^A(1, 2)>=1$, normalized. If at time t the system is in the state φ, so its dynamical properties at t are given by the set of probabilities given by $\omega=|<X|\varphi>|^2$-------------------- (33).

In reality, the two fermions are part of a system of N fermions. Hence, N-2 fermions are left out of the investigation.

Now let us see whether the dynamical properties are the same as those found before if the N-2 fermions are taken into account.

The normalized antisymmetrical wave function representing the dynamical state of N-2 fermions is (3, 4, 5,---N). If 1 and 2 were not identical to fermions 3, 4, 5, 6,--N, so the state of the total system would be represented by $[\varphi(1, 2)\psi(3, 4,--N)]$.

This representation is based on the assumption that no interaction possible between the particles 1 and 2 and the rest represented by N-2 particles. On this basis, let $|\phi>$ be the state of the whole system, so $1 \phi > \propto A | \varphi\psi >$ where $A|\varphi\psi >$ is antisymmetric, and A is the antisymmetrization operator for N particles, $[A = \frac{1}{N!}\sum_p(-1)^p P]$.

By the hypothesis φ *and* ψ do not overlap, the two fermions are inside D, but the N-2 are outside D (as proposed), so the problem concerns only the dynamical properties of the two fermions inside D. Let $\mathbb{F}(3, 4, 5,\text{---}N)$ be antisymmetric wave function of norm 1, which vanishes when any of the N-2 position vectors $r^3,\text{------}r^N$ are inside the domain D; hence, $\mathbb{F}$ represents the state of the system of N-2 fermions, which is outside D. By hypothesis ψ is their function, suppose that $\mathbb{F}_1$, $\mathbb{F}_2,\text{---}\mathbb{F}_i$ are a complete orthonormal set of such functions, then $\psi = \sum_i \mathbb{F}_i < F_i | \psi >$------------------(34).

Let $\mathcal{X}(1, 2)$ be any normalized antisymmetric wave function of 1 and 2 electrons, where $\mathcal{X}(1,2) = 0$ if r^1 and r^2 are outside D because $\mathcal{X}$ represents fermions 1 and 2 inside D only.

Now let φ, by hypothesis be a such function, and then the permutation of the N particles may be put into two categories, depending upon their action on the vector $1\mathcal{X}\varphi >$. These categories are:

1. Denote category 1 by f which can, at most, change the sign of $1\mathcal{X}\varphi >$. They are the 2!(N-2)! permutation that exchanges 1 and 2 and/or exchanges 3, 4, 5, ----N, among themselves.

So $F|\chi\theta > = (-)^F|\mathcal{X}\theta >$----------------------------------(35).

2. Denote all other permutations by G, which exchange at least one of the particles 1 2 with one of the N-2 particles.

Thus, G $|\mathcal{X}\theta\rangle$ represents the state in which at least one of the particles 1, 2 is definitely outside D. Therefore, $G|\mathcal{X}\theta\rangle$ is orthogonal to any vector of the type

$|\mathcal{X}\theta\rangle \Rightarrow \langle \mathcal{X}^-\theta^-|G|\mathcal{X}\theta\rangle = 0.$

Going back to the definition of A, where $A = \frac{1}{N!}\sum_p (-1)^p P$, it is possible to deduce the following identity:

$\langle \mathcal{X}^-\theta^-|A|\mathcal{X}\theta\rangle = \frac{1}{N!}\sum_p (-1)^p \langle \mathcal{X}^-\theta^-|P|\mathcal{X}\theta\rangle =$

$\frac{1}{N!}\sum_f (-1)^f \langle \mathcal{X}^-\theta^-|F|\mathcal{X}\theta\rangle =$

$\frac{2!(N-2)!}{N!} \langle \mathcal{X}^-\theta^-|\mathcal{X}\theta\rangle$ ---------- (36)

It is noticeable that the norm of $A|\mathcal{X}\theta\rangle$ is $\langle \mathcal{X}\theta|\mathcal{X}\theta\rangle = \frac{2!(N-2)!}{N!}$.

The probability that the particles 1, 2 inside D are in the state $|\mathcal{X}\rangle$ can be found as follows, if N-2 fermions are distinguishable from 1 and 2 fermions! then the state of the system is represented by $|\varphi\psi\rangle$. Hence, $\omega = \sum_i \left|\langle X\theta_i|\varphi\psi\rangle\right|^2$

$= |\varphi\psi|^2 (\sum_i \langle |\theta_i|\psi\rangle|^2 = |\langle \mathcal{X}|\varphi\rangle|^2$ ------------------ (37).

It is the probability.

Since N fermions are identical, the state of the system is given as $|\Phi\rangle = \sqrt{\begin{bmatrix} 2 \\ N \end{bmatrix}} A|\varphi\psi\rangle$) ------------ (38).

$\begin{bmatrix} 2 \\ N \end{bmatrix} = \frac{N!}{2!(N-2)!}$

So the probability of finding the system in any one of the states represented by the orthonormal antisymmetric vector,

$$|X_i> = \sqrt{\left[\frac{2}{N}\right]} A|X\theta_i>, \text{ is given by}$$

$$\omega = \sum_i <|X_i \Phi>|^2 = \left[\frac{2}{N}\right]\sum_i |<X\theta_i|A|\varphi\psi>|^2 \text{ using equation 35}$$

and 36, then $\omega = |<X|\varphi>|^2$ as it is in equation 37, which is just the one in equation 33. This implies that it is possible to ignore the existence of the N-2 fermions outside the domain D, and still it is possible to obtain the correct result to the probability. So the question is positively answered.

6.3.4. Nuclear System in Quantum Mechanics

As a basic knowledge, the nuclear system constitutes Z protons and N neutrons. The protons and the neutrons are in general similar particles; their common name is the nucleon.

The proton is a positively charged particle, but the neutron is a neutral particle. Both particles are fermions with spin of half integer. Therefore, they obey the Fermi-Dirac statistics.

The mass of the neutron m_n is bigger than the mass of the proton m_p, i.e., $m_n > m_p$, $m_n - m_p \approx 0.8$ MeV; it is $> m_e$ (electron mass = 0.511 MeV). When the nucleon is in the proton state, it behaves as proton; but if it is in the neutron state, it behaves as neutron. It behaves according to the charge or the spin space.

So the nucleon has space of two degrees of freedom. This space historically has the names, charge space, isobaric space, or isotopic spin space (isospin space). This will be classified in the next sections. This nuclear system, as any physical system, has a Hamiltonian (H) which is the kinetic energy (KE) plus the potential energy (PE).

To distinguish between the protons and the neutrons, the protons must be numbered from 1 to Z, and the neutrons must be numbered from Z+1 to Z+N. Let A_Z denotes the antisymmerizing operator for the first Z particles (protons) and A_N is for the last N (neutrons). Therefore, the wave function representing a possible state of this system in the symmetric representation $\{Q\}$ is given by $(q^{(1)}, q^{(2)}, q^{(3)}, -- q^{(Z)}, q^{(Z+1)}, --)$ and is subjected to the antisymmetrization condition, i.e., $A_N\varphi=\varphi$, $A_Z\varphi=\varphi$ -------------------------- (39).

This formalism is quite ordinary.
So far, the protons and the neutrons are treated as two different particles. But as it was pointed out before, each of these particles physically represents a different state of the nucleon particle. Hence, the nuclear system is a system constitutes of $\mathcal{N}$ nucleons. Therefore, it is a system of identical particles (nucleons), and they are fermions; the state of this system is, as usual, antisymmetrical with respect to the permutation of the $\mathcal{N}$ identical particles (fermions).

Now with this useful information, a new formalism will be described to show that it is equivalent to the known ordinary one. According to this new formalism, there are two states for the nucleon, as shown in the following diagram,

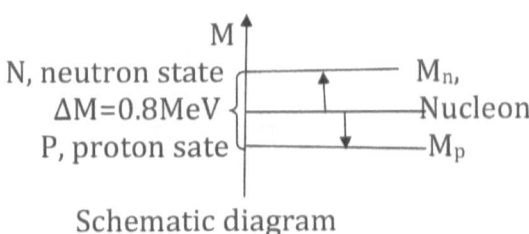

Schematic diagram

The required in this problem is how to distinguish between these two states. It is required that to each nucleon state, it has to be assigned a new dynamical variable. This variable is found to be two possible values of its charge. So let $\mathcal{W}$ be the proton state and $\mathcal{V}$ is the neutron state. This leads to the proposal of the so-called the charge space for the nucleon of two dimensions, likes the spin space (up or down). The charge space is called isotopic spin-space or isobaric spin-space, or in general, isospin-space, as mentioned before. So like any physical phenomenon in quantum physics, it is possible to define operators to its dynamical variables. It is analogous to those of the spin-space with the same mathematical properties.

The Pauli matrices play the main roles in the mathematical manipulation here because the nucleons are fermions obeying the Pauli Exclusion Principle. The vector operator representing the charge space is defined as $\tau \equiv (\tau_1, \tau_2, \tau_3)$, which can be represented by Pauli matrices $\sigma \equiv (\sigma_1, \sigma_2, \sigma_3)$ in isospin-space.

So τ is analogous to the vector (σ), where $t=(1/2)\tau$ is analogous to $s=1/2\ \sigma$, in the spin-space, where t is called the isotopic spin of the nucleon. Go back to the defined basis vectors $W\ and\ V$ for the proton and the neutron respectively; it can be written as

$$t_3 W = (1/2)\tau_3 W = 1/2\ W$$
$$t_3 V = (1/2)\tau_3 V = -1/2\ V \qquad \text{---------------- (40).}$$

Now define Π_W, as a projector onto the proton state, where,
$\Pi_W = (1/2)\ (1+\tau_3)$ ----------------------------------- (41).

and Π_V is the projector onto the neutron state where,
$\Pi_V = (1/2)(1-\tau_3)$ ----------------------------------(42).

Hence, the operator which is representing the charge of the nucleon is given by, $e\Pi_W = (1/2)(1+\tau_3)\ e$ --------- (43).

Where e is the electron charge unit.
Now the product of $\mathcal{N}$ individual charge spaces is the charge space $\mathcal{E}_c$ for the system of the $\mathcal{N}$ nucleons. So the total charge of the system is represented by the operator:
$C = \sum_{i=1}^{\mathcal{N}} e\Pi_W\ [(1/2)\mathcal{N} + T_3]$ ----------------------- (44).

Here, T_3 is the third component of the total isospin (T). It is given by $T=\sum_{i=1}^{N} t_i$ --------------------------------------- (45).

The orthonormal basis in $\mathcal{E}_c$ is obtained by taking the product of $\mathcal{N}\mathcal{W}$ or $\mathcal{N}\mathcal{V}$ vectors. In particular, the basis vector is given by $\zeta = \mathcal{W}^1\mathcal{W}^2 \text{-----} \mathcal{W}^z\mathcal{V}^{z+1} -- \mathcal{V}^N$, which represents the charge state in which the first Z particles are protons and the last are N neutrons. For simplicity, consider the states of the charge C=Ze only, i.e., with $T_3=1/2[Z\text{-}N]$.

Therefore, it is possible to construct $(\mathcal{N}!/Z!\ N!)$ basis vectors corresponding to the eigenvalue T_3. A typical one of these is $\zeta_\alpha = \mathcal{W}^{\alpha_1}\mathcal{W}^{\alpha_2}\text{---}\mathcal{W}^{\alpha_z}\mathcal{V}^{\alpha_z+1}\text{--}\mathcal{V}^N)$. This is the state in which the Z particles $\alpha_1, \alpha_2, \text{-------}\alpha_z$ are protons, and the rest are neutrons.

The ket vectors for the system of $\mathcal{N}$ nucleons are the vectors of the space which is formed by taking the tensor product of $\mathcal{E}_c$ with the space $\mathcal{E}_0$ of the other dynamical variables. When the permutation is applied to these nucleons, the charge variables and other variables are similarly affected. Let p_c be the permutation of the charges and p_o is the permutation of the others. Hence, the overall permutation is the product of the these two permutations, i.e., $P=p_op_c$; therefore, the antisymmetrization operator for the system of $\mathcal{N}$ nucleons is given as $A=\frac{1}{\mathcal{N}!}\sum_p(-1)^pP=\frac{1}{\mathcal{N}!}\sum_p(-1)^pp_o\ p_c$ -------- (46).

The states of the system corresponding to the charge Ze are those vectors Φ of the space $\mathcal{E}=\mathcal{E}_c\otimes\mathcal{E}_0$, which meet the antisymmetrization condition, $A\Phi=\Phi$ ------------- (47).

And it satisfies $T_3\Phi =(1/2)(Z\text{-}N)\ \Phi$ --------------------(48).

Where Z-N is the eigenvalue of T_3. It can be shown that there is a one-to-one correspondence between the vectors subjected to the conditions (46, 47), and the vectors of the space subjected to the condition (39), namely:

$$|\Phi\rangle = \sqrt{\frac{\mathcal{N}!}{N!Z!}} A|\varphi\rangle|\zeta\rangle \text{ --------------- (49).}$$

$$|\varphi\rangle = \sqrt{\frac{\mathcal{N}!}{N!Z!}} \langle\zeta|\Phi\rangle \text{ ------------------ (50).}$$

It conserves the scalar product.

It is possible to take a vector $|\varphi\rangle$ which obeys the condition (39). The corresponding $|\Phi\rangle$ which is obtained from equation 48 clearly satisfies (46, 47). Since T_3 commutes with A, and $T_3|\zeta\rangle = 1/2(Z-N)|\zeta\rangle$. Now it can be shown that the partial scalar product (49) gives back the vector $|\varphi\rangle$. Using equations 45 and 48, it is possible to get:

$$\sqrt{\frac{\mathcal{N}!}{N!Z!}}\langle\zeta| A|\Phi\rangle = \frac{1}{N!Z!}\Sigma_p(-1)^p p_0 |\varphi\rangle\langle\zeta|p_c|\zeta\rangle\text{-----(51).}$$

Now it is useful to put $\mathcal{N}!$=two groups of permutation:

1. F is thefirst group). It permutes the first Z particles among themselves and/or the last N particles that permute among themselves. The (N!Z!) permutations of F leave the vector $|\zeta\rangle$ unchanged. Multiply the vector $|\varphi\rangle$ by $(-)^f$, then $\langle\zeta|F_c|\zeta\rangle = 1$, $F_0|\varphi\rangle = (-)^f|\varphi\rangle$.
2. G is the second group. It takes the vector $|\zeta\rangle$ over into another vector $|\zeta_\alpha\rangle$; hence, $\langle\zeta |G_c|\zeta\rangle = 0$. It leads to the sum of the right hand side of equation 50 which has (N!Z!) nonzero terms, each one equals to 1, giving the required result (49). It can easily be seen that the correspondence (48) conserves the scalar product, for if $|\Phi\rangle$ and $|X\rangle$ correspond to $|\varphi\rangle$ and $|\chi\rangle$ respectively, then since

$$|X\rangle = \sqrt{\frac{\mathcal{N}!}{N!Z!}} A|\chi\rangle|\zeta\rangle,$$ and since $|\Phi\rangle$ obeys equations 46 and 48;

therefore, φ to one nature. Let $|\Phi\rangle$ be a vector subjected to conditions that $A\Phi = \Phi$, and $[T_3\Phi = (1/2)(Z-N)\Phi]$; it is given by equation 47. In the light of equation 44, $|\Phi\rangle$ is a linear combination of the vectors $|\zeta_\alpha\rangle$, where the coefficients are being vectors of ζ_0. Hence, $|\Phi\rangle$ can be written as:

$$|\Phi\rangle = \sqrt{\frac{N!}{N!Z!}}\Sigma_\alpha|\varphi_\alpha\rangle|\zeta_\alpha\rangle.$$ Hence, $|\varphi_\alpha\rangle$ is coefficient of in the sum of the RHS. Now, a permutation of the type F acting on Φ

gives $F|\Phi\rangle = \sqrt{\frac{N!}{N!Z!}}\Sigma_\alpha \langle f_0|\varphi_\alpha\rangle\langle f_c|\zeta_\alpha\rangle$ ------------- (52).

Also, $F|\Phi\rangle = (-)^f|\Phi\rangle = \sqrt{\frac{N!}{N!Z!}}\Sigma_\alpha(-1)^f|\varphi_\alpha\rangle|\zeta_\alpha\rangle$ --(53).

Note that the action of f_c on one of the $|\zeta_\alpha\rangle$ leads to another of the $|\zeta_\alpha\rangle$ and in particular, it leaves $|\zeta_\alpha\rangle$ invariant. Therefore, the coefficient of $|\zeta\rangle$ in the expansion of $F|\Phi\rangle$ is the coefficient of $f_c|\zeta\rangle$ on the RHS of the equation 52. Equating this to the coefficient of $|\zeta_\alpha\rangle$ in equation 52; hence, $f_0|\varphi_\alpha\rangle = (-)^f|\varphi_\alpha\rangle \Rightarrow |\varphi\rangle$ is corresponding to $|\Phi\rangle$ has indeed the antisymmetrical properties ($A\varphi=\varphi$). This implies that $A_N\varphi=\varphi$, $A_Z\varphi=\varphi$.

So the one-to-one correspondence is a natural case.

Chapter Seven
Scattering Theory in Quantum Mechanics

7.1. *Introduction*

The scattering is a physical phenomenon which always takes place in nature. It behaves as a simplest collision processes. It is a type of interaction. Usually it occurs in the atomic and nuclear interactions. As an example, the famous scattering experiment done by Rutherford in 1911, which discovered the nucleus of the atom, is an interaction between alpha particle and the atom nucleus.

In addition to the discovery of the positively charged nucleus, it showed that the main atomic mass is concentrated in this nucleus (more than 99 percent). Also by the scattering processes, the constituents of the nucleus were discovered (the protons and the neutrons). Then in 1964, the constituents of the nucleon (proton or neutron) were discovered by the scattering experiment to be the quarks, as it is well known in nuclear physics. This discovery showed that the nucleon is not fundamental particle as it was thought before. As it is known in Rutherford scattering experiment, a particle was used as projectile, and a a thin foil of gold was considered as a target. Alpha particle is equivalent to the helium nucleus; its size in terms of the radius is about 10^{-15} m. Usually the force acting between alpha particle (positively charged) and the gold nucleus (positively charged) is the known Coulomb force; it is a repulsive in this case, given by

$F_c = k \frac{q_\alpha q_n}{r^2}$, where the potential V(r) is related to the force such as, $-\nabla V(r) = F_c = k \frac{q_\alpha q_n}{r}$.

Where, q_α and q_n are the charge of alpha particle and the target nucleus respectively, k is the interaction constant. Since r is the

167

shortest distant possible at the point of the alpha particle deflection under the effect of the nucleus field, it was considered as the nuclear radius. The measurement gave the approximate radius of the nucleus is about 10^{-14}m. Therefore, in general, the scattering experiment is an projectile particle moving with certain energy, i.e., with a velocity (v) and momentum (P) and mass (m) toward a certain target either fixed or moving too, according the required setup and the objective of the experiment.

The theory of the scattering in quantum mechanics is of very important. It deals with forces, fields, and the probabilities as quantum physics's best tools to tackle main problems in quantum physics of fields. In the scattered particles in different directions, the probability is used to find the scattered particle somewhere with certain percentage. It is clear that such probabilities can be found by the potential field or the relative interaction $V(r_{ab})$ between the interacting particles A and B.

7.2. Scatterin Cross –Section:

The mathematical theory of scattering enables to calculate the relative probabilities of scattering that takes place. It shows the possible various directions of the scattered particles (incident particles beam). It is very well known that the potential $V(r)$ or the relative potential (V_{ab}) have a main role in this calculations.

In addition to that, the experimentally measured cross sections data can be used to find out these possible different directions. Accordingly, the cross section of the scattering (interaction) is defined as the probability that the incident projectile particle will be scattered as it passes through a given thickness (dx) of the target. So consider the following simple example of incident particle (p) on a target of thickness d and then calculate the cross section.

The cross section is usually given by the area $\sigma = \pi d^2$ which represents the probability that the interaction between the incident particle and the target occurs. There are two points to be noted:

1. For a very thin foil target, the area is $(A_{thi.}) = \sum \sigma$.
2. For a very thick foil target, the area is $(A_{thic.}) < \sum \sigma$.

Actually, the used target foil is quite thin enough.

On this basis, let A=area; dx is the target thickness, and ρ is the density of the incident particles. Therefore, the particles that are contained in the target are ρAdx). Hence, the effective target area is given by $A_{eff.} = \rho A\sigma dx/A$. The fraction area of the target blocked by A is given as $A_{fr.} = A_{eff.}/A = \rho\sigma dx$; it is the *probability that incoming particles collide with the target. Hence,* $dP = \rho\sigma dx$. The cross section (X-section) is a function of the scattering angle (θ). The schematic diagram bellow illustrates the simple scattering case,

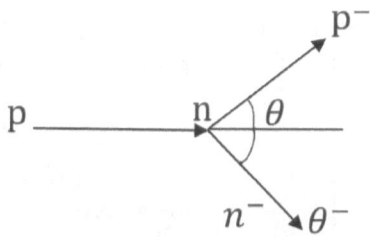

Take the following example

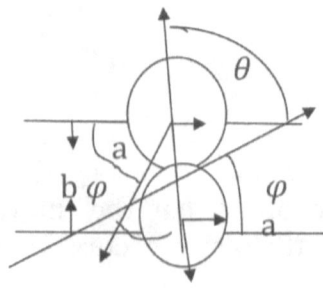

$\theta=2\varphi$, $\cos\varphi = b/2a$, $\rightarrow \theta = 2 \cos^{-1}b/2a$, b is the impact parameter which is the least distance, where the scattering occurs.

7.2.1. *Experimental Types of Collisions (Scatterings)*

As mentioned before, the scattering is a type of interactions between the colliding particles or the incident particles and the targets. There are two types of the experimental setup; it depends on the states of the projectiles and the targets. They are

1. The Laboratory System (LS) of scattering, where the targets are fixed while the incident particles (projectiles) are moving toward the targets with a velocity V_L and a momentum P_L, where $P_L=mV_L$. This is illustrated in the following schematic diagrams,

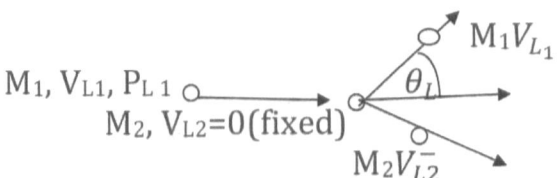

2. The Center of Mass System (CMS), where both the incident particles and the targets are moving toward each other. This $M_2V_{c2}^-$ can be illustrated in the following schematic diagrams:

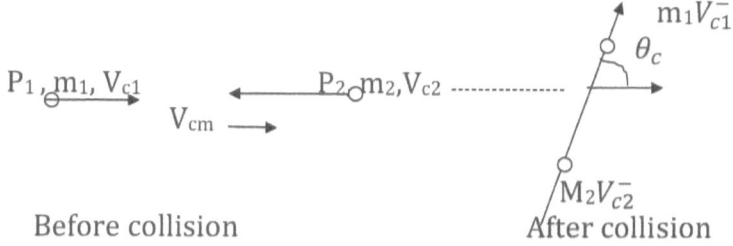

P_1, m_1, V_{c1}

$V_{cm} \longrightarrow$

P_2, m_2, V_{c2}

$m_1 V_{c1}^-$

θ_c

$M_2 V_{c2}^-$

Before collision   After collision

V_{cm} is the velocity of the center of the mass (CM) frame with respect to the laboratory frame. It is known that in the CM frame, $P_1+P_2=0$. It is the total momentum of the system after the collision. Hence, $P_1=-P_2$ are directed through the CM scattering angle without changing the magnitudes of P_1 and P_2. Thus, $V_{c1}^-=V_{c1}, V_{C2}=V_{c2}^-$ ---------------------------- (1).

An equivalent description to the process in CM may be obtained by replacing m_1 by μ with a velocity equals to the relative velocity of the two particles (due to the fact that r, the position of μ, is $(r_1 -r_2)$, the distance between m_1 and m_2. Consider m_2 to be very large approaching the infinity ($m_2 \to \infty$). This is schematically represented as follows:

$\mu \multimap V_{\mu c}$

$m_2 \to \infty$

$V_{\mu c}^- \mu_{sc}$

θ_c

, CM feame

after collision

Before collision

Therefore, $V_{\mu c} =V_{L1}, V_{\mu c}^- =V_{\mu c}$ ------------------- (2).

To determine V_{CM} with respect to the laboratory frame, the classical velocity transformation equation might be used, so $V_{c1}=V_{L1}-V_{CM}, V_{c2}=V_{CM}$ -------------------------- (3).

Using the conservation law ($P_1+P_2=0$) in the CM frame, the following relation is obtained: $m_1V_{c1}=m_2 V_{c2}$ -------(4).

Hence, $m_1(V_{L1}-V_{CM})=m_2V_{CM} \longrightarrow m_1V_{L1}= (m_1 +m_2)V_{CM}$, so

$$V_{CM}=(\frac{m_1}{m_1+m_2})\,V_{L1} \text{------------------------------------- (5).}$$

Equation 5 relates V_{CM} to V_{L1} (the velocity of m_1 in the laboratory frame). It is also useful to find the relation between θ_c(CM) and θ_L(laboratory), using the following sketching diagram:

$$\overrightarrow{V_{L1}}=\overrightarrow{V_{cM}} +\overrightarrow{V_{c1}} \text{-------------------------------(6)}$$

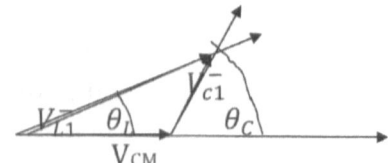

V_{CM}

Transformation of the velocity from CMS to LS.
Hence, by investigating this diagram, the following relations can be found:

$$\overrightarrow{V_{L1}} =\frac{\overrightarrow{V_{c1}}\,sin\theta_c}{sin\theta_l} \text{-------------------------- (7).}$$

Where $V_{l1}^{-}sin\theta_L =V_{c1}^{-}sin\theta_c$--------(8).
(Transverse components)

And $V_{L1}^{-}cos\theta_L =V_{CM}+V_{C1}^{-}\cos\theta_C$ ------------(9).
(Longitudinal components)

Using equations 7 and 9, the following is obtained:

$$Tan\theta_L=\frac{sin\theta_L}{\frac{m_1}{m_2}+cos\theta_c} \text{----------------(10).}$$

7.2.2. *The Relation between the Scattering Cross Sections and the Differential Scattering Cross Sections for the CMS and the Laboratory. System*

Let I be the incident beam of particles (flux) at the target as

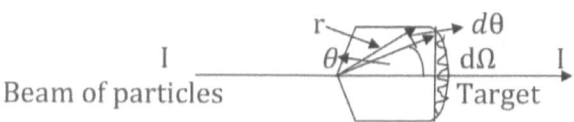

$d\Omega$ is solid angle between θ and $d\theta$ is given by,
$d\Omega=2\pi r^2\sin\theta d\theta/r^2=2\pi\sin\theta d\theta$

Let $\mathbb{I}$ be the probability per second that the incident particle will be carried out by the beam a cross 1cm² area, normal to the direction of the beam. Now define S as the scattered fluxes, which is the probability per second that the incident particle will be scattered into all angles, and $S(\theta)$ is the probability per second that the incident particle will be scattered into a unit solid angle at angle (θ) from the incident beam direction. So S and $S(\theta)$ are both related to the scattering cross section σ as follows:

$S=\sigma\mathbb{I}$--- (11).
$S(\theta)=\dfrac{d\sigma}{d\Omega}\mathbb{I}$-------------------------------------- (12).

To differentiate between the CM frame and the laboratory frame, let S_C for the CM frame and S_L for the laboratory frame. For illustration, let μ be the mass of the particle in the CM frame which is scattered under the effect of the potential V(r) associated with a central force at the origin.

The scattering cross section (σ) in this case can be calculated by using the solution of the specified Schrödinger equation, as it will be shown later on for the CM frame. Due to the relations between the two frames, it is easy to transform the (cross section in CM frame) to the laboratory frame for comparison with the experimental data. It is notable that the probability per second for the scattered incident particle is the same in both frames, i.e., $S_L=S_C$. Using equation 11, then $\sigma_L \mathbb{I}_L=\sigma_c \mathbb{I}_c$

but $\mathbb{I}_L=\mathbb{I}_c \Rightarrow \sigma_L=\sigma_c$ ------- (13).
Similarly, $S_L(\theta_L)d\Omega=S_C(\theta_c)\ d\Omega$

From these relations or by differentiating equation 13, it can be found that $\dfrac{d\sigma_L}{d\Omega_L}=\dfrac{d\sigma_c}{d\Omega_c}$ --------------------------------- (14).

{As an exercise, use equation 9 to evaluate equation 14 and

show that: $\dfrac{d\sigma_L}{d\Omega_L}=\dfrac{d\sigma_c}{d\Omega_c}=\dfrac{\left[(1+\frac{m_1}{m_2})^2+2\left(\frac{m_1}{m_2}\right)Cos\theta_c\right]^{3/2}}{1+\frac{m_1}{m_2}\cos\theta_c}$ $----(15)$.

7.2.3. The Scattering by the Central Force V(r)

This type of scattering physically occurs under the effect of the central force field. It can be illustrated in the following schematic diagram:

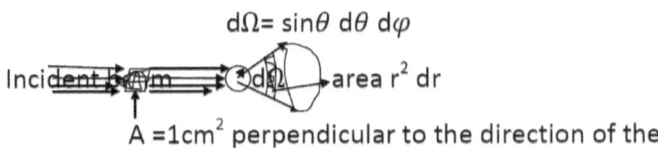

$$d\Omega= \sin\theta\ d\theta\ d\varphi$$

Incident beam → area $r^2\ dr$

$A =1cm^2$ perpendicular to the direction of the

Incident particles beam

The cross section (X-section) is either a differential cross section by integrating it; the total cross section can be obtained, i.e.,

1. $\sigma(\Omega)=d\sigma(\Omega)/d\Omega$ -differential cross section.
2. $\sigma_t=\int (d\sigma(\Omega)\ /d\Omega)d\Omega=\int d\sigma(\Omega)=$total cross section.

Accordingly, the total cross –section can be defined as

$$\sigma_t = \frac{Number\ of\ scattered\ particles\ into\ solid\ angle\ per\ unit\ time.}{Number\ of incoming\ particlrs\ per\ unit\ area\ per\ unit\ time}$$

If the incoming particles are just one (1), then $d\sigma(\Omega)$ is the number of the scattered particles that represents the probability of scattering.

It is quite useful to notice the following remarks:

1. $d\sigma(\Omega)/d\Omega$ is angle dependent. (It is a function of angle.)
2. The dimension of the differential cross section $\sigma(\Omega)$ is an area dimension (squared meter). Its unit is barn (1 barn=10^{-24}cm^2).
3. σ_t (total cross section) approaches infinity($\sigma_t \rightarrow \infty$) in the case of coulomb force ($\sim 1/r^2$).

The question now is, how is it possible to compute σ for a given central force? As indicated before, this can be done through solving the Schrödinger equation for a moving particle under the effect of a central force. To clarify this problem, consider the stationary scattering wave in solving Schrödinger equation with a potential V(r). This Schrödinger equation is given by [$\frac{-\hbar^2}{2m}\nabla^2+V(r)]\psi(r)=E\ \psi(r\ )$-------- (16).

The necessary conditions for the scattering are V(+∞)=0, E>0, i.e., the system is unbound.

$E=\frac{\hbar^2 k^2}{2m}$, $P=\hbar k$ (momentum of incident particle), where $\hbar$ is the reduced Plank constant and k=1/λ=wave number. $P=\frac{h}{\lambda}$ or $\lambda=\frac{h}{P}$ (de- Broglie wavelength).

Suppose that V(r) $\longrightarrow$ 0 (as r $\longrightarrow$ ∞) faster than $\frac{1}{r}$.

Therefore, according to these conditions, $\psi(r)$ is the solution of equation 16 for free particle (V=0). Hence, it is a plane wave, i.e., $\psi(r) \longrightarrow$ (as $r \longrightarrow \infty$) $e^{i\vec{k}.\vec{r}}$.

Also, the solutions of equation 16 for a potential of central force (F=-∇V) are known to be:

$$1 - j_{\ell(kr)}Y_{\ell m}(\theta, \varphi) \longrightarrow \frac{Cos\ kr}{kr}Y_{\ell m}(\theta, \varphi)$$

$$2\text{-}n_\ell(kr)\ Y_{\ell m}(\theta, \varphi) \longrightarrow \frac{sinkr}{kr}Y_{\ell m}(\theta, \varphi)$$

Where the phase shift δ is zero. (Why?)

$j_\ell(kr)$ is Bessel spherical function which represents the radial solution of (16).($Y_{\ell m}$) is the spherical harmonic function represents the angular solution of equation 16. n_ℓ is the Neumann function; it is related to $j_\ell(kr)$. (See Q. M. I. Schiff). These solutions lead to

$\psi(r) = \frac{e^{\pm ik.r}}{r}$, where $e^{+ik.r}/r$ is the spherical outgoing wave after scattering (divergent), and $e^{-ik.r}/r$ is the spherical incoming wave (convergent). Therefore, $\psi(r)$, is given by:

$$\psi(r)=e^{+ik.r}+f(\theta, \varphi)\frac{e^{+ik.r}}{r}\text{------------------(17)}, r \longrightarrow \infty.$$

f (θ, φ) is the scattering amplitude where

$$\frac{d\sigma(\Omega)}{d\Omega} = \text{lf } (\theta, \varphi\text{l}^2 \text{ --------------------(18)}$$

Equation 18 indicates that if the scattering amplitude $f(\theta, \varphi)$ is found, the total cross section (σ_t) can be easily calculated.

7.3. *Partial Wave Analysis(Phase-Shift Analysis)*

This is so important in understanding in depth the scattering theory. This can be illustrated schematically as follows:

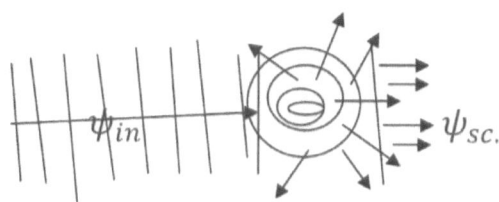

$\psi(r)=\psi_{in}+\psi_{sc.}$. The total wave function is composed of the incident wave ψ_{in} and the scattered wave $\psi_{sc.}$. From equation 17, the total wave function is $\psi(r)=e^{ikz}+f(\theta)\frac{1}{r}e^{ik.r}$, e^{ikz} is incident plane wave along the Z-direction, and $\frac{e^{ik.r}}{r}$ is the outgoing scattering spherical wave function moving in the direction of increasing r. It represents the possible ways through which the scattered particles can move.

Since the potential V(r) is symmetrical one, the scattering amplitude f(θ) is a function of θ only. Note that waves might be so tangled which makes quite difficult to distinguish between the incident and the scattered waves. But as r→ ∞ (at large r), this complications can be overcome. The (1/r) provides the concept that the necessary inverse square law leads to a decreasing in the intensity of the scattered wave. (Also, equation 17 shows that the f(θ) is a solution of Schrödinger time independent equation in the region r for free particle [V(r)=0].

As it is known that the intensity I of the incident particles is

$$I = v\psi^*\psi = ve^{-ikz}e^{ikz} = v \text{ --------------- (19)}.$$

Where v is the velocity=P/m=$\hbar$ k/m.
so I =v, intensity of incident beam.

P is the momentum of the particle, and m is its mass.
Similarly, the intensity of the scattered particles can be found, such as, $I_{sc.} = v[f^*(\theta)][f(\theta)]\frac{1}{r^2} = \frac{v}{r^2}f^*f.$

The(v/r^2) $f^*(\theta)$ $f(\theta)$ represents the probability per second that the scattered particle cross section of 1cm² area normal to its direction of motion. Hence, the probability per second to cross dA cm² is vf^*fdA/r^2.

By definition, $(dA/r^2)=d\Omega$, so $(vf^*fd\Omega)$ is the probability into the solid angle $d\Omega$; dividing by $d\Omega$, the probability per second for scattering into a unit solid angle is obtained. This in fact is the scattered flux $S(\theta)$. It is given by

$S(\theta)=vf^*(\theta)f(\theta)$ --(20).

Using equations 19 and 20, it is obtained that:

$\frac{d\sigma}{d\Omega}=|f(\theta)|^2$ -- (21)

so as it was shown in equation 18. Again, the calculation of the differential cross section $\frac{d\sigma}{d\Omega}$ is a matter of calculating the scattering amplitude $f(\theta)$. It is done by matching the incident wave with the scattered wave in a way to obtain a solution to Schrödinger equation for a central potential V(r) that behaves like equation 17 for r→ ∞. The first requirement is to find out how to decompose the plane wave $e^{ikz}=e^{ikr\cos\theta}$, [. K.z=kr $\cos\theta$, z=r $\cos\theta$]. ------------------------------22.

It is associated with a momentum $P=\hbar k$ into a set of partial waves. Each is associated with orbital angular momentum L; its value is given by $\sqrt{\ell(\ell+1}$ $\hbar$.

Going back to the solution of Schrödinger equation for the central force, it is given by ψ (r,θ,ϕ)=R(r)$\Theta(\theta)\Phi(\phi)$ ----(23).

V(r) is symmetric; the ψ (r, θ,ϕ) is independent of ϕ. This leads to Φ_0 (ϕ=0)$\Rightarrow$ $\frac{1}{\sqrt{2\pi}}$ $e^{\pm im\phi}=1$, m=0.

$\Theta(\theta)$ is an acceptable solution of Schrödinger equation, never divergent. These functions are $\Theta_{\ell 0}$ (θ), where m=0, and ℓ is the angular momentum quantum number. They are mathematically called Legendre Polynomial denoted by $P_\ell(\cos)$, where $P_0 = 1$, $P_1 = \cos$, $P_2 = (3\cos^2 - 1)/2$.,------.

Their integral properties (-1 to +1) is given by
$\int_{-1}^{+1} P_{\ell-} (\cos) P_\ell (\cos\theta\,) d\,(\cos\theta) = \frac{2}{2\ell+1}\delta_{\ell-\ell}$. This is the angular solution to Schrödinger equation. While R(r) is the radial solution to the radial part of Schrödinger equation given by
$(\frac{1}{r^2})\frac{d}{r}(r^2\frac{dR}{dr}) + \{-\frac{\ell(\ell+1)\hbar}{r^2} + \frac{2\mu}{\hbar^2}E\}R(r) = 0$. ---------------- (24).

V(r)=0.
In this case where E>0, E is not quantized, and the quantum number (n) does not arise. The solution of equation 24 is known to be the well-familiar spherical Bessel functions denoted by $j_\ell(kr)$. Where,

$j_0(kr) = \sin kr/kr$, $j_1(kr) = \sin kr/(kr)^2 - (1/kr)\cos kr$,
$j_2(k\,r) = [3/(kr)^3 - 1/k\,r]\sin kr - 3/(kr)^2 \cos kr$,---etc.

7.3.1. Calculation of the Scattering Amplitude F(θ)

It is important to illustrate this with a simple diagram as shown below:,

$$\psi(r,\theta,\phi) = \sum_{\ell m} R(r)\, Y_{\ell m}(\theta,\phi)$$

For a spherically symmetric (m=0) central potential, $P_{\ell 0} \propto P_\ell(\cos\theta)$; therefore,

$$\psi(r,\theta) = \sum_{\ell=0}^{\infty} R(r)\, P_\ell(\cos\theta)$$ ------------------------- (25).

Since the potential V(r) is spherically symmetric, the scattering must occur symmetrically around the direction of the incidence. Now, let the radial solution R(r) be as $\frac{u_\ell(r)}{r}$, so the equation 25 takes the form

$$\psi(r,\theta)= \sum_{\ell=0}^{\infty}\frac{u_{\ell(r)}}{r} P_\ell(\cos\theta) \text{ -------------------------- (26).}$$
$$=\sum_{\ell=0}^{\infty} b_\ell R(kr)P_\ell(\cos\theta), \text{ in general.}$$

So the radial Schrödinger equation reduces to

$$[\frac{d^2}{dr^2} +(k^2 - \frac{2m}{\hbar^2}V(r)-\frac{\ell(\ell+1)\hbar}{r^2})u_\ell(r)=0, \text{----------------(27).}$$

Applying the boundary condition, $u_\ell \longrightarrow 0$ as r $\longrightarrow 0$.

So $u_\ell(0)=0$ at B.C.

The solution of Schrödinger equation for a central force gives u_ℓ (r)(as r → 0) $\propto r^{\ell+1}$or r^ℓ(Q. M. I. Schiff); hence, the solution of $u_\ell(r)$ is uniquely fixed. So the solution is (asymptotic) as r $\longrightarrow \infty$; hence,

$u_\ell(r)\rightarrow a_\ell \sin (kr-(1/2) \ell\pi +\delta_\ell).$

Where δ is a real phase shift due to the potential V(r).

If V(r)=0, free particle, then of course δ consequently is zero, then $u_\ell \longrightarrow \sin[kr -(1/2)\ell\pi].$

But in case V(r)≠ 0, the wave function which is given in equation 26 becomes

$$\psi(r,\theta) \longrightarrow \sum_{\ell=0}^{\infty} a_\ell \frac{\sin (kr-\frac{\ell\pi}{2}+\delta_\ell)}{r} P_\ell(\cos) \text{---------------- (28).}$$

From equations 17 and 22, where
$$e^{ik.z} =e^{ikr\cos\theta}=\sum_{\ell=0}^{\infty}(2\ell + 1)i^\ell j_i(kr) P_\ell (\cos\theta) \text{------- (29).}$$

Hence, the total wave function $\psi(r, \theta)$ takes the form

$$\psi(r, \theta) = \sum_{\ell=0}^{\infty}(2\ell + 1)\, i^{\ell} j_{\ell}(kr)\, P_{\ell}(\cos \theta) + f(\theta, \varphi) + \frac{e^{ikr}}{r}. \quad (30)$$
$$r \longrightarrow \infty$$

Equation 30 can be described by the linear combination (26) since it contains all possible orbital angular momentum. In addition, it is a general solution to Schrödinger equation for a free particle (V(r)=0). So equation 30 is verified. For a free particle (V(r))=0,

$$\text{Now } j_{\ell}(kr) \longrightarrow \frac{\sin\left(kr - \frac{\ell\pi}{2}\right)}{kr}, \; \delta_{\ell}=0, \text{for } V(r)=0\text{------------}(31)$$
$$r \longrightarrow \infty$$

Here, it is considered that V(r) is symmetric, so $f(\theta, \varphi) \longrightarrow f(\theta)$ is the scattering amplitude.

Now equation 30 is the desired decomposition of the plane wave associated with the linear momentum $P(r)=\vec{k}\hbar$ into a set of partial waves. Each is associated with orbital angular momentum of value $\sqrt{\ell(\ell + 1)}\hbar$. The Bessel function $j_{\ell}(kr)$ and Legendre polynomial $P_{\ell}(\cos)$ specify the radial and the angular dependence of ith *partial wave*, respectively. Therefore, the plane wave is obtained by mixing together the totality of partial waves, giving each wave amplitude $(2\ell + 1)$, with a phase factor i^{ℓ}. It is quite necessary to notice that $R_{\ell}(kr)$ is essentially different from $j_{\ell}(kr)$ only for small r, where V(r)≠0. But for the free particle (V=0) and large r, both satisfy the same differential equation. So for large r, these functions essentially are the same and particularly for the r $\rightarrow\infty$ case, (asymptotic case). Hence, $R_{\ell}(kr)=\frac{Sin\left(kr - \frac{\ell\pi}{2} + \delta_{\ell}\right)}{kr}$, as $r \rightarrow \infty$ ------------------(32).

For r $\rightarrow \infty$, each of $R_{\ell}(kr)$ at most can differ from the corresponding $[j_{\ell}(kr)=[\frac{Sin\left(kr - \frac{\ell\pi}{2}\right)}{kr}]$ by a phase shift δ_{ℓ}. This phase arises from the different of r dependence of the two functions in the region where V(r)≠0, by equating the right sides of equations 17 and 26, obtained

$$\sum_{\ell=0}^{\infty}(2\ell+1)\, i^{\ell} j_{\ell}(kr)\, P_{\ell}(\cos)+f(\theta)\frac{e^{ikr}}{r} =$$
$$\sum_{\ell=0}^{\infty} b_{\ell}\, R_{\ell}(kr)\, P_{\ell}(\cos\theta),\ \text{as}\ r \to \infty$$

(Remember that $R_{\ell}(kr)$ only differs from $j_{\ell}(kr)$ by the phase shift δ_{ℓ}, as pointed out before.)

Now from equations 31 and 32, it is obtained that

$$\sum_{\ell=0}^{\infty} b_{\ell}\frac{Sin\left(kr-\frac{\pi\ell}{2}+\delta_{\ell}\right)}{kr}P_{\ell}(\cos\theta) \equiv$$
$$\sum_{\ell=0}^{\infty}(2\ell+)1)\, i^{\ell}\frac{Sin(kr-\frac{\pi\ell}{2})}{kr}P_{\ell}+f(\theta)\frac{e^{ikr}}{r},\ \text{as}\ r \to \infty.$$

From this equation, both $f(\theta)$ *and the entire of* b_{ℓ} can be determined in terms of δ_{ℓ}. With the aid of the complex variables, $\sin(kr\text{-}\ell\pi/2) = \frac{e^{i\left(kr-\frac{\ell\pi}{2}\right)}-e^{-i(kr-\frac{\ell\pi}{2})}}{2i}$. Multiply both side of the equation by (kr), grouping the coefficients of e^{ikr} and e^{-ikr}, it is obtained that

$$\sum_{\ell=0}^{\infty}\left\{[(2\ell+1)\, i^{\ell}/2i\{e^{-i\pi\frac{\ell}{2}}\}-\frac{b_{\ell}}{2i}e^{-i\pi-\frac{\ell}{2}e^{\delta_{\ell}}}]P_{\ell}(\cos\theta+k\, f(\theta)e^{ikr}\right\}$$

$$+\sum_{\ell=0}^{\infty}\left\{\frac{-(2\ell+1)}{2i}\left[i^{\ell}e^{i\pi\frac{\ell}{2}}+\frac{b_{\ell}}{2i}e^{-i\delta_{\ell}}e^{\frac{i\pi\ell}{2}}P_{\ell}(\cos\theta)\right]e^{-ikr}\right\} \text{------ (33).}$$

As $r \to \infty$

Equation 33 can be satisfied for arbitrary (r) only if the coefficients of both e^{ikr} and e^{-ikr} vanish separately. It implies that

$$\sum_{\ell=0}^{\infty}[-\frac{(2\ell+1)}{2i}i^\ell e^{\frac{i\pi\ell}{2}} +\frac{b_\ell}{2i}e^{\frac{i\pi\ell}{2}}e^{-i\delta_\ell}]P_\ell(\cos\theta)=0.------------(34).$$

From equation 34, it is demanding the coefficients $of\ P_\ell$ vanish, and then it is obtained that

$$\frac{2\ell+1}{2i}i^\ell =\frac{b_\ell}{2i} \Rightarrow [\ b_\ell = (2\ell +1)i^\ell e^{i\delta_\ell}]---------------------------(35).$$

By introducing equation 35 into equation 33, it is easy to find that the scattering amplitude f(θ) is given by

$$f(\theta)=\frac{1}{2ik}\sum_{\ell=0}^{\infty}(2\ell + 1)(e^{2i\delta_\ell} -1)P_\ell(\cos\theta),\ \text{leads to}$$
$$f(\theta)=\frac{1}{2ik}\sum_{\ell=0}^{\infty}(2\ell + 1)\ e^{i\delta_\ell}\sin\delta_\ell P_\ell(\cos\theta\ )\ ---------------(36).$$

Hence, since f(θ) is known, the total cross section can be calculated as follows, from equation 20 the differential scattering cross section is given as $\frac{d\sigma}{d\Omega}$=f*(θ)f(θ) =lf(θ)l². Using equation36, remembering that $\int_{-1}^{+1}P_\ell\text{-}\ P_\ell d(\cos\theta)=\frac{2}{2\ell+1}\delta_{\ell\text{-}\ell}$, where $\delta_{\ell\text{-}\ell}$=1 if ℓ^-=ℓ or 0 if $\ell^-\neq\ell$.

(The use ofℓ and ℓ^- is to indicate specifically that the series are to be summed independently.) As it was early shown, the scattering cross section $\sigma=\int\frac{d\sigma}{d\Omega}$ dΩ $=\int\frac{d\sigma}{d\Omega}(2\pi\sin\theta d\theta)$ =2π $\int_{-1}^{+1}\frac{d\sigma}{d\Omega}$d($\cos\theta$).

So it can be evaluated by multiplying the two series (36) and integrating each of the resulting terms over $\cos\theta$ using the known properties of $\cos\theta$ such as

d($\cos\theta$)=-$\sin\theta d\theta$

Then the scattering cross section will be:

$$\sigma = \frac{4\pi}{k^2} \sum_{\ell=0}^{\infty} (2\ell + 1) \sin^2 \delta_\ell \quad \text{------------------------------ (37).}$$

By investigation of equation 37, it is clear if k is given, which represents the energy of the incident particle; and also the phase shift (δ) for a given ℓ is known, the total cross section (σ) easily can be calculated. In addition, the differential scattering cross section ($\frac{d\sigma}{d\Omega}$) can be computed too. The energy of the incident particle is known according to the nature of the designed experiment for the required measurement. The phase shift can be evaluated by solving the well-known radial equation with $V(r) \neq 0$, for each ℓ.

Now, the scattering problem in view of the radial equation might be represented graphically as shown below:

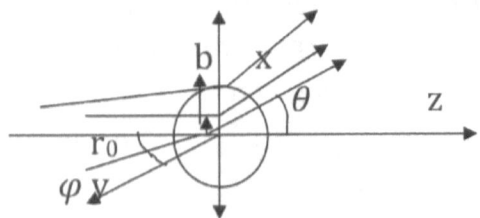

$$b = \text{impact parameter} = \frac{\ell\hbar}{P} = \frac{\ell}{k}$$

This can be illustrated more clearly by drawing the energy verses the potential range (r) as shown below:

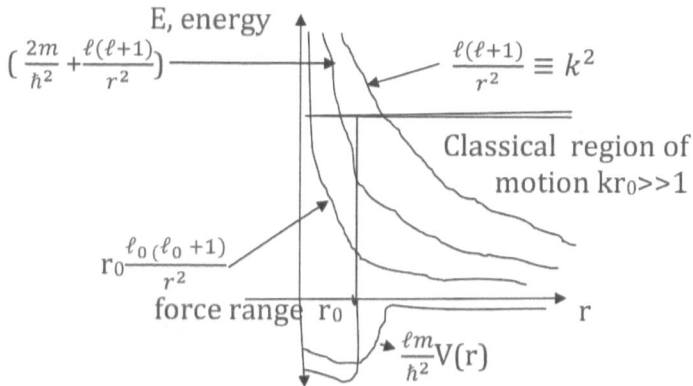

Sketch illustration of partial wave analysis scattering.
To give this its physical meaning, it is necessary to approximate the wave shift analysis to be useful. Here, $V(r)=0$ for $r>r_0$, beyond the force range. So $k^2 = \frac{\ell_0^2}{r_0^2}$ or

$kr_0 = \ell_0$, $\hbar k \longrightarrow \hbar\ell_0 \to pr_0 = L$
For $\ell > kr_0$, small δ_ℓ, $\ell \leq \ell_0$, large δ_ℓ.

[In the case of Coulomb potential ($V \propto 1/r$), the wave phase shift analysis is less of use.] Now assume the condition that $V(r)=0$, $r>r_0$, which implies δ_ℓ is energy or k dependent; it behaves just as $\delta_\ell \propto k^{2\ell+1}+n\pi$ (as $k \longrightarrow 0$), so if $k=0$, $\delta_\ell = n\pi$, n is integer. This means the phase shift measures only the amount by which the phase of the radial wave function for any angular momentum differs from the nonscattering case ($\delta_\ell=0$). Each phase shift is, of course, a function of the energy or k. [As an exercise: show that $\int u_{\ell k}(r)u_{\ell k^-}(r)dr = \frac{\pi}{2k^2}\delta(k-k^-)$, where $\frac{u_{\ell k}}{r}=R_{\ell k}(r)=\sin(kr-\frac{\ell\pi}{2}+\delta_\ell)$.]

Now if the impact parameter $b=\frac{\ell\hbar}{p}=\lambda$ accedes r_0 or when $\ell > kr_0$, no appreciable scattering takes place, i.e., $\delta_\ell=0$. But if $kr_0 \gg 1$, then the classical argument is applicable, so it is expected, for $\ell > kr_0$ is vanishingly small. But for the case $kr_0 \ll 1$ (low

185

energy), the incident waves are long; and at given instant, the phase of the wave changes very little across the scattering region. So all special senses of direction are lost. This implies that $f(\theta)$ has to be independent of (θ)!!.

By equation 36, all phase shifts vanish, except for corresponding to $\ell = 0$. This means only the S-wave scattering occurs, and it is isotropic in characteristic because $P_0(\cos\theta) = 1$. The conclusion is: the partial wave sum (36) is a particularly useful representation of the scattering amplitude $f(\theta)$.

On a physical ground, only a few angular momenta are expected to contribute significantly. In fact, if $V(r)$ is not known before hand, one may attempt to determine as a function of k empirically by comparing scattering data at various energies with the formula (37).

In the differential scattering cross section (36), or angular distribution, as it is often called, contributions from the different partial wave (different angular momenta) interact with each other because $f(\theta)$ is a sum of terms with different values. This sum is often said to be a coherent superposition of different angular momenta. But such interference does not occur in σ given by equation 37. This implies that σ constitutes in a coherent sum of partial waves contribution, σ_ℓ.

Hence, $(\sigma_\ell)_{\max} = \frac{4\pi}{k^2}(2\ell + 1)$ ------------------- (38).

This equation is a contribution of each angular momentum value, at most the partial cross-section; the total scattering cross-section is given by equation 37. It is possible to compare with the classical scattering cross section (σ_c) per unit $\hbar$ of angular momentum. σ_t is of the same value (as an order of magnitude) as the maximum σ_c (the domain of the classical physics, $(\hbar \rightarrow 0)$. To see this comparison, a formulation to σ_c

will be performed next. Consider the following diagram for beam of particles incident on the target A (classical case).

Let b the impact parameter; it is the least distance between the point at which the incident particle is scattered.

Classically, $d\sigma$=b db dφ, and the scattering is through the angle (θ) as shown in the diagram bellow:

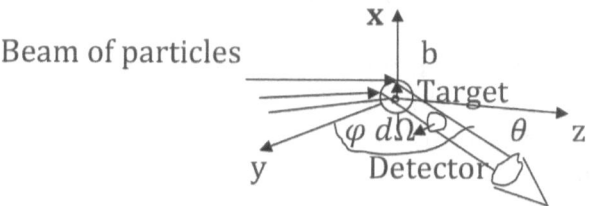

It is a cylinder bounded by b,--b+db, and φ--+dφ where φ is the azimuthal angle.

General Theory: $\delta_\ell \propto k^{2\ell+1}+n\pi$, n is integer.

δ_ℓ =t_ℓ +ρ_ℓ, where t_ℓ is the kinematical (energy dependent only), and ρ_ℓ is the dynamical part. The behavior of the scattering amplitude f(θ) is given as:

$$f(\theta) \xrightarrow[k \longrightarrow 0]{} 1/k\sum_{\ell=0}^{\infty}(2\ell + 1)\, e^{i\delta_\ell}\sin\delta_\ell P_\ell(\cos)\text{-------(39)}.$$

δ_ℓ is the difference in the phase shift between the asymptotic form of the actual radial function R_ℓ (r) and the radial functionj_ℓ(kr), where V(r)=0. Hence,δ_ℓ completely determines the scattering where each of δ_ℓ =0^0, 180^0 which implies (no scattering,$\sigma = 0$). It is necessary to note that the derivation of equation 35 is valid whether or not there exists the assumed radius; beyond it, V(r) is negligible, with the condition V(r)=0, faster than 1/r.

Therefore, the conclusion is the method of partial waves is most useful at low energies. Also, as it was shown before, the closest distance b is given by $\frac{\ell}{k'}$, it is equivalent to the saying that the classical particle is not scattered if its angular momentum is quite enough to help it not entering the region of the affected potential field ($r<r_0$).

7.3.2. Determination of δ_ℓ and Scattering Resonances

For more understanding of this important subject, it is advisable to go further to find out how phase shift can be calculated and how its properties are related to the potential $V(r)$. To do so, it is needed to solve Schrödinger equation for the given potential and obtain the asymptotic form of the solution (as $r \to \infty$). There are two wave functions to the solution, as it is known. They are exterior and interior given by

$$R_{\ell k}(r)=A_\ell j_\ell(kr)+B_\ell n_\ell(kr)=\frac{u_{\ell k(kr)}}{r}, \text{ } r>r_0 \text{ -------(40).}$$

Where $j_\ell(kr)$ is the spherical Bessel function, and $n_\ell(kr)$ is the spherical Neumann function. Let $\psi_1= j_\ell(kr) Y_\ell^m(\theta,\varphi)$, where (Y_m^ℓ) is the spherical harmonic function.

And let $\psi_2=n_\ell (kr)(Y_\ell^m)$. Remember that

$$j_\ell(kr)=j_\ell(z)=\frac{z^\ell}{2^{2\ell+1}}\int_{-1}^{+1} e^{izs} (-S^2)ds ----(41).$$

And $j_\ell(z) = \sqrt{\frac{\pi}{2z}}J_{\ell+\frac{1}{2}}(z)$ ---------------------- (42).

(See QM by Merezbacker, 1994 Edition.)

Also, it is known that the Bessel functions are either even or odd functions, depending on the parity which it depends on the values of ℓ. So $j_\ell(kr)$ is the regular solution of the Schrödinger equation, $\frac{d^2 R}{d\rho^2}+\frac{dR}{d\rho}+[1-\frac{\ell(\ell+1)}{\rho^2}]R=0$ ----- (43).

Where $\rho=kr=\sqrt{\frac{2mE}{\hbar^2}}.r$, $V(r)=0$, free particle.

The regular solution is

$$j_\ell(\rho)(\rho \to \infty) = \frac{Cos(\rho-\frac{\pi(\ell+1)}{2}}{\rho}, \quad \rho \gg \ell.$$

The irregular solution at the origin is

$$n_\ell(\rho)(\rho \to \infty) \frac{Sin(\rho-\frac{\pi(\ell+1)}{2}}{\rho}, \quad \rho \gg \ell.$$

$\left.\begin{array}{c} \\ \\ \\ \\ \\ \\ \end{array}\right\}$ ------- (44)

Another two solutions of equation 43 (singular solutions) are the well familiar for mathematicians and mathematical physicists. They are the so-called the spherical Hangle functions, $h^{(1)}(z)$, and they are defined as

$h_\ell^{(1)}(z)=j_\ell(z)+i\,n_\ell(z)$ -------------------------- (45).

And, $h_\ell^{(2)}(z)=j_\ell(z)-in_\ell(z)$ --------------------(46).

Where, $h_\ell^{(1,2)}(z)=\mp\frac{i(2\ell)!}{2^\ell\,\ell!}Z^{-\ell-1}+\mathbb{O}^{(1,2)}(Z^{-\ell+1})$-----(47).

So for asymptotic solutions it is possible to find

$h_\ell^{(1)}(z) \cong \frac{1}{z}exp.\,[i\{z-(\ell+1)\frac{\pi}{2}\}]$ ----------- (48).
$r \to \infty\ z\gg\ell$

$h_\ell^{(2)}(z) \to \cong \frac{1}{z}exp.[-i\{(z-(\ell+1)\frac{\pi}{2}\}]$------------(49).
$\quad r \to \infty,\ z=\ell$

To determine δ_ℓ, it is necessary to calculate the interior wave function and make it to join smoothly the exterior solutions

(40) at $r=r_0$ (applying boundary conditions), and then δ_ℓ can be expressed in terms of the logarithmic derivatives at $r=r_0$; hence, $B_\ell=(\frac{r_0}{R_\ell}\frac{dR_\ell}{dr})r=r_0,$

From equation 40, $\frac{B_\ell}{A_\ell}=$-tan δ_ℓ, using this, it leads to

$$B_\ell(k)=k\ r_0\frac{\overline{j}_\ell(kr_0))cos\delta_\ell-\overline{n}_\ell(kr_0)sin\delta_\ell}{j_{\ell(kr_0)}\ cos\ \delta_\ell-n_\ell(k\,r_0)\ sin\delta_\ell}\ \text{-----------}\ (50)$$

Conversely, $let\ S_\ell=e^{2i\delta_\ell}=\frac{j_\ell-in_\ell}{j_\ell+in_\ell}\times\frac{B_\ell-kr_0\frac{\overline{j}_\ell-i\overline{n}_\ell}{j_\ell-in_\ell}}{B_\ell-\frac{\overline{j}_\ell+ij_\ell}{j_\ell+in_\ell}}\text{-------}\ (51).$

All j_ℓ and n_ℓ and their derivatives are to be evaluated at the argument (kr_0). Now let equation 51 represents the real phase angle (ξ_ℓ). It is a phase shift at $B_\ell\longrightarrow\infty$, which means that $R_\ell(r_0)=0$; hence, $R_\ell\longrightarrow0$ at $r=r_0$ for all values of ℓ. If V(r) is a type of a hard sphere core of radius, the wave function cannot penetrate into the interior at all. It indicates that ξ_ℓ is the hard sphere shift. This concept is useful to be worked out. For that, let the parameters $\Delta\ell$ and δ_ℓ introduced such that $\Delta\ell+iS_\ell=$ $kr_0\frac{\overline{j}_\ell+i\overline{n}_\ell}{j_\ell+in_\ell}$ ----------------- (52).

Then the following simple relation can be found (show it as exercise), exp. [2i $(\delta_\ell-\xi_\ell)$] $=\frac{B_\ell-\Delta\ell+iS_\ell}{B_\ell-\ell\Delta-iS_\ell}$ ------------- (53).

And $e^{i\delta_\ell}\ sin\ \delta_\ell=e^{2i\xi_\ell}[\ \frac{S_\ell}{B_\ell-\Delta\ell-iS_\ell}+e^{i\xi_\ell}sin\xi_\ell\ ]$-----------(54).

Equation 54 shows explicitly that the partial wave contributions to the scattering amplitude depend on B_ℓ.

7.3.2.1. Application on Spatial Case (Square Well Potential)

Take a square potential well of range (a) and depth V_0. Consider the solution that $R(kr) = \frac{u_{\ell(r)}}{r} = j_\ell(kr)$, where $k = \sqrt{\frac{2m(E+V(r))}{\hbar^2}}$, $V(r) = -V_0$. Hence, $B_\ell = k^- a \frac{j_\ell^-(ka)}{j_{\ell(ka)}}$, $r=a$, j^- is a derivative of j_ℓ. Now for $\ell = 0$, (S-wave). Then $j_0(z) = \frac{\sin z}{z}$, $n_\ell(z) = -\frac{\cos z}{z}$.

These data leads to $\xi_0 = -ka$, $\Delta_0 = -1$, $S_0 = ka$, $B_0 = k^- a \cot an k^- a -1$, (show that as an exercise). Applying these data to equation 54, where the scattering amplitude for S-wave is given by:

$$f_0 = \frac{1}{k} e^{i\delta_0} \sin\delta_0 = [\frac{k}{k^- cotonk^- a -ik} - e^{ika}\sin ka] \text{ ------ (55)}.$$

In the limit, when $k \longrightarrow 0$, $E \longrightarrow 0$, then

$$\sigma_0 = 4\pi a^2 [\frac{\tan k_0^- a}{k_0^- a} -1]^2 \text{ ----------------------------------- (56)}.$$

This is the nonvanishing isotropic S-wave scattering cross section.

7.3.2.2. Resonances of Scattering

It might happen that in a small range of energy, a rapid change in the logarithmic derivative can be expressed as $B_\ell(E) = C+bE$. It is a linear approximation, ------------- (57).

While ξ_ℓ, $\Delta\ell$, and S_ℓ, which are characterizing the external wave function, very slowly and smoothly with energy, they

191

may be considered constants. Substituting these approximations (53) lead to $e^{2i(\delta_\ell-\xi_\ell)}=\dfrac{E-E_0-i\frac{\Gamma}{2}}{E-E_0+i\frac{\Gamma}{2}}$ ------- (58),

where $E_0=\dfrac{\Delta\ell-c}{b}$, $\Gamma/2=\dfrac{-S_\ell}{b}$----------------------(59).

Then $\tan(\delta_\ell-\xi_\ell)=\dfrac{\Gamma}{2(E_0-E)}$---------- (60).

(As an exercise, prove equation 60.) Since B_ℓ is a decreasing function E, and then b must be negative, S_ℓ is positive, so Γ is positive; therefore, equations 58 and 60 are useful if E_0 and Γ are reasonable constants and if equation 57 is accurate relation over an energy range compared with Γ. From equation 58, it can be seen that the phase $2(\delta_\ell-\xi_\ell)$ changes by 2π as E is given by $E_0+\dfrac{\Gamma}{2}\geq E\geq E_0-\dfrac{\Gamma}{2}$, and if ξ_ℓ is also nearly constant in this interval of E, then δ_ℓ changes by π and $\sigma_\ell \propto \sin^2\delta_\ell$ changes abruptly. Such sudden variations in the phase shifts are called resonances, where E_0 is the resonance energy, and Γ is the resonance width. In the case δ_ℓ is near the resonance, the contribution of the corresponding partial wave to the scattering amplitude $f(\theta)$ can be written as (according to 57, 58, 59):

$$f_\ell=\frac{2\ell+1}{k}e^{2i\xi_\ell}[(\frac{\frac{\Gamma}{2}}{E_0-E+i\frac{\Gamma}{2}})_r+(e^{-i\xi_\ell}\sin\xi_\ell)_{nr}]P_\ell(\cos).\text{---------} (60).$$

This gives a neat separation of the resonant part (r) of the partial wave amplitude from the nonresonant part (nr), which only depends on (ξ_ℓ). For low energy or high angular momentum (L) the phase shifts ($\xi_\ell \approx 0$). This reduces equation 60 to

$$\tan\delta_\ell=\frac{\Gamma}{2(E_0-E)}\text{----------------------------}(61).$$

Then the resonant terms in equation 61 predominates the other and contributes to σ_t an amount $\frac{d\sigma}{d\Omega}=|f_\ell|^2$ to become

$$\sigma_t=\int \frac{d\sigma}{d\Omega}\,\Omega=\int |f_\ell|^2 d\Omega=\frac{4\pi}{k^2}(2\,\ell+1)\sin^2\delta_\ell=$$

$$\frac{4\pi(2\ell+1)}{k^2} \quad \frac{\Gamma^2}{4(E-E_0)^2+\Gamma^2}\text{--------------------}(62).$$

[Equation 62 is Briet-Wigner cross section in nuclear interaction.]

[Remember $\int_{-1}^{+1} P_\ell P_\ell d(\cos\theta =\frac{(2\ell+1)}{4\pi}]$.

For small Γ (width of resonance), a sharp maximum peak is obtained. It is centered at E_0 with a symmetrical shape similar to that of the transmission peaks shown in the figures below,

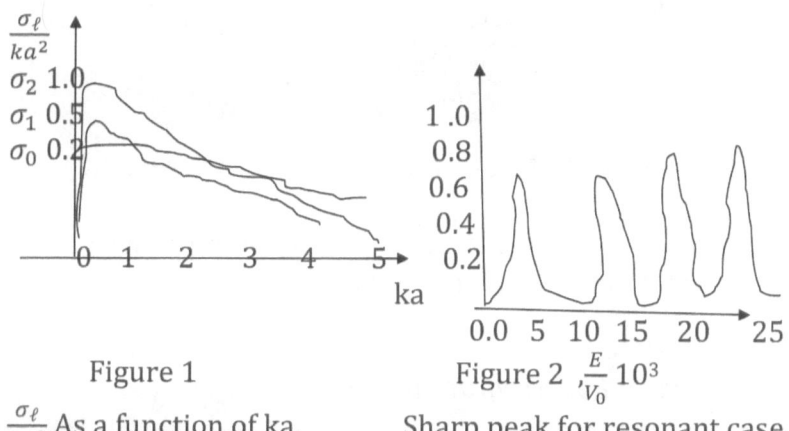

Figure 1

$\frac{\sigma_\ell}{ka^2}$ As a function of ka.

Figure 2 , $\frac{E}{V_0} 10^3$

Sharp peak for resonant case.

Or this might be demonstrated more clearly as

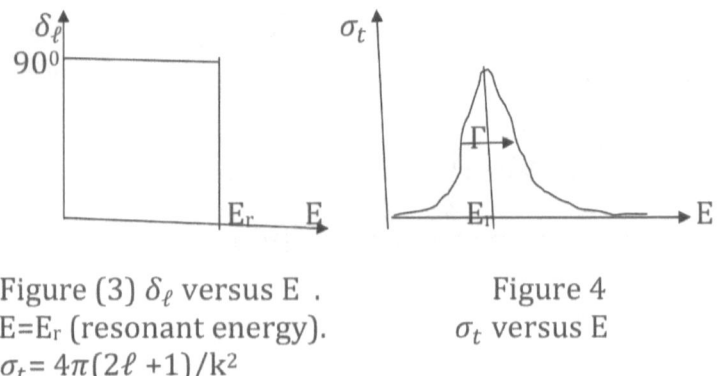

Figure (3) δ_ℓ versus E .
If E=E$_r$ (resonant energy).
$\sigma_t = 4\pi(2\ell +1)/k^2$

Figure 4
σ_t versus E

So one might plot σ_t versus the energy E to show the resonant cross section as in figure 5 below,

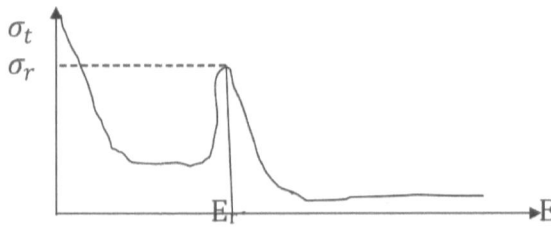

Figure 5 σ_t versus E (Notice the resonant cross section.)

In the case of going down (phase shifts going down), it needs only to put (-) in front of Γ (peak width).

7.3.2.3. Integral Representation of δ_{ℓ}.

Let $u_\ell(r) \rightarrow \sin(kr - \frac{\ell\pi}{2} +\delta_\ell)$, --corresponds to V(r).
As r$\rightarrow \infty$

And $u_\ell(r) \rightarrow \sin(kr - \frac{\ell\pi}{2} +\delta_\ell^-)$, - corresponds to V(r$^-$).
As r $\rightarrow \infty$

It is possible to show that:

$k \sin (\delta_\ell^- - \delta_\ell) = -\frac{2m}{\hbar^2} \int_0^{+\infty} [V(r) - V^-(r)]\, u_{\ell^-}(r)\, u_\ell(r) dr.$

If $V^-(r) = 0$, $\delta_\ell^- = 0$, $u_\ell^-(r) \rightarrow kr\, j_\ell(kr)$.

Hence, $\sin\delta_\ell = -\frac{2m}{\hbar^2} \int_0^{\infty} V(r)\, j_\ell(kr) u_\ell(r)\, r\, dr.$

But if $V(r)$ is sufficiently weak, $u_\ell(r) = kr j_\ell(kr)$.

Because δ_ℓ is very small, $\sin\delta_\ell = \delta_\ell$; therefore δ_ℓ takes the integral form, $\delta_\ell = -\frac{2mk}{\hbar^2} \int_0^{\infty} V(r)\, j_\ell(kr) u_\ell(r)\, r^2\, dr$ ------- –(63).

This equation is called Born approximation for the phase shifts. As was shown before, if kr goes to zero, $j_\ell(kr)$ takes the form: $j_\ell(kr) \propto (kr)^\ell$, and $\delta_\ell \propto k^{2\ell+1}$

As $kr \rightarrow 0$

Now suppose $\Delta V = V^- - V$ is very small, then $\Delta\delta_\ell = \delta_\ell^- - \delta_\ell$, so considering u and u^- are identical, it is possible to write that $\sin\Delta\delta_\ell = -\frac{2m}{\hbar^2 k} \int_0^{\infty} \Delta V(r) u_\ell^-(r) u_\ell(r) dr.$

Since, u and u^- are identical, then $\sin\Delta\delta_\ell = -\frac{2m}{\hbar^2 k} \int_0^{\infty} \Delta V(r)\, u_\ell^2(r) dr$ ------------------(64).

Suppose $V(r) > 0$ for all (r), then $V(r)$ is more repulsive. It means $\Delta\delta_\ell < 0$. But if $V(r) < 0$, it is more attractive. It means that $\Delta\delta_\ell > 0$. From this, the following can be concluded:

1. $\delta_\ell > 0$, for attractive potential $V(r) < 0$.
2. $\delta_\ell < 0$, for repulsive potential $V(r) > 0$

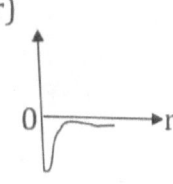

This is the reason why the resonance passes up to 90^0 with δ_ℓ(figure 3, page 194), where suddenly, the attractive potential increases, leading to $\delta_\ell > 0$.

7.3.3. Scattering Length and Effective Range for Short-Range V(r)

Take the S-wave ($\ell = 0$), where Schrödinger equation takes the form, $\frac{d^2 u}{dr^2} + (k^2 - U(r)) u(r) = 0$ ---------------------(65).

Where $U(r) = \frac{2m}{\hbar^2} V(r)$. Suppose $u_{1(r)}$ and $u_2(r)$ are the two solutions of different energy k_1^2 and k_2^2 respectively. They satisfy the condition, $u_1(0) = 0, u_2(0) = 0$ ------------- (66).

Also, they are normalized such that asymptotically, they take

the forms $u_1(r) \rightarrow \frac{1}{\sin\delta_1} \sin(k_1 r + \delta_1)$ $\left.\begin{array}{c}\\\\\end{array}\right\}$ ---------------(67).

$u_2(r) \rightarrow \frac{1}{\sin\delta_2} \sin(k_2 r + \delta_2)$

Using equation 64 and equation 67, the following is obtained:

$[u_2 \frac{du_1}{dr} - u_1 \frac{du_2}{dr}]_0^R = (k_2^2 - k_1^2) \int_0^R u_1 u_2 dr$ ------------------(68).

(As an exercise, show equation 68, where R is an arbitrary radial distance. For simplification, take two free particles solutions, $\mathcal{V}_1(r) = \frac{1}{\sin\delta_1}\sin(k_1 r + \delta_1)$, --------------------(69).

$\mathcal{V}_2(r) = \frac{1}{\sin\delta_2}\sin(k_2 r + \delta_2)$ --------- ----------(70).

For equation 64, where U(r)=0, as it was the case for equation 65, with the aid of equations 68 and 69, it will be obtained that $[\mathcal{V}_2 \frac{d\mathcal{V}_1}{dr} - \mathcal{V}_1 \frac{d\mathcal{V}_2}{dr}]_0^R = (k_2^2 - k_1^2)\int_0^R \mathcal{V}_1 \mathcal{V}_2 dr$ -- (71).

The scattering length is defined as

$$-\frac{1}{a}=\text{limit } [\text{k cot}\delta(k)] \text{ -------------------- (72)}.$$
$$K \to 0$$

From equation 72 substitute in equation 70, as $R \to \infty$, the following is obtained:

$$k_2 \cos\delta_2 - k_1 \cot\delta_1 = (k_2^2 - k_1^2)\int_0^R \mathcal{V}_1 \mathcal{V}_2 - u_1 u_2)dr \text{-------(73)}.$$

If $k_1 \to 0$, denote k_2 by k and δ_2 by δ equation 73 can be written as $\text{k cot}\delta = -\frac{1}{a} + \frac{b}{2} k^2 \text{ ------------------(74)}.$

Where, $b = 2\int_0^\infty \mathcal{V}_0 \mathcal{V} - u_0 u) dr \text{---------------(75)}.$

The factor 2 in equation 75 is introduced so that b (or r_0 as it will be seen later on) has the meaning of the potential range.

From equations 67, 68, and 69, it is noticed that the integrand of (75)$\neq$0 only in the region where U(r) is appreciable. In this region, the wave function u(r) will not depend very much on the energy k^2 if $|U(r)| >> k^2$. This is making the approximation of replacing u by u_0, (for zero energy) in equation 73, i.e., the first two terms will be taken in a power series expansion in k^2 of $\text{k cot}\delta$, $\text{k cot}\delta = -\frac{1}{a} + \frac{r_0}{2} k^2 + O(k^4) + \text{--(76)}.$

Where $r_0 = 2\int_0^\infty (\mathcal{V}_0^2 - u_0^2) dr \text{ ------------------------- (77)}.$

This is defined as the effective range of the potential V(r). According to equations 67, 69, 70, the zero energy has the asymptotic form:

$$u_0(r) \to \mathcal{V}_0(r) = \lim_{k \to 0} (\cos kr + \cot\delta \, \sin kr) = 1 - \frac{r}{a} \quad ----(77).$$

Also, from equations 67, 69, 70, $(\mathcal{V}_0^2 - u_0^2)$ vanishes for $r > r_0$ outside the range of U(r). It is that both (a) and (r_0) are determined by U(r), so they are sensitive to the exact form of U(r). But they depend on some, integrated or average, property of U(r). The physical meaning of(a) can be clarified by the following considerations:

1. For $k^2 \to 0$, then from equations 35 and 74, (let $\ell = 0$ in equation 35) where $f(\theta) = \frac{1}{2ik}(e^{2i\delta_0} - 1)$ and $\sigma = \frac{4\pi}{k^2}\sin^2\delta_0$, then

$$\sigma = \frac{4\pi}{[k^2(1 + \cot^2\delta)]} = 4\pi a^2 ----------------- (78).$$

2. From equation 63 for δ_0 and for $k \to 0$, a takes the form.

$$a = \int_0^\infty rU(r)\, u_0(r)\, dr ------------------------ (79).$$

Where $u_0(r)$ is normalized by the asymptotic behavior,
$u(r) \to \frac{1}{k} \sin (kr - \ell\frac{\pi}{2} + \delta_\ell)$; hence, it is different from that in the

equation 77, it is, $u_0(r) \to \lim(k \to 0) = (\frac{\sin(kr+\delta)}{k}) =$

$$(r - a)\cos\delta -------- ----------------------------(80)$$

Where $\cos\delta$ has the values $+1$ or -1.

Now if Born approximation is valid, then u_0 is replaced by the

field free solution $r\sqrt{\frac{\pi}{2kr}}j_{\frac{1}{2}}(kr) = \frac{1}{k}\sin kr$ (as $k \to 0$) $\to r$. Hence,

equation 79 becomes, $a \cong \int_0^\infty U(r)r^2\, dr --------(81).$

(Note: Born approximation is not good for the proton-neutron and e-He scattering.) Also, it can be noticed that there are more than three relations(71, 79, 81) which are characterizing the scattering length (a).

3. The sign of a depends on U(r). The variation of a with U(r) can be brought out explicitly by the following example:

Take a rectangular well potential such that:

$$U(r) = \begin{vmatrix} -\beta^2 \ r<R \\ 0 \quad r>R \end{vmatrix} \text{----------------------------- (82)}$$

If the energy $(E)=-\gamma^2$, $\gamma>0$, a discrete state, then Schrödinger equation for this physical system is

$$\left. \begin{array}{l} [\frac{d^2}{dr^2}-\gamma^2+\beta^2\ ]u_\ell=0, r<R, V(r)=-\beta^2 \\ [\frac{d^2}{dr^2}-\gamma^2]u_\ell(r)=0, r>R, V(r)\ 0 \end{array} \right\} \text{------------(83).}$$

The solution of equation 83 is very well known; it is

$$u_\ell(r) = \begin{vmatrix} A\sin(\sqrt{\beta^2-\gamma^2}\ r, r<R \\ \\ Be^{-i\gamma r}, r>R \end{vmatrix}$$

With the aid of the boundary conditions, it is obtained that

$$[\tan(\sqrt{\beta^2-\gamma^2}\ R) = \frac{-\sqrt{\beta^2-\gamma^2}}{\gamma}], \text{at r=R ------- (84).}$$

The conditions which are imposed on βR for the existence of 1, 2, or more discrete states are:

1. $\frac{\pi}{2}<\beta R<\frac{3}{2}\pi$, for only one discrete state.
2. $\frac{3\pi}{2}<\beta R<\frac{5}{2}\pi$ for only two discrete states.

And so on.

To calculate the scattering by the potential described by equation (82), again using Schrödinger equation for U(r)=0, r>R, and U(r)≠0 for r<R, but E>0, i.e.,

$$[\frac{d^2}{dr^2}+k^2+\beta^2]u_k=0, \ r<R,$$

$$[\frac{d^2}{dr^2}+k^2]u_k=0, \ r>R,\} \ E>0 \ \text{------------------(85);}$$

Unbound System

The solution is given by

$$u_k = \begin{cases} C \sin\sqrt{\beta^2+k^2}\ r, & r<R \\ \\ D \sin(kr+\delta), & r>R \end{cases} \quad \text{----------- (86).}$$

The continuity condition at the boundary (BC) at r=R yields
$k \cot (kr+\delta) = \sqrt{\beta^2+k^2} \cot\sqrt{\beta^2+k^2}\ R.$

Or $k \cot\delta = \dfrac{k \tan\left(\sqrt{\beta^2+k^2}\ R\right)\tan kr+\sqrt{\beta^2+k^2}}{\tan \left(\sqrt{\beta^2+k^2}\ R-\frac{1}{k}\sqrt{\beta^2+k^2} \tan kR\right)}$ ------------------(87).

With the definition equation 72 for (a), equation 87 gives
$a = R -\dfrac{1}{\beta} \tan\beta R$ --- (as exercise verify this) ----- --- (88).

As $0<BR<\frac{\pi}{2}$, it means U(r) has no discrete state, then a decreases, a<0 as $\beta R \to \frac{\pi}{2}$, (a) becomes negatively infinite. The cross section (σ) given in equation 78 becomes infinity (as $\beta R \to \frac{\pi}{2}$), a resonance is there at zero energy.

7.3.4. Integral Equation of Scattering, Green's Function:

So far, exact theory of scattering of a particle by a potential field effect has been dealt with. It is based on the solution of Schrödinger (differential) equation subject to the asymptotic conditions, corresponding to the physical situation of the incident wave and the outgoing spherical wave. In the following, this problem will be treated on the basis of transforming Schrödinger equation into an integral equation in a form that automatically ensures the correct prescribed asymptotic conditions. Up till now, $\psi_a^+(r)$ is defined as a solution of Schrödinger equation satisfying certain boundary conditions.

Next, ψ_a^+ will be expressed as a solution of a certain integral equations, where it will be written as corrections to Born approximation. To introduce Green function G(r), Schrödinger equation, $[\Delta + k^2 - U(r)] \psi(r) = 0$, $\Delta = \nabla^2$, can be written as $[\mathcal{H}_0 - E]\psi(r) = \mathcal{X}(x)$- ---------------------- (89).

Where $\mathcal{H}_0 = -\Delta$, $E = k^2$, and $\mathcal{X} = -U(r)\psi(r)$. ------------(90).

The asymptotic condition (as stated previously) is

$$\psi^+(r) \rightarrow e^{ikz} + f(\theta)\frac{e^{ikz}}{r}$$ --------------------------- (91).

Now according to the theory of the differential equation, equation 89 is a nonhomogeneous equation; therefore, it has two solutions:

$$\psi(r) = \phi(r) + \int G(r, r^-)\mathcal{X}(r^-)dr^-$$ ------------------- (92),

where $\phi(r)$ is the particular solution to the homogeneous equation, $[\mathcal{H}_0 - E]\phi_k(r) = 0$, $\phi_k = e^{i\vec{k}.\vec{r}}$ --------------- (93),

and $G(r, r^-)$ is the Green's function for the nonhomogeneous solution, i.e., $[\mathcal{H}_0 - E] G(r, r^-) = -\delta(r - r^-)$ ---------------(94).

So the general solution to equation 92 is the sum of the two solutions, (referred to any advanced calculus book).

$\delta(r - r^-)$ is the known Dirac delta function (refer to appendix A). Hence, the problem is to prove equation 92, then find G(r, r^-) explicitly. To do that, let a particular integral of the nonhomogeneous equation 89 be expanded in terms of the complete orthonormal set in equation 93. Hence,$\psi(r)$ takes the form, $\psi(r) = \int C_k \text{-} \phi_k \text{-} dk^-$, where C_k- to be determined from equation 89, using the simple form of equation 94.(Notice equation 90), i.e.,

$$(\Delta + k^2)G(r,r^-) = -4\pi\delta(r) \text{-----------------------------(95)}.$$

By introducing Fourier integral:
$$G(r) = \int g(k^-) e^{ik^- \cdot r} d^3k^- \text{----------------------------- (96)}.$$

And the Fourier representation of δ function. (See appendix A.) Substituting equation 96 into equation 95, one gets

$$g(k^-) = \frac{1}{2\pi^2} \frac{1}{k^{(-)2} - k^2}, \text{ so G will be,}$$

$$G(r) = \frac{1}{2\pi^2} \int \frac{e^{ik^- \cdot r}}{k^{(-)2} - k^2} d^3k^- \text{--------------------(97)}.$$

By integrating over the angles, with an algebraic trick, the following convenient form is obtained:

$$G(r) = \frac{1}{i\pi r} \int_{-\infty}^{+\infty} \frac{e^{ik^- \cdot r}}{k^{(-)2} - k^2} k^- \, d \, k^- \text{--------------------(98)}.$$

The singularity of equation 98 is at $k^- = \pm k$. It implies that this integral does not really exist. Again, it indicates that the representation of the solution of equation 95 as a Fourier integral is failed. This problem, mathematically, can be solved using a certain trick to let this integral of useful use. This can be done by adding an infinitesimal positive quantity (imaginary) such as $i\epsilon$ to k^2, so $k^{(-)^2}$ -$(k^2+i\epsilon)$ =$k^{(-)^2}$ -k^2-$i\epsilon$=$i\epsilon$, if $k^- =\pm$ k.

This gets rid of singularity. Therefore, equation 98 takes the form $G_{+\epsilon}(r)=\frac{1}{i\pi r}\int_{-\infty}^{+\infty}\frac{e^{ik^-.r}}{k^{(-)^2}-(k^2+i\epsilon)}k^-dk^-$ ------------ (99).

The new definition of $G_{+\epsilon}(r)$ also exists, but it is no longer a solution of the equation 94. But certain mathematical manipulations (some tricks) is required to evaluate equation 99 for$\epsilon\neq0$. Then the quantity ε is considered zero, where $G_{+\epsilon} \rightarrow G$. This step usually takes place after the integration is done. It is also important to observe that a unique solution to equation 94 does not exist because if it does exist, it would then be the inverse of $[-\frac{1}{4\pi}(\nabla^2+ k^2)]$, but this operator in fact has no inverse. Because the homogeneous equation $(\nabla^2 +k^2)\psi=0$ does have nontrivial solutions. However, the inverse of $[-\frac{1}{4\pi}(\nabla^2 +k^2+i\epsilon)]$ does exist; it is the Green function $G_{+\epsilon}(r)$ in equation 99).

From the complex variable theory, it is known that to deal mathematically with this problem, the use of the so-called complex plane is required to compute (or evaluate) the integral in equation 99. This complex plane can be shown in the following diagram, where there are two poles at the points $(k^2+i/2k)$ and $-(k^2 +i/2k)$ along the real axis.

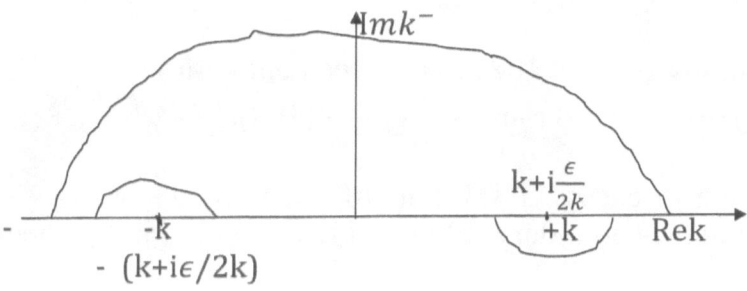

Path of integration in the complex plane k⁻ (referred to the book, complex variables by Churchill).

The path of integration leads along the real axis from (-∞) to (+∞). Since r is necessarily positive, the closed contour may be used if the complete path is performed by a semicircle of a very large radius through the upper half of k⁻ plane. It encloses the pole in the right half plane. The result of the integration is not altered by introducing the detours to avoid the two points k⁻=+k, k⁻=-k, as shown in the above diagram.

If this is done, then the limit $\epsilon \longrightarrow 0$ can be taken prior to the integration; therefore, $G_{+(r)}$ can be written as

$$G_{+(r)} = \text{limit} G_{+\epsilon}(r) = \frac{1}{i\pi r} \oint \frac{e^{ik^-.r}}{k^{(-)^2} - k^2} k^- dk^- \text{ -------- (100)}.$$
$$\epsilon \longrightarrow 0$$

Using the residue theorem (referred to Churchill) at k⁻=k, it becomes, $G_+(r) = \frac{e^{ikr}}{r}$ --------------------------- (101).

If ϵ is replaced by(- ϵ), in equation 99, a second solution to G(r) is obtained. It is given by

$$G_-(r) = \text{limit } G_{-\epsilon}(r) = \frac{e^{-ikr}}{r} \text{ ------------- (101)}.$$
$$\text{As } \epsilon \rightarrow 0$$

Equations 101 and 102 are in agreement with
$G_\pm(r) = e^{\pm ik(r-r^-)}/|r - r|$ ------------- (103).

Which represents a solution to equation 94. Using Cauchy integral (see boundary value problems by Churchill), one finds

$$G_1(r) = 1/2 \ G_+(r) + 1/2 \ G_-(r) = \frac{\cos kr}{r} \text{ ------------(104)}.$$

Green functions (101, 103, 104) might be identified as equation 101 is outgoing wave, equation 103 is incoming wave, and equation 104 is standing wave. To appreciate these terminologies, these might multiplied by the time factor exp. (-iEt/ℏ). Furthermore, the scattering description in terms of wave packets suggests $G_+(r)$ is retarded Green function, and $G_-(r)$ is advanced Green function; therefore, equation 94 might be written as

$$(\nabla^2 + k^2)G(r, r^-) = -4\pi\delta(r - r^-)$$ ---------------- (105).

This has a solution in terms of $G(r, r^-)$, which is the particular solution of equation $(\nabla^2 + k^2)\psi = U\psi$ ----- (106).

Using the well-known properties of $\delta\ function$, the solution of equation 106 is $\psi_p(r) = -\frac{1}{4\pi}\int G(r, r\text{-}) U(r\text{-})\psi_p r\text{-})d^3r^-$.

Also, the solution of the homogeneous equation $(\nabla^2 + k^2)\psi = 0$ is $\psi_h(r) = Ne^{ikr}$, where N is a normalization factor. Therefore, the total wave function $\psi_k(r)$ is given by $\psi_p + \psi_h$; hence,

$\psi_k(r) = Ne^{ikr} - \frac{1}{4\pi}\int G(r, r\text{-}) U(r\text{-})\psi_p r\text{-})d^3r^-$ ------ (107).

Substituting the special forms for $G_\pm$ into equation 107, it takes the form

$$\psi_k^{(\pm)}(r) = Ne^{ikr} - \frac{1}{4\pi}\int \frac{e^{(\pm ik(r - r^-))}}{|r - r^-|} U(r^-)\psi_k^{(\pm)}(r^-) d^3r^-$$ ---- (108).

Equation 108 represents Schrödinger equation rewritten in a form of which is particularly convenient for the use in the scattering theory. To show that in the asymptotic limit $(r \to \infty)$, where the right hand side of equation 108 takes the simple form $\psi_k^{(+)}(r) = N[e^{ikr} + f_k\frac{e^{ikr}}{r}]$ ------(large r)------ (109).

For large (r), the integrand in equation 109 can be closely approximated in the view of the fact that U(r)≠0, only for values of the case r<a. So it is possible to expand the exponent klr-r⁻l in power series of r^-.

Therefore, klr-r^-l≅k[r²-2r.r^-+ $r^{(-)^2}$]$^{1/2}$ =kr-kr^.r^- +neglected terms. If r is so large that $\frac{ka^2}{r}$<<1 and lr - r^-l ≅ r, then,

$$\psi^\pm(r)=N(e^{ik^\rightarrow.r^\rightarrow}+ f_k^\pm(r^\wedge)\frac{e^{\pm k^\rightarrow.r^\rightarrow}}{r})----------(110).$$

Where, $f_k^\pm(r^\wedge)=-\frac{1}{4\pi N}\int e^{\mp k^-.r^-}U^-(r)\psi^\pm(r-)d^3r^-$ ------ (111).

Since $U^-(r^-)=\frac{2mV(r)}{\hbar^2}$, hence, equation 111 takes the form

$$f_k^\pm(r^\wedge)=-\frac{m}{2\pi\hbar^2 N}\int e^{\pm k^-.r^-}V(r^-)\psi_k^\pm(r)d^3r^- -----------(112).$$

This is the scattering amplitude, where $\sigma =\int f^* f d\Omega$ and m is the mass of the incident particle.

7.3.5. Born Approximation and Applications

In physics, all physicists know most physical problems are complicated; that is due to the fact that the physical systems more likely to be complex nature, especially the nuclear systems where the force is still not very well defined; and consequently, the corresponding potential may take different shapes, as shown before. Therefore, the approximation methods play main roles to tackle such complicated problems. Hence, Born approximation is one of these methods. It is usually used in the scattering problems. For illustration, the mathematical formalisms in the preceding sections will be used.

Let a beam of particles with a momentum P=ℏK, scattered by the effect of a potential V(r) as shown below:

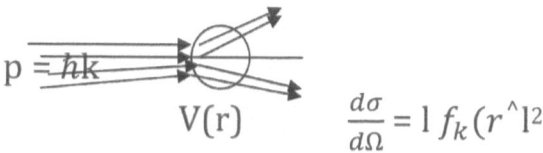

$$\frac{d\sigma}{d\Omega} = 1 f_k (r^\wedge) ^2$$

$f_k(r^\wedge)$ is the coefficient of the outgoing spherical wave represents the scattering amplitude for the asymptotic solution given by equation 112 for large r. It is the solution of Schrödinger equation $(\nabla^2 + k^2)\psi(r) = (\frac{2m}{\hbar^2})V(r)\psi$. Where for elasting scattering in the direction $^\wedge k^-$, f_k is given by:

$$f_k(^\wedge k^-) = -\frac{m}{2\pi N \hbar^2} \int e^{-ik^-.r^-} V(r-) \psi_k^{(+)}(r^-)d^3r- \text{--------} (113).$$

$\psi_k^{(+)}$ appears in equation 113 is not an explicit expression; however, it can be used to obtain an estimated f_k if it is replaced by a normalized wave function in the right hand side of the equation 108, neglecting the scattered wave $\psi_{sc.}$.Hence, this approximation yields

$$f_k(\cdot k^\wedge) = -\frac{m}{2\pi\hbar^2} \int e^{-ik^-.r^-} V(r^-)e^{ik.r^-} d^3r- \text{------------------}(114).$$

This is the scattering amplitude in the Born first approximation.

By noticing equation 114, it is clear that f_k is proportional to the matrix elements of the potential V(r), where $V_{k-k}(r) =$ $<k^- lV(r^-)k> \Rightarrow f_k = -\frac{m}{2\pi\hbar^2} [V_{k-k}]$ ------------------ (115).

The approximation used is the scattered wave $(\psi^{(+)})$ in equation 113 is replaced by the plane wave $(e^{-ik.r^-})$, which in fact represents the main approximation. Physically, it means as if the outgoing wave is very little affected by the potential field, so it is approximated to a plane wave.

Examples for Applications: Take the central force potential V(r), the Born approximation for the scattering amplitude is given by the equation 114), can be written as

$$f_{Born} = f_{B.} = -\frac{m}{2\pi\hbar^3} \int V(r^-) e^{-i(k^- - k)r^-} d^3r^-$$
$$= \frac{-m}{2\pi\hbar^3} \int V(r\text{-})e^{-q \cdot r\text{-}} d^3r\text{-}, \quad [q=k\text{--}k)], \text{------------ (116)}.$$

Where q is the transferred momentum in ℏ unit or momentum transfer during the scattering processes (interaction) due to the effect of the potential V(r), $d^3r\text{-}=r^2d\Omega = r^2\sin\theta\, d\theta d\varphi$.

By integrating equally 116 with respect to the angles (θ, φ), $f_{B.}$ is found to be $f_{B.} = -\frac{2m}{\hbar^2} \int_0^\infty V(r^-) \frac{\sin qr\text{-}}{qr\text{-}} r\text{-}^{(2)} dr\text{-}^2$, so,

$$f_{B.} = \frac{-2m}{\hbar^2} \int_0^\infty V(r\text{-})j_\ell(qr^-) r^{-(2)} dr^{-(2)} \text{------------- (117)}.$$

Here, j_ℓ is the spherical Bessel function. θ is the scattering angle between k and k- (see the figure above). For elasting scattering k- =k, so q=2k sin θ/2 ----------------(118).

For application of equation 117, consider Yukawa (or screened Coulomb) potential, where $V(r)=V_0\frac{e^{-ar}}{ar}$, $(1/\alpha)$ might be considered as the range of V(r). So to find σ_t, f(θ) has to be found from equation 117. Using equation 118 and integrating equation 117, f will be

$$f(\theta)=\frac{-2m}{\alpha\hbar^2} V_0 \left(\frac{1}{q^2+a^2}\right)=\frac{-2m}{\alpha\hbar^2}V_0\left(\frac{1}{4k^2\sin^2\theta/2 + a^2}\right)\text{-----(119)}.$$

(Show equation 119). Then σ_t=? if $I(\theta)$ $l^2d\Omega$ is to be found, using equation 119. Find it as an exercise.)

Problem1: For unscreened Coulomb potential due to two charges, q_1 and q_2, where $\alpha \to 0$, and $V_0 \to 0$ with $\frac{V_0}{\alpha} = q_1 q_2$ and $V_0 = \frac{q_1 q_2}{a}$, $a = \frac{1}{\alpha}$ (potential range), using Born approximation, find $\frac{d\sigma}{d\Omega}$ (differential scattering cross section). Show that it is in agreement with the classical Rutherford scattering cross section and the quantum mechanical evaluations of Rutherford scattering cross section (RSCS). (Answer: $\frac{d\sigma}{d\Omega} = \frac{q_1^2 q_2^2}{16E^2} \frac{1}{\sin^4 \theta/2}$).

Problem 2: Show that for $1S \to 2p$ state, $\frac{d\sigma}{d\Omega}$ is given by

$$\frac{d\sigma}{d\Omega} = \frac{m^2}{\hbar^3} \cdot \frac{p_2}{p_1} \left(\frac{16e^4}{q^5 a^8} \right) \left[\frac{Z^2 (a^3 - 2aq^2)}{(a^2 + q^2)^3} \right] - \frac{2\alpha}{(q^2 + \alpha^2)^2} \right]^2$$

Hint: Excitation $1S \to 2p$, it means $n=1$, $\ell = 1$, $m = 1, 0, -1$. Hence,

$$\psi_{100} = \frac{1}{\sqrt{\pi a_0^3}} e^{-R/a_0}, \psi_{210} = \frac{1}{\sqrt{32\pi a_0^3}} \frac{R}{a_0} \cos \alpha e^{-R/2a_0}$$

$$\psi_{2,\pm 1} = \mp \frac{1}{\sqrt{64\pi a_0^3}} \frac{R}{a_0} \sin \alpha e^{\pm i\phi} e^{-R/2a_0}$$

Choose q as polar axis, then $\psi_{2,\pm 1}$ gives no contribution, and $\int_0^\pi e^{\pm i\varphi} d\varphi = 0$, it implies that

$$\psi_i(r, \theta) = \frac{1}{\sqrt{T}} e^{i k_1 \cdot r} \psi_i(R^-)$$

$$\psi_f(r, \theta) = \frac{1}{\sqrt{T}} e^{i k_2 \cdot r} \psi_f(R^-),$$

where $\frac{1}{\sqrt{T}}$ is the normalization factor. Then find the transition probability $(\omega_{if}) = \frac{2\pi}{\hbar} |(\psi_f, V\psi_i)|^2 \rho(E)$.

Also, $\rho(E) = \frac{mkT}{(2\pi)^2\hbar^2} d\Omega$, remember that $j_1 = P_1/mT$, then $\frac{d\sigma}{d\Omega} = \frac{\omega_{if}}{j_1} = ?$

Relation between $\frac{d\sigma}{d\Omega}$ and ω_{if} for Coulomb scattering:

This is another example where an electron is scattered by a nucleus (Nuclear Rutherford Scattering), as shown in the following diagram:

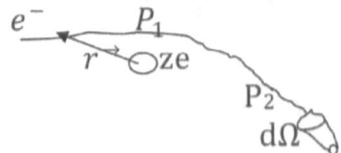

$$d\sigma = \frac{number\ of\ particles\ scattered\ per\ second\ into\ solid\ angle\ d\Omega}{number\ of\ incident\ particles\ per\ sequared\ centemeter\ per\ sec.}$$

Transition rate (ω_{fi}) can be calculated if the initial state wave function φ_1 and the final state wave function φ_2 are known and also the potential affecting the electron is known. Then using the second golden rule, the transition rate is found, such as $\omega_{21} = \frac{2\pi}{\hbar} \rho(E) \, | \, (\varphi_2, V_1\varphi_1) |^2$.

$\rho(E)$=Energy density for the final state, it is, as shown earlier, $\rho(E) = \frac{mkT}{(2\pi)^3\hbar^2} d\Omega$. T is normalization factor. V_1 is the potential due to the Coulomb force between the nucleus and the electron. It is$(-ze^2/r)$, the potential energy. But the kinetic is given as $P^2/2m_e$. Therefore, Hamiltonian of the system is $H = P^2/2m_e - Ze^2/r = H_0 + H_1$. $\varphi_1 = \frac{1}{\sqrt{T}} e^{i\vec{k_1}.\vec{r}}$, $\varphi_2 = \frac{1}{\sqrt{T}} e^{i\vec{k_2}.\vec{r}}$

$k_1 = p_1/\hbar$, $k_2 = p_2/\hbar$. Now $d\sigma/d\Omega = \omega_{fi}/j_1$, $j_1 = \frac{v_1}{\tau} = |A|^2 v_1$, $A = 1/|T|^{1/2}$, $v_1 = p_1/m_e$, then $j_1 = p_1/m_e T$.

$\rho(E_2)$ is the density of the state per unit energy (final states); it is given by $\rho(E_2)=\frac{T}{(2\pi)^3}d\Omega k^2\frac{dk}{dE}$, the energy for nonrelativistic system is $E=\frac{P^2}{2m}=\frac{\hbar^2 k^2}{2m} \Rightarrow \frac{dE}{dk}=\frac{\hbar^2 k}{m}=(\frac{\hbar k}{m})\hbar=(\frac{p}{m})\hbar=v\hbar \Rightarrow \frac{dk}{dE}=\frac{1}{VK}$

Hence, $\rho(E_2)=\frac{T}{(2\pi)^3}\cdot\frac{mk}{\hbar^2}\,d\Omega.$

The matrix elements of the potential V_1 is by

$(\varphi_2, V_1\varphi_1)=\frac{1}{T}.Ze^2\int\frac{e^{i\vec{q}\cdot\vec{r}}}{r}\,d\tau=\frac{1}{T}.Ze^2\,[\frac{4\pi}{q^2}]$, q as defined before.

Therefore, $\omega_{21}=\frac{2\pi}{\hbar}|V_1|^2\rho(E).$

Using the data for ρ and the matrix elements of V_1 and that q is $\vec{k_1}-\vec{k_2}$, where $q^2=k_1^2+k_2^2-2k_1k_2\cos\theta$, then

$\omega_{21}=\frac{4mkT}{\hbar^3}(\frac{Z^2 e^4}{q^4 T^2})$, then $d\sigma=\frac{\omega_{21}}{j_1}=\frac{4m^2 k}{p_1}(\frac{Z^2 e^4}{q^4})d\Omega$

Now, $\vec{k_1}$ ————————→ $\vec{k_1}=\vec{k_2}$- for elasting
                    θ scattering.
                   ↘$\vec{k_2}$

So, $q^2=2k^2(1-\cos\theta)=4k^2\sin^2\theta/2$

Remember that $\sin\theta/2=(1/2(1-\cos\theta)$, now substituting for q^2 into $d\sigma$, d σ will take the form $d\sigma=(\frac{Ze^2}{4E}\cdot\frac{1}{\sin^2\theta/2})\,d\Omega$. Find σ_t, as an exercise.

7.3.6. Application of Partial Waves Analysis Method on the Nucleons Scattering (N-P)

The scattering of the nucleons (n-n, n-p, p-p) is the most probable interaction in nuclear physics. From this type of interaction, many features of nuclear force, which is still ambiguous, will be clarified. So it is too important to study the nucleon-nucleon scattering using the partial wave analysis method. Starting with the total scattering, $\sigma_t = \int \sigma(\theta)d\Omega$ $= \frac{4\pi}{k^2} \sum_{\ell=0}^{\infty}(2\ell + 1) \sin^2\delta_\ell$.

Let $\ell \le a$ be considered as case of study, where ℓ is the orbital angular momentum, and a is the potential range. δ_ℓ is the phase shift due to the potential.

Consider the energy value is E=5 MeV, and accordingly, the energy in the laboratory frame is E_l=10 MeV. The energy is given as E=$\frac{\hbar^2 k^2}{2m}$, where m is the nucleon mass. Also, it is known the reduced mass M of two nucleons is given as

$$\frac{1}{M} = \frac{1}{m_p} + \frac{1}{m_n}$$

M= $\frac{m_p m_n}{m_p + m_n}$. For $m_n \approx m_p$, M=$\frac{m_n}{2}$=$\frac{m}{2}$, [let m_n=m]. This implies that $k^2 = \frac{ME}{\hbar^2}$, this gives $k^2 = 12 \times 10^{24}$ cm^{-2}, so $k \approx 3.5 \times 10^{12}$ cm^{-1}.

Since $a \approx 2 \times 10^{-13}$cm, hence, ka=0.7$\Rightarrow \delta_0$ wave and $\delta_1 = \delta_2 = \delta_3 = 0$. Therefore, this has to be restricted to the S-wave, $f(\theta) = \frac{1}{2ik}(e^{2i\delta_0} -1)$. Also, it is known from the complex variable that $\sin\delta_0 = \frac{e^{i\delta_0} - e^{-i\delta_0}}{2i}$. This implies that $f(\theta) = \frac{e^{i\delta_0} \sin\delta_0}{k}$.

Then $d\sigma(0) = \frac{1}{k^2} \sin^2\delta_0$, so $\sigma_t = \int_0^\pi \frac{1}{k^2}\sin^2\delta_0 d\Omega = \frac{4\pi}{k^2} \sin^2\delta_0$

E_L <10 MeV. In fact, this σ_t can be obtained directly from the equation $\sigma_t = \frac{4\pi}{k^2} \sum_{\ell=0}^{\infty}(2\ell + 1)\sin^2\delta_\ell$. For S-wave, $\ell = 0$, which leads to $\sigma_t = \frac{4\pi}{k^2} \sin^2\delta_0$, as found before. So if σ_t is known experimentally, it is quite possible to find the phase shift for

S-wave. If the phase shift for the p-wave $(\ell = 1)$ δ_1 is very small, then the scattering amplitude $f(\theta)$ takes the form

$$f(\theta)=\frac{e^{i\delta_0}\,sin\delta_0}{k} - \frac{3}{k}\delta_1 cos\theta, \text{ (As an exercise find } \sigma_t \text{])}.$$

For low energy $k \to 0$ [very low energy], $f(\theta) \to f_0(0)$, which represents the scattering length (a), i.e., $f_0(0)=a$, then

$\sigma_t=4\pi a^2$ -- (120).

This is the total scattering cross section of incident particle with a very low energy. It is equivalent to an area of sphere with a radius (a). Think of such sphere of dimension $a \sim 0.3\,bn(1barn = 10^{-24}$ cm^2; it is within the nucleus dimension $\sim 10^{-13}$. This is the total scattering cross section of incident particle with a very low energy. It is equivalent to an area of sphere with a radius $(a = 10^{-13}$ cm$)$.

Square well potential, as another example V(r):
It is the simplest representation of a nuclear potential. It can be represented graphically as follows:

$$\begin{cases} V(r)=-V_0 & r \leq a \\ V(r)=0 & r \geq a \end{cases}$$

The Schrödinger equations for this system are as follows:

1. $\frac{d^2 u(r)}{dr^2}+(k^2 + \beta^2)u(r)=0$, for $r \leq a$, and $\beta=\frac{MV_0}{\hbar^2}$

2. $\frac{d^2 u(r)}{dr^2}+k^2 u(r)=0$, for $r \geq a$, and $k^2=\frac{2ME}{\hbar^2}$

The solutions of 1 and 2 are very well known to physicists; they are, respectively,

1. $u(r) = A \sin\sqrt{k^2 + \beta^2}\, r$, $r \leq a$
2. $u(r) = B \sin(kr + \delta_0)\, r$, $r \geq a$

Now applying the boundary conditions such as

$[\frac{u^-}{u}]_a = [\frac{u^-}{u}]_0$, $[u^- = du/dr]$, then by simplifying this relation, it leads to, $\sqrt{k^2 + \beta^2}\, \cot\sqrt{k^2 + \beta^2} = k \cot(ka + \delta_0)$ ----- (121).

β is known from the bound state if it is solved for δ_0, then

$$\delta_0 = -ka + \cot^{-1}[\sqrt{1 + \frac{\beta^2}{k^2}}\, \cot^{-1}\sqrt{k^2 + \beta^2}\, a]$$

Suppose that $(\beta^2 - \alpha^2)^{1/2} \cot(\beta^2 - \alpha^2)^{1/2} a = -\alpha$ --------(122).

Where $\alpha = \sqrt{\frac{-ME}{\hbar^2}}$. For low energy of bombarding particle less or equal to 1 MeV, so $k^2 \ll \beta^2$. It means k^2 in equation 120 might be neglected; therefore,

$\beta \cot \beta a = k \cot(ka + \delta_0)$ ----------------------------(122).

If $\alpha^2 \ll \beta^2$, then from equation 121, $\beta \cot \beta a = -\alpha$. Therefore, k cot $(ka + \delta_0) = -\alpha$, but for very low energy, $ka \ll \delta_0$, so k cot$(ka + \delta_0) \cong k \cot \delta_0 =$

$-\alpha = \sqrt{\frac{-ME_B}{\hbar^2}}$, where E_B is the binding energy E_0. So

$\cot\delta_0 = -\sqrt{\frac{E_0}{E}}$ ----------------------- (123).

Now $\sigma_T = \frac{4\pi}{k^2} \sin^2\delta_0 = \frac{4\pi}{k^2} \cdot \frac{1}{1+\cot^2\delta_0}$, also, $1+\cot^2\delta_0 = 1 + \frac{\cos^2\delta_0}{\sin^2\delta_0} = \frac{1}{\sin^2\delta_0}$,

from (123), $1+\cot^2\delta_0 = 1 + (-\sqrt{\frac{E_0}{E}})^{1/2})^2 = 1 + \frac{E_0}{E}$.

Hence, $\sigma_T = \frac{4\pi}{k^2} \frac{1}{1+|E_0/E|}$, as k $\rightarrow$ 0, $\sigma_T = \frac{4\pi\hbar^2}{M} \cdot \frac{1}{E_B} \sim 2$ barns

So $\sigma_T = \frac{4\pi\hbar^2}{M} \cdot \frac{1}{E_B}$

As k, $E \rightarrow 0$. Let the versus between σ_T and the energy E is drawn as in the figure below, where the bound state of p.n is triplet for s=1 (parallel) and singlet for s=0 (anti-parallel), where σ_T =(3/4) σ_t+(1/4) σ_s. The measured σ_T is 80 barns. Hence, 80 barns=(3/4) σ_t (small)+(1/4)σ_s(large). Here σ_s is the dominant part.

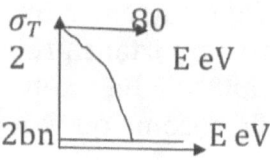

Chapter Eight

Approximation Methods, Theories, and Applications

8.1. *Introduction*

In natural sciences in general and in physics specifically, there are some problems so complicated; it is not easy to deal with, unless some approximations are taken. In classical physics, it might be quite possible to solve the problem in general; but in quantum physics, where the systems are microscopical in nature, it is difficult to solve the physical problems exactly, so the approximations impose themselves to solve the problems.

Even in the physics of systems of microscopical nature, there are certain problems which cannot be solved exactly. Therefore, some approximation methods have to be used, such as the well-known Wigner-Kremer-Brillion (WKB) methods. Hence, in general, most important problems in classical or in quantum physics cannot be solved exactly. So the approximation methods become quite necessary to be applied. The important approximation methods to be used in this book as quantum physics are

1. The Perturbation Method
2. The Variation Method

But in quantum physics, it is important that the nature of the system under study to be understood cleverly. So the suitable method is chosen. Usually, the physical systems are atoms, subatomic, nuclearic, and subnuclearic microscopic systems. The interactions in quantum physics are fields type of interactions; therefore, the system under study is described by a wave function (ψ), where the dynamical variable to be measured, is associated with an operator, which is usually Hermitian. This operator operates on the eigenfunction, its eigenfunction, the measured value, eigenvalue, is obtained. Therefore, for a dynamical variable A, there is an operator $A_0 \equiv A^\wedge$ (remember for the momentum P the operator is P_0 is given by $-i\hbar \frac{\partial}{\partial q}$]).

So $A_0 \psi_\alpha = \alpha \psi_\alpha$. This is an eigenvalue equation (as mentioned earlier). ψ_α is the eigenfunction, and α is the eigenvalue (measured value). The eigenfunction is either time dependent or time independent (stationary). In case it is time dependent, the time evolution of ψ is too necessary to be known. The time evolution operator $U(t, t_0)$, as mentioned before, is given as $U(t, t_0) = e^{-i\frac{H(t-t_0)}{\hbar}}$.

So the properties of $U(t, t_0)$ are directly related to those of the Hamiltonian H. If H is time dependent, this implies that it is a complicated case. Then the eigenfunctions and eigenvalues which are time dependent will be so important to determine the properties of the time evolution operator $U(t, t_0)$.

In the next sections, the following subjects will be tackled briefly:

1. Perturbation for time-independent (stationary) system.
2. Perturbation for time-dependent system.
3. Degenerate and nondegenerate systems.
4. Variation method.

8.2. Stationary Perturbation (Time Independent)

Introduce $H = H_0 + \lambda V$------------------------------------ (1).
Where λ is a real parameter $0 \leq \lambda \leq 1$.

H_0 is unperturbed Hamiltonian, its eigenvalue equation can be easily solved, where

$H_0 | E_{i,\alpha}^{(0)} >= E_i^{(0)} | E_{i,\alpha}^{(0)} >$ ---------------------------------- (2).

α is used to stand for degeneracy. The eigenfunction $|E_{i,\alpha}^{(0)}>$ can be represented by Dirac ket (|>) in terms of the quantum numbers (n, ℓ, and m), such as $|n, \ell, m>$, where n is the principal quantum number, ℓ is the orbital angular momentum quantum number, and m is the magnetic quantum number (magnetic projection of ℓ). The basic assumption is H|E>=E|E>, exact eigenfunction, exact eigenvalue.

Or one might write it as $H | \psi> = E|\psi>$ ---------------------- (3).

Also, from the previous mathematical formalisms can be written that $<0|\psi>=<0|0>=1$ due to normalization property.

Now the energy E and the eigenfunction |E> can be expanded in a series of λ, as $\lambda \to 0$, so E in equation 3 takes the form $E_i^{(0)}$, then

$|E>=|E_{i,\alpha}^{(0)} > \longrightarrow \Sigma_{i,\alpha} C_\alpha^{(0)} | E_{i,\alpha}^{(0)} >$, where $|E_{i,\alpha}^{(0)}>$ is the 0th order. Hence, for $\lambda \neq 0$. E takes the form,

$E=E_i^{(0)}+\lambda \mathcal{E}_1 +$ higher orders terms. ------------------- (4).

And $|E>=\Sigma_\alpha C_\alpha^{(0)} | E_{i,\alpha}^{(0)} >+\lambda \Sigma_{i,\alpha} C_{i,\alpha} | E_{i,\alpha}^{(0)} > +\lambda^2 \Sigma_{i,\alpha} C_{i,\alpha}^2 | E_{i,\alpha}^{(0)} >+$ higher orders.-- (5).

The coefficients C_α^0, $C_{i,\alpha}^1$, and $C_{i,\alpha}^2$ can always be determined. The basic difficulty of the perturbation theory is if the expansion exists for $\lambda = 1$. It may not converge rapidly enough.

8.2.1. *Nondegenerate Case*

There are two possible situations in dealing with the perturbation theory. Either it is degenerate case where there are more than one eigenfunction for a state of defined energy (E) or nondegenerate case where there is only one eigenfunction for certain defined energy (E). The nondegenerate case will be dealt with here. The eigenstate (referred to Q.M. by Mesieh, Volume II) is given by

$$|E>=|0> +\lambda|1> +\lambda^2|2> +---+ \text{higher order terms} ------(6).$$

Therefore H can be written as
$$H=H_0 +\lambda V +\lambda^2 V+---, --------- (7).$$

For illustration, take the following example: particle of mass m and momentum p with charge e are moving in an electromagnetic field (EM) of a vector potential $\mathcal{A}$. The Hamiltonian H of this moving particle is given by

$$H=\frac{p^2}{2m} -\frac{e}{2m}(p\mathcal{A}+\mathcal{A}P)+\frac{1}{2m}(\frac{e}{c})^2\mathcal{A}^2+--=H_0-\lambda V+\lambda^2 V^- +---,----(8).$$

Hence, the 0th order equation is already known as,
$$H_0|0>=E_a^{(0)}|0> --(9).$$

The first order equation is
$$[H_0-E_a^{(0)}]|1>+(V-\mathcal{E}_1)|0>=0, ----------------------------(10).$$

The second order equation is:
$$(H_0 -E_a^{(0)})|1> + (V-\mathcal{E}_1)|1> -\mathcal{E}_2|0>=0,------------(11).$$
-

-

Nth order equation is given by:

$((H_0 - E_a^{(0)})|n> + (V - \mathcal{E}_1)|n-1> ---- + \mathcal{E}_{2n}|0> = 0 --------(12).$

The energy E is expanded as:

$E = E_a^{(0)} + \lambda \mathcal{E}_1 + \lambda^2 \mathcal{E}_2 + ----- + \lambda^n \mathcal{E}_n + --,$ -------------- (13).

Equation 5 can be written as:

$|\psi> = |E> = |0> + \lambda|1> + \lambda^2|2> + ---- + \lambda^n|n> +--.$ ------- (14).

Starting with equation 10 to work out its solution,

$(H_0 - E_a^{(0)})|1> + (V - \mathcal{E}_1)|0> = 0$, multiply by $|E_a^{(0)}>$ from the left, the following form is obtained:

$<E_a^{(0)}|H_0|1> - <E_a^{(0)}|E_a^{(0)}|1> + <E_a^{(0)}|V|0> - <E_a^{(0)}|\mathcal{E}_1|0> = 0$; hence,

$\underline{<E_a^{(0)}|E_a^{(0)}|1>} - \underline{<E_a^{(0)}|E_a^{(0)}|1>} + <E_a^{(0)}|V|0> - <E_a^{(0)}|\mathcal{E}_1|0> = 0 \longrightarrow$

$<E_a^{(0)}|V|0> = <E_a^{(0)}|\mathcal{E}_1|0>$, from equation 6 $|E_a^{(0)}> = |0>$, which implies that $<0|\mathcal{E}_1|0> = \mathcal{E}_1<0|0> \Rightarrow$

$\mathcal{E}_1 = <E_a^{(0)}|V|E_a^{(0)}> = V_{aa}$ ------------------ (15).

This is the first order correction to the energy due to perturbation.

So $E_a^{(1)} = E_a^{(0)} + \mathcal{E}_1$. if $|E_a^{(0)}> = |\psi>$ is known for the system, and also the potential $V(r)$ is known, then $\mathcal{E}_1$ is found explicitly.

Similarly, it is possible to determine the successive expansion coefficients in equations 13 and 14. Now by substituting equations 1, 13, and 14 into both sides of equation 3, then it will be an equal equation between two power series in λ. (Do it as an exercise.) Then the coefficients of each power of λ must separately be equal so that the equality between two power series is satisfied. Then equations 9, 10, 11, and 12 are obtained. According to the condition $<0| \psi>=<0|0>=1$, the following can be concluded:

$<0|1>=<0|2>=<0|3> = $ -------$<0|n>=0$---------------- (16).

Consider $\psi_n =\psi_0,\psi_1,\psi_2,$ ------,$(n=0,1,2,3,$-----$)$.
Equation 9 determines the eigenvalue and the eigenfunction (eigenvector) for the 0th order. With equation 15, equation 10 determines the first order corrections, and equation 11 determines the second order corrections to the eigenvalues and the eigenvectors. Equation 12 determines the nth order corrections to E and $|E>$. It is useful to show that equation 12 effectively determines $\mathcal{E}_n$ and $|n>$ in terms of the lower corrected terms.

To show this, one has to project onto the basis vectors of H_0 (0th order). This means the projection takes place onto the 0th order eigenvector $|0>$. If it is done, and with the aid of equation 15, the nth energy is obtained as

$\mathcal{E}_n=<0|V|n-1>$ ---------------------------- (17).

By projecting onto the other basis vectors of H_0, the corresponding components of $|n>$ along each of them are obtained, i.e., $<E_a^{(0)}|n> = \frac{1}{E_a^{(0)}-E^{(0)}} [< E_{a,\alpha}^{(0)}|(V-\mathcal{E}_1)| n-1 >-$
$\mathcal{E}_2<E_{a,\alpha}^{(0)} |n-2>--- \mathcal{E}_{n-1}<E_{a,\alpha}^{(0)}| 1 >], E_a^{(0)} \neq E_0$-------------- (18).

Since $<0|n>=0$, orthogonal, $|n>$ is thereby completely determined.

For convenience, define $P_0=|E_a^{(0)} >< E_a^{(0)}|$ as an operator projects into space, and Q_0 as an operator projects out of the specified space, where $P_o + Q_o=1$; therefore, $Q_0 = 1 - P_0=$
$1- E_a^{(0)} >< E_b^{(0)}=\sum_{a\neq b} \left| E_a^{(0)} >< E_b^{(0)}\right|$ ------(19).

Problem: Let the harmonic oscillator be subjected to external force $V(r)=1/2\ ma^2\omega^2 q^2$. The Hamiltonian is given as $H=H_0+V(r)$, where $H_0 = P^2/2m+(1/2)m\omega^2 q^2$, so $V(r)$ is the perturbation term. Also, $E_0 =(1/2)\ \hbar\omega$, ground state energy $(n=0)$. $\mathcal{E}_1=<0|V|0>\equiv<E_a^{(0)}|V|E_a^{(0)}>$, from equation 15. Find the exact eigenvalue of E, $E=1/2\ \hbar\omega+\mathcal{E}_1$. If H is written as $H=P^2/2m + (1/2)m\{\omega^2+\alpha^2\}q^2$, then, $E\equiv E(\omega,\alpha)$. Hence, E can be expanded in terms of α, considering α as a very small. Using the so far developed methods from equation 15,

$\mathcal{E}_1=<0|V|0>=<E_a^{(0)}|V|E_0>$, which leads to:

$E=<0|H|0>+0(\lambda^2)$--(20).

Where E is just the average value of the Hamiltonian H that is calculated using the eigenvector of the unperturbed situation, and H is H_0, i.e., $\lambda=0$ in equation 1. Equation 19 gives the first order correction to the eigenvector, so

$|1>=\frac{Q_0}{a}(V-\mathcal{E}_1)|0>$, but since $Q_0|0>=0$, then $|1>=\frac{Q_0}{a}V|0>$ ---(21).

Therefore, in general, the eigenvector $|n>$ can be written as

$|n>=\frac{Q_0}{a}V(n-1)^1= (\frac{Q_0}{a}V)^2(n-2)=---(\frac{Q_0}{a}V)^n|0>$. -------- (22).

Hence, $|\psi>=(1+\lambda\frac{Q_0}{a}V)|0>+0(\lambda^2)$, for first order ------(23).

The norm of $|\psi>$ is given by $<\psi|\psi>=\cong<0|0>+\lambda^2<1|1>=$
$1+\lambda^2<1|V\frac{Q_0^2}{a^2}V|0>= 1+\lambda^2<1|V\frac{Q_0}{a}.V\frac{Q_0}{a}|0>$, by equation 21.

0

$<\psi\ |\ \psi>\cong1+0(\lambda^2)$, so the components of the first order correction to $|0>$ along the other basis vectors of H_0 is given by

$$\lambda <E_\alpha^{(0)}|1> = \frac{<E_\alpha^{(0)}|v|0>}{E_\alpha^{(0)}-E^0}, \quad [E_\alpha^{(0)} \neq E^0]. \text{-----------------} (24).$$

Hence, the component along $|E_\alpha^{(0)}>$ is the perturbation matrix elements linking the eigenstate $|0>$ to the eigenstate $|E_\alpha^{(0)}>$ divided by the energy difference between these two unperturbed eigenstates. As the parameter quantity λ is small, the power series rapidly converges. Now the next steps will deal with the normalization of the eigenstate $|E>$.

1. $<E_\alpha^{(0)}|E_\alpha^{(0)}>=1$; it is a well-known property to $|\psi>$ or $|E>$.
2. $<E_\alpha^{(0)}|E>=1$ but $\neq<E|E>$.
3. Hence, $<E_\alpha^{(0)}|E>=1\cong1+\lambda<E_\alpha^{(0)}|1>+\lambda^2<E_\alpha^{(0)}|2>$ $+----$, it leads to $\lambda<E_\alpha^{(0)}|1+\lambda^2<E_0^{(0)}|2>+---=0$, since $\lambda\neq0$, for perturbation case, $\Rightarrow$

$<E_\alpha^{(0)}|1>=0$, $<E | E>=1+0(\lambda^2) \approx 1.\rightarrow|E>$ is the first order eigenstate is normalized for small λ.

$<E_\alpha^{(0)}|1>=0$ $\rightarrow$ $<E|E>=1=1+0\lambda^2 \approx 1$

$<E_\alpha^{(0)}|2>=0$ $\rightarrow$ $|E>$ is the first order eigenstate, normalized forsmall λ

$<E_\alpha^0|n>=0$

Suppose it is possible to solve this problem to n-1 order, but the problem is still how to find the solution for the order n. The answer is $|n>=\sum_{b\neq a} E_a^{(0)}><E_b^{(0)}|n>$. (Show that as an exercise.)

Let us look more at first-order perturbation from equation 15.

$\varepsilon_1 = <E_a^{(0)}|V|E_a^{(0)}>$, then

$<E_b^{(0)}|1>=(E_b^{(0)}-E_a^{(0)})<E_b^{(0)}|1>+<E_b^{(0)}|V-\varepsilon_1|E_a^{(0)}>=0.$

Also, it is known that $<E_b^{(0)}|\varepsilon_1|E_a^{(0)}>=0$ if $a \neq b$, then

$<E_b^{(0)}|1> =(E_b^{(0)} -E_a^{(0)})<E_b^{(0)}|1>+<E_b^{(0)}|V|E_a^{(0)}>=0$

$\qquad = -(E_a^{(0)}-E_b^{(0)})<E_b^{(0)}|1> + <E_b^{(0)}|V|E_a^{(0)}>$

Of course, $<E_b^{(0)}|1>=0$; therefore, $(E_a^{(0)} -E_b^{(0)})<E_b^{(0)}|1>=$

$<E_b^{(0)}|V|E_a^{(0)}>$. This gives that $<E_b^{(0)}|1>=\dfrac{<E_b^{(0)}|V.E_a^{(0)}>}{E_a^0 -E_b^{(0)}}$

Now multiplying from the right side by $|E_b^{(0)}>$, the following is obtained: $|1> =\sum_{a \neq b} \dfrac{V_{ab}}{(E_a^{(0)}-E_b^{(0)})}| E_b^{(0)}>$----------(25).

As an exercise, find the ground state energy first order correction for ${}_2^4He$ atom, where the Hamiltonian H is given by

$H= [\dfrac{P_1^2}{2m} +\dfrac{P_2^2}{2m} -\dfrac{Ze^2}{r_1} -\dfrac{Ze^2}{r_2}] +\dfrac{e^2}{r_{12}^2} =H_0 +V$

$r_{12} = |r_1 - r_2|$

Hint: $E_a^{(0)} =-2zE_H$

$\psi(r)=<r_1 r_2|E_a^{(0)}>=\dfrac{e^{-(r_1+r_2)}}{\pi a^2}$, (spin singlet, nondegenerate).

$\dfrac{1}{r_{12}} =\dfrac{1}{|r_1-r_2|}$, to be expanded as it is well known to do it.

For $\dfrac{r_1}{r_2}, r_1<r_2$

$\qquad \dfrac{r_2}{r_1}, r_2<r_1$

$\qquad \varepsilon_1 =<E_a^{(0)} |V|E_a^{(0)} >$, equation 15.

Results information, $\varepsilon_1 = \frac{5}{4} Z E_H \Rightarrow E = E_a^{(0)} + \varepsilon_1 = -2Z E_H + \frac{5}{4} Z E_H$

=-108+34=-74.8 eV. If it is done for $_3^6 Li$, then E=-193 eV compared to experimental result -197.1 eV. If Be(z=4), then E=-365.5 eV compared to $E_{expt.}$=-370.1 eV. It is clear that the perturbation calculations are close to the measured ones.

Therefore, it is necessary to go to higher order terms of perturbation to find the correction to the energies and the eigenstates. The concept of the projection operator, as defined previously, is to be used. So let the first order be worked out to find ε_1 and $|1>$ as they are in equations 15 and 25, respectively. Now the projection operator, as defined earlier, is

$P_0 = |E_a^{(0)}><E_a^{(0)}|$ --- projection in------------------ (26).

And $Q_0 = 1 - P_0 \equiv$

$\sum_{b \neq a} E_b^{(0)}><E_a^{(0)}|$ --|(projection out)------(27).

From the previous section, $Q_0 \frac{1}{E_a^{(0)} - H_0} =$

$\sum_{b \neq a} E_b^{(0)}><E_a^{(0)}| \frac{1}{E_a^{(0)} - E_B^{(0)}}$, $[\frac{1}{H_0}|E_b^{(0)}> = \frac{1}{E_b^{(0)}} | E_b^{(0)}>$, but $\frac{1}{E_a^{(0)} - H_0}$ and Q_0 commute, which yield to $(Q_0 \frac{1}{E_a^{(0)} - H_0}) --(\frac{1}{E_a^{(0)} - H_0}) Q_0 = 0$

As shown before, $Q_0^2 = Q_0$ which yields that $Q_0 = \pm 1$, it represents the main property of the projector operator; therefore,

$\frac{Q_0}{E_a^{(0)} - H_0} = (\sum_{b \neq a} (|E_b^{(0)}><E_a^{(0)}|)(\frac{1}{E_a^{(0)} - H_0})$ ------- (28).

Equation 28 represents the operator which is so important in formulating the perturbation theory. Hence, it is also necessary to be acquainted with. Thereafter, let the first order correction be tackled on the basis of the projection operator applications.

For the first order, one has equation 10:

$P_0(H_0 - E_a^{(0)})\,|1> + (V - \mathcal{E}_1)|E_a^{(0)}> = 0$, $|E_a^{(0)}> = |0>$. Hence,

$H_0\,|1> = E_a^{(0)}\,|1>$, and $P_0\,(E_a^{(0)} - E_a^{(0)}) + P_0(V - \mathcal{E}_1)|E_a^{(0)}> = 0$, so

$P_0\,(V - \mathcal{E}_1)|E_a^{(0)}> = 0 \Rightarrow VP_0|E_a^{(0)}> = \mathcal{E}_1 P_0|E_a^{(0)}> \Rightarrow$

$V\psi(P_0\,E_a^{(0)}) = \mathcal{E}_1\psi(P_0\,E_a^{(0)})$, which is an eigenvalue equation, where $\mathcal{E}_1$ is the eigenvalue of V, and $P_0\,|E_a^{(0)}> \equiv \psi(P_0, E_a^{(0)}>$ is the eigenvector of V. Now to find $\mathcal{E}_1$. Take P_0; as defined before, it is possible to write $|E_a^{(0)}><E_a^{(0)}|(V - \mathcal{E}_1)|E_a^{(0)}> = 0$, which implies that, $|\ E_a^{(0)}>(<E_a^{(0)}|V|E_a^{(0)}> = |E_a^{(0)}>(\mathcal{E}_1 <E_a^{(0)}|\ E_a^{(0)}>) = \mathcal{E}_1|E_a^{(0)}>$. Therefore, $V_{aa}\,|E_a^{(0)}> = \mathcal{E}_1|E_a^{(0)}>$; hence,

$\mathcal{E}_1 = V_{aa}$ as it is in equation 15.

To find $|1>$, follow the same procedures using Q_0, so

$$Q_0|1> = \frac{Q_0 \ V\,|E_a^{(0)}>}{E_a^{(0)} - H_0} = \frac{Q_0}{E_a^{(0)} - H_0}\ V|E_a^{(0)}> -\!-\!-\!-\!-\!-\!-\!-\!-\!-\!-\!-\!-\!-\!-(29).$$

Also, using the relation between P_0 and Q_0 where $P_0 + Q_0 = 1$, the same result (equation 29) will be obtained. Using equation 29 and the definition of Q_0, it can be obtained that

$<E_b^{(0)}|1> = \dfrac{<E_b^{(0)}|V\,E_a^{(0)}>}{E_a^{(0)} - H_0}$, multiply by $|E_b^{(0)}>$ from the right, then

$<E_b^{(0)}|\ E_b^{(0)}>|1> = |1> = \sum_{b \neq a} \dfrac{<E_b^{(0)}|V\,E_a^{(0)}>}{E_a^{(0)} - E_b^{(0)}}|E_b^{(0)}> -\!-\!-\!-\!-\!-\!-\!-\!-\!- (30).$

This is equation 25 derived directly.

Next, consider the second order equation:

$$(H_0 - E_a^{(0)})|2> + (V - \mathcal{E}_1)|1> - \mathcal{E}_2|E_a^{(0)}> = 0.$$

Apply the projection operator P_0 to this equation then

$$P_0((H_0 - E_a^{(0)})|2> + P_0(V - \mathcal{E}_1)|1> - P_0\mathcal{E}_2|E_a^{(0)}> = 0 \Rightarrow$$
$$P_0(V\mathcal{E}_1) - P_0\mathcal{E}_2|E_a^{(0)}> = 0.$$

Since $P_0 + Q_0 = 1$, so $P_0 (V - \mathcal{E}_1)(P_0 + Q_0)|1> - P_0\mathcal{E}_2|E_a^{(0)}> = 0.$
With some mathematical manipulations, it is obtained that

$$\mathcal{E}_2 = \Sigma_{E_a \neq E_b} \frac{\Sigma_{a \neq b} <E_a^{(0)}|V|E_b^{(0)}><E_b^{(0)}|V|E_a^{(0)}>}{E_a^{(0)} - E_b^{(0)}} \text{---------------(31)}.$$

Now the correction for the eigenstate $|E_a^{(0)}>$ can be easily calculated with the aid of equation 19

$$|n> = \frac{Q_0}{a}[(V - \mathcal{E}_1)|n-1> - \mathcal{E}_2|n-2> \text{-------} \mathcal{E}_{n-1}|1>, \text{ where } |2> \text{ can be found.}$$

So for n=2, this equation becomes

$$|2> = \frac{Q_0}{a}[(V - \mathcal{E}_1)|1> - \mathcal{E}_2|0>], \ Q_0 = 1 - P_0 = 1 - |0><0|, \ a = \frac{E_a^{(0)} - H_0}{Q_0}.$$

So $Q_0|1> = 1$ and $Q_0|0> = 0$ show that. Also, $Q_0 (\mathcal{E}_2|0>) = \mathcal{E}_2(Q_0 |0> = 0)$. Taking all these data, the eigenstate ($|2>$) takes the form

$$|2> = \frac{Q_0}{E_a^0 - H_0}[(V - \mathcal{E}_1)\frac{Q_0}{E_a^{(0)} - H_0}(V - \mathcal{E}_1)|0>] \text{ or}$$
$$|2> = \frac{Q_0}{E_a^{(0)} - H_0}[(V - \mathcal{E}_1)]|1>.$$

This can be implied to be $12>=\frac{Q_0}{E_a^{(0)}-H_o}$ $(V-\mathcal{E}_1)Q_0 l1>$, using

equation 29 for $Q_0 l1>$, then Q_0 $12>=\frac{Q_0^2}{[E_a^{(0)}-H_o]^2}$ $[V-\mathcal{E}_1]lE_a^{(0)}>$

With some algebraic manipulation, it is possible to get (do it)

$$12>=\frac{\sum_{a\neq b}<E_a^{(0)}\left|V\right|E_b^{(0)}><E_b^{(0)}\left|V\right|E_a^{(0)}>}{[E_a^{(0)}-H_o]^2}--------------(32).$$

$lE_b^{(0)}><E_b^{(0)}>$ is the intermediate state.

Problems:

1. Consider $V=\sum_{i<j}\frac{e^2}{r_{ij}}$ as a perturbation term due to Coulomb force, find Coulomb energy E_C
2. Consider rigid rotator in an electric field $\vec{E}$ where $H_0=\frac{L^2}{2I}$, where L is the angular momentum, and I is the moment of inertia, $V=-d. \vec{E}=-dE\cos\theta=H^-$(perturbation term), find the energy correction term.

8.2.2. Pertubation of a Degenerate Level

As stated before, the degenerate level is a level of energy E described by more than one eigenfunction (ψ). If it is subjected to an external force, it is quite possible this degeneracy is removed. It is well known that the eigenvalue equation is

Hl E>=ElE>, where H= H_0 +λV, $0\leq\lambda\leq1$.

$lE>=lE_a^{(0)}>+\lambda lE_1>+\lambda^2 lE_2>+----=l0>+\lambda\ l1>+\lambda^2 l2>+-------(33).$

And E=$E_a^{(0)}$+$\lambda\mathcal{E}_1$+$\lambda^2\mathcal{E}_2$ +---------------------------------- (34).

Consider $E_a^{(0)}$ as the eigenvalue of the unperturbed Hamiltonian (H_0), which is of g_a-fold degenerate. Let $\mathcal{E}_a^{(0)}$ be the subspace corresponding to $E_a^{(0)}$, which can be obtained by the aid of the projector P_o (as defined before), so there may be more than one eigenvalue of H tending to $E_a^{(0)}$ when $\lambda \to 0$.

These eigenvalues are E_1, E_2, E_3,------E_n; their orders of deg. are g_1, g_2, g_3, ----g_n, and their respective subspaces $\mathcal{E}_1$, $\mathcal{E}_2$, $\mathcal{E}_3$,----$\mathcal{E}_n$, respectively. Hence, $g_a = g_1 + g_2 + g_3 +-----+ g_n$------------- (35).

And $E_{(a)}^{(0)} \to \mathcal{E}_1 + \mathcal{E}_2 + \mathcal{E}_3 +-----+\mathcal{E}_n$ as $\lambda \to 0$------- (36).

Now if P is the projection onto $\mathcal{E}_a^{(0)}$, it is a continuous function of λ and $P \to p_o$, as $\lambda \to 0$.It is clear that in order to determine E_i and $| E_i>$ of Hamiltonian H by the perturbation theory, in the case of a degenerate level case, it will be too complicated compared to the problem of nondegenerate level.

For the illustration, the lowest orders will be discussed here.

From the last section, it was found that $[H_0 - E_a^{(0)}] \, |0> = 0$, where $|0> = |\psi>$ as $\lambda \to 0$. The energy E is one of E_1, E_2,--- E_n.

Also, it is known that:

$(H_o - E_a^{(0)}) |1> + (V - \mathcal{E}_1) |0> = 0$------------------------(a).
$(H_o - E_a^{(0)}) |2> + (V - \mathcal{E}_1) |1> - \mathcal{E}_2 |0> = 0$ ----------------(b).

Now let $|0>$ be expanded in terms of $|E_a^{(0)}, \alpha >$, where α stands for the degeneracy. Hence,

$|0> = \sum_\alpha E_a^{(0)}, \alpha > < E_a^{(0)}, \alpha \, |0>$ ---- (37).

Where $\sum_\alpha | E_a^{(0)}, \alpha > < E_a^{(0)}, \alpha | = P_o$, the well-defined projection operator. Its value is 0 or1. It is into projection. Where Q_o is defined as the out projector given as $Q_o = 1 - P_o$ ------ (38).

Using the well-defined properties of these projection operators, it helps a lot, such as $P_0 + Q_0 = 1$, $P_0 = 0$ or 1; hence, $Q_0 = 1$ or 0.

Hence, $P_0 |0> = |0>$. The projection of a onto the subspace $\mathcal{E}_a^0$ gives $P_0(V-\mathcal{E}_1)|0> = 0$ -------------------------------- (39).

And the projection onto the complementary spaces gives

$Q_0|1> = \frac{Q_0}{a} V|0>$-------------------------------------- (40).

This is obtained by the use of previous relations such as $Q_0 = 1 - P_0$

And $\frac{Q_0}{a} = \sum_{E_a^0 \neq E_0} \frac{\sum_a |E_a^0><E_a^0|}{E_a^0 - E_0}$ (as shown before).

Equation 39 is eigenvalue equation in the subspace $\mathcal{E}_a^{(0)}$, $\mathcal{E}_1$ is the eigenvalue of the operator $P_0 V P_0$ in $\mathcal{E}_a^{(0)}$ and $|0>$ is the corresponding eigenvector $[|\psi> \rightarrow |0>$ as $\lambda \rightarrow 0]$. In the $|E_a^{(0)}>$representation, equation 39 takes the form

$\sum_{\alpha-} < E_{a,\alpha}^0 |V|E_{a\alpha}^0->|0>=\mathcal{E}_1|0>, \alpha = \alpha^-$------------(41).

Thus, the first order correction $\mathcal{E}_1$is obtained by diagonalyzing the $g_a \times g_a$ matrix whose elements are

$V_{\alpha\alpha-} \equiv <E_{a,\alpha}^0 |V|E_{a,\alpha}^0->$ ----------------------------(42).

Where the possible values for $\mathcal{E}_1$are the eigenvalues of this matrix. If there are g_a different eigenvalues, it means they are nondegenerate, which indicates the degeneracy is completely removed by the perturbation. But if there are less than g_a different eigenvalues, it means the degeneracy is partially removed. Using the projection operator P_0, the following is obtained:

$\sum_{\alpha}^{g_a} \sum_{\alpha} - |E^0_{a,\alpha}><E^0_{a,\alpha}|V - \varepsilon_1|E^0_{a,\alpha}-><E^0_{a,\alpha}-|0>=0$

By certain algebraic elaborations, it is easy to find

$\sum_{\alpha} - <E^0_{a,\alpha}|V - |E^0_{a\alpha}-><E^0_{a,\alpha}-|0>=0$, $\alpha = 1, 2, 3, \text{-----} g_a$.

The solution is Slater determinant, given by

$det.[<E^0_{a,\alpha}| V - \varepsilon_1| E^0_{a,\alpha}->]=0$, which yields

$$\begin{pmatrix} V_{11} - \varepsilon_1 & V_{12} & V_{13} & V_{1g_a} \\ V_{21} & V_{22} - \varepsilon_2 & V_{23} & V_{2g_a} \\ V_{31} & V_{32} & V_{33} - \varepsilon_3 & V_{3g_a} \\ \text{---} \\ V_{g_a1} & V_{g_a2} & V_{g_a3} & V_{g_ag_a} - \varepsilon_g \end{pmatrix} = 0 \text{--------(43).}$$

This gives g_α different ε_1 correspond to $<E^0_{a,\alpha}|0>$. If all ε_i are different, then the degeneracy is completely removed.

8.2.3. *Important Notes*

1. If the first-order correction ε_1 is nondegenerate eigenvalue, the eigenstate $|0>$ is completely determined to the zeroth order; it is defined within constant by the equations 9 and 10.
2. The projection of $Q_0|1>$ of the first order correction to $|\psi>$ onto the component of ε^0_a is given by equation 40.
3. The projection of $Q_0|1>$ onto the subspace ε^0_a remains undetermined, except for the condition $<0|\psi> = <0|0>=1$ where $|E^0_a> = |0>$ as $|\psi> \rightarrow |0>$, when $\lambda \rightarrow 0$.
4. If ε_1 is g_1-fold degenerate, equations 9 and 10 show only the $|0>$ belongs to the corresponding g_1-dimensional subspaces.

5. If it is required to measure $|0>$ more precisely, it is needed to go for higher orders.
6. If one of the values of $\mathcal{E}_1$ has been chosen, it can obtain the second-order correction $\mathcal{E}_2$ by projecting (equation 11) onto the subspace of $\mathcal{E}_1$,where this subspace $(\mathcal{E}_a^{(1)})$ is contained in $\mathcal{E}_a^{(0)}$, and the corresponding projector is $P^{(1)}$. The projector onto its complement in $E_a^{(0)}$ is P^-, where $P_0 = P^{(1)} + P^{(-)}$. Therefore, $P^{(1)} + P^{(-)} + Q_0 = 1$, using these relations, the projection of equation 11 gives $P^{(1)}$ $VQ_0|1> - \mathcal{E}_2 P^{(1)}|0>$. With the aid of equation 40, it is obtained that

$$P^{(1)}[(V\frac{Q_0}{a} V) - \mathcal{E}_2]|0> = 0 \text{ -----------------------(44).}$$

This is the second-order correction; it is analogous to equation 39 for the first-order correction by the same way as $\mathcal{E}_1$ is the eigenvalue of $P_0 V P_0$ in the space $\mathcal{E}_a^0$, and $\mathcal{E}_2$ is the eigenvalue of $P^{(1)}V(\frac{Q_0}{a})P^1$ in the space $\mathcal{E}_a^{(1)}$ and the corresponding eigenvector is $|0>$. If $\mathcal{E}_1$ is nondegenerate eigenvalue $(g_{1=1})$, then $|0>$ is determined by the equation of lower order, i.e., $\mathcal{E}_2 = <0|V\frac{Q_0}{a}V|0>$ as in the nondegenerate case (equation 31). But if $g_1 > 1$, the calculation of $\mathcal{E}_2$ will involve the finding of eigenvalues of a $(g_1 \times g_1)$ matrix. Also, it should be noted that

1. If these eigenvalues are quite different, the degeneracy is completely removed in the second order. If it is not so, and it is needed, then it has to proceed to higher orders.
2. It happens, sometimes, that the degeneracy remains in all orders (as in the case of Coulomb energy E_C and the Stark effect in the rigid rotator).

Problem: Take $^1H_1$ atom under the effect of the electric field $\vec{E}$, where the Hamiltonian $H = H_0 - e\vec{E}.\vec{r} = H_0 + V$

And E_n^0 is the ground state eigenvalue. Let $\psi_{n\ell m}(n\ell m_\ell s_z) \equiv$ ln$\ell m_\ell s_z>$. The contribution of the permutation is given by

$<n\ell m s_z|V|$ln $\ell^- m^- s_z^->$. Carry on the calculation for n=2 to find the effect of $E^\rightarrow$ on the level l2ℓm S$_z$>.

Problem: Atoms in a central force field, where $\mathbb{H}$ is given by

$$\mathbb{H} = \sum_{i=1}^z \left(\frac{p_i^2}{2m} - \frac{Ze^2}{r_i}\right) + \sum_{i<j}\frac{e^2}{r_{ij}} + \sum_{i=1}^z (\vec{\ell_i}.\vec{s_i})$$

$\mathbb{H}$ is the Hamiltonian of the system, under the effect of the central force field, neglecting the nuclear reaction. If the electrons are so close to the nucleus, so it is advised to take the average of the nucleus charge that interacts with the surrounding electrons, then with neglecting $(\ell.s)$ term, Hamiltonian of the system becomes

$$\mathbb{H} = \sum_{i=1}^z \left(\frac{p_i^2}{2m} - U_c(r_i)\right) + \sum_{i=1}^z \left[(U_c^-(r_i) - \frac{ze^2}{r_i}\right] + \sum_{i<j}^z \frac{e^2}{r_{ij}} =$$

H$_C$ +V$_1$,.where U_c^- is the coulomb potential due to nuclear charge effect, and H$_C$=$\sum_{i=1}^z (\frac{p_i^2}{2m}$ -U$_c$(r$_i$)), and V$_1$ is the other three terms. Consider V$_1$ as a perturbation term, work it out to deduce the spectrum of H atom.

Worked Example: Take the carbon atom ($_6^{12}C$), find its ground state. The configuration of the carbon ground state is 1S^{2}2S^{2}2P^2 where 1S^{2}2S^2 are complete shells, but 2P^2 shell is not complete.

It has only two electrons. It is $\binom{6}{2}$ fold-degenerate, i.e., $\binom{6}{2}$=15-fold degenerate.

For the purpose of finding the possible values for the pair (LS) and their degeneracy, that is to say the number of the corresponding series of (2L+1) (2S+1) vectors, the closed shells may be ignored, taking only the nonclosed one such as the 2p electrons. Now if Pauli principle considered neglected, the different spectral terms can be formed with 2P electrons, which are: 3S3P3D1S1P1D. This can be illustrated as

$1S^2$ (n=1, $\ell=0$) two electrons occupy 1S level.
$2S^2$ (n=2, $\ell=0$) two electrons occupy 2S level.
$2P^2$ (n=2, $\ell=1$, $m_\ell=-1, 0, +1$, $S_z=\pm1/2$; m_ℓ takes ($2\ell+1$) values.

The degeneracy is 15-fold as shown before. V_1 (perturbation term in Hamiltonian) is invariant under coordinate rotation, which means that [V, L]=0, where L is the generating function of rotation as shown previously. For the same reason, [V,S]=0.

$1S^2 \equiv |1, 0, 0>|1, 0, 0>[\frac{1}{\sqrt{2}}(\alpha\beta - \beta\alpha)]$, where S=0 L=0.

If 2P is considered where there are two electrons of spins and orbital angular L for each one, then there two possible combinations. =1 (parallel spins); it is triplet, (2S+1) values of m_s or S=0 (anti parallel), it is a singlet because (2S+1)=1, one state.

If L=2, 1, 0 due to the combination of L_1 and L_2 for electron one and two respectively, then

S=1, 0
L=2, 1, 0

If S=1 is considered with L=2, 1, 0, it means S3D3P3S. These are quite clear from that 3D(s=1, l=2)3P(s=1, l=1) and 3S(s=1, l=0), which is a triplet state. But if S=0 is a singlet state (2s+1)=1, is taken with L=2, 1.0, it leads to 1D1P1S. In the spin exchange, S=1 is symmetric and S=0 is antisymmetric.

It is important to notice S and D states are symmetrical in the interchange of the orbital variables, while P state is antisymmetrical for such interchange. This character of antisymmetrical for P, 1S, 1D states imply that 3P1S1D satisfy Pauli principle in all $(2l_p+1)_p$ $(2S+1)_p+(2l_D+1)_D$ $(2s+1)_D$ $+(2s+1)_s(2l+1)_s=9+5+1=15$ linearly independent antisymmetrical states.

In LSM$_L$M$_S$ basis, $V_1=V$ is necessary diagonal. In over all, there are three different eigenvalues: $\wp(3P)$, $\wp(1S)$, and $\wp(3D)$ which are of 9, 1, and 5 fold of degeneracy respectively, as pointed above. So if the individual state wave functions are known, these eigenvalues can be numerically calculated. As a representative example, the 2P state might be taken, where

$$L=1 \Rightarrow \begin{cases} m = 1 \\ m = 0 \\ m = -1 \end{cases} \times 2, \text{ 2 is for the possibility of spinning up or}$$

down, where there are two different combination (as shown before). Also, $[V, L]=0$, $[V, S]=0$; V commutes with both L and S.

It had been shown in the chapter of identical particle the wave functions (as antisymmetrical) are represented by Slater determinant, and the solution for the eigenvalues also given by the Slater determinant (equation 43). The wave functions are chosen such that the off diagonals vanish because $[V, L]=0$, and $[V, S]=0$. Hence,

$$\begin{pmatrix} V_{11} - \mathcal{E} & 0 & 0 \; 0 \\ 0 & V_{22} - \mathcal{E} & 0 \; 0 \\ - & - & - \\ \cdot & \cdot & \cdot \\ V_{n1} \text{---} V_{n2} \text{--} V_{nn} \text{-} \mathcal{E} \end{pmatrix} =0$$

The wave functions are represented by lL, S, M$_L$, M$_S$>. Take 2P^2:3P, 1D, 1S, where:

3P of (L=1, S=1)→ 3 of M_L, 3 of M_S, $(2l+1)(2s+1)= 9$
1D of (L=2, S=0)→ 5 of M_L, $1M_S$, $(2l+1)(2s+1) =$ 5
1S of (L=0, S=0)→ $1M_L$, $1M_S$, $(2l+1)(2s+1) =$ 1

So there are fifteen different states $=(9)_p+(5)_D+(1)_s 15= \begin{bmatrix} 6 \\ 2 \end{bmatrix}$.

To solve this kind of problem, the calculation of the matrix elements of V is necessary. Consider the wave function given as lL, S, M_L, M_S>, the matrix elements are given by:

<L, S, M_L, M_S lVlL⁻, S⁻, M_L^-, M_S^- >=$<V>_{LS}\delta_{LL}-\delta_{SS}-\delta_{M_LM_L^-}\delta_{M_SM_S^-}$.

Then the details of the calculations are known according to the conditions of the problem.

Problem: An atom is subjected to a magnetic field ($\mathbb{H}^{\rightarrow}$), where the Hamiltonian H is given by H=H_0 $-\frac{e}{2mc}\mathbb{H}^{\rightarrow}.(L+2S)$

<E_0, J, M_Jl$\frac{-e}{2mc}\mathbb{H}^{\rightarrow}.(L_Z+2S_Z)$l E_0J, M_J>=$-\frac{-e}{2mc}\mathbb{H}.M_J\hbar.g$

Where g= $\frac{3J(J+1)+S(S+1)-L(L+1)}{2J(J+1)}$, itis g-Lande factor.

Compute g for $^{22}_{11}Na$; its configuration (GS) is $1S^2 2S^2 2P^6 3S^1$.

8.3. Time-Dependent Perturbation Theory

So far, the time-independent perturbation for degenerate and nondegenerate levels has been dealt with, formations and some applications. Since the physical systems usually are subjected to events which, in fact, of time character in general, where the time is physically related to these events. So the wave function that is describing the physical dynamical variables during the evolution of the time is, of course, time dependent.

This physical situation is quite well expressed by the time-dependent Schrödinger equation which is

$$i\hbar \frac{d}{dt}|\Psi(t)\rangle = H(t)|\Psi(t)\rangle \ \text{--------------(45)}.$$

Hence, in fact, even the dynamical variable such as H in equation 45 is time dependent. This might be clarified by considering a quantum system in a certain dynamical state at time (t_0), and it is requested to find it during going from t_0 to t. In this case, the time evolution operator $U(t, t_0)$ has to be used; therefore, $|\Psi(t)\rangle = U(t, t_0)|\Psi(t_0)\rangle \ \text{-------}$ (46).

where $U(t, t_0)$ is given by $U(t, t_0) = e^{-iH(t-t_0)/\hbar}$ (as shown before).

So the problem is to determine $U(t,t_0)$ as exactly as possible, which describes the time evolution of the dynamical state of the system in Schrödinger equation representation. Therefore,

$$i\hbar \frac{\partial}{\partial t}U(t, t_0) = H(t)U(t, t_0), \text{ where } U(t_0, t_0) = 1 \ \text{--------}(47).$$

Using the integral representation of $U(t, t_0)$, which is given by

$$U(t, t_0) = 1 - \frac{i}{\hbar} \int_{t_0}^{t} H(\tau)U(\tau, t_0)d\tau \ \text{-------------------} (48).$$

From equation 47, where $U(t_0, t_0) = 1$, as initial condition,

$U(t, t_0)$ can be determined once $H(t)$ is known. Also, it is known that $U(t, t_0)$ is a unitary, i.e., $U^+U = U\ U^+ = 1$, which can be easily proved. Hence, $U^+U = UU^+ = 1 \ \text{--------------}$ (49).

In addition to that, it obeys the time composition law, so

$$U(t, t^-)=U(t, t^=)U(t^=, t^-) \text{-----------------------------(50)}.$$
$$\text{Whence } U^+(t, t^-)=U(t^-, t) \text{ ----------------------------(51)}.$$

Equation 48 is an approximation solution for $U(t, t_0)$.

Using equation 51, $U(t, t_0)$ might take the form:
$$U(t, t_0)=1-\frac{i}{\hbar}\int_{t_0}^{t} U(t, \tau)H(\tau)d\tau \text{ --------------(52)}.$$

This formalism is based on the fact that:
$$H(t)=H_0(t)+V(t). \text{ ---------------------------------- (53)}.$$

Hence, the philosophy of the perturbation theory is based on knowing H_0 (t) and $U(t, t_0)$ first. So if $U(t,t_0)$ is the time evolution operator of $H_0(t)$, then it can be written that

$$i\hbar\frac{\partial}{\partial t}U_0(t, t_0)=H_0(t)\, U_0(t, t_0), \ U(t_0, t_0)=1,\text{-------(54)}.$$

If $U_0(t, t_0)$ is known, then U will be determined if it is possible to form the unitary operator

$$U_I(t, t_0)=U_0^+(t, t_0)U(t, t_0)\text{----------------------(55)}.$$

$U_I(t, t_0)$ is the time evolution operator for the state in the intermediate representation that is derived from Schrödinger equation representation by the unitary transformation $U_0(t, t_0)$.

Now the interaction picture (intermediate representation) can be obtained by applying U_0^+ to $|\Psi(t)>$, such as:

$$|\Psi_I(t)>=U^+(t, t_0)|\Psi(t)>\text{----------------------------------(56)}.$$
$$\text{Where } U^+|\Psi(t) > =|\Psi_0(t)>, \text{ Heisenberg picture)-------(57)}.$$
$$\text{Now, } i\hbar\frac{d}{dt}|\Psi_I(t)>=V_I(t)|\Psi_I(t)> \text{ or } i\hbar\frac{\partial}{\partial t}U_I(t, t_0)=V_I(t)U_I(t, t_0)$$

Also, $U_I(t_0, t_0)=1$, and $V_I(t)=U_0^+(t, t_0)\, V(t)\, U_0(t, t_0)$ ---------(58).

So the time-evolution operator is

$$U_I(t, t_0)=1-i\hbar^{-1}\int_{t_0}^{t} V_I(t)U_I(\tau, t_0)d\tau \text{ --------------------(59).}$$

It has all the properties of an evolution operator, and in particular, it satisfies the equations 48–54, where H(t) is replaced by $V_I(t)$ everywhere.

Now
$|\Psi_I(t)>= U^+(t, t_0)|\Psi(t)>=U_0^+(t, t_0)U(t, t_0)|\Psi(t_0)>=U_0^+(t, t_0)|\Psi(t)>$
At t=t_0, $U_0^+(t_0, t_0)\, U_0(t_0, t_0)=1$, it implies that $|\Psi_I(t_0)>= |\Psi(t) >$.

It is initial condition. Therefore:

$|\Psi_I(t) >=U_0^+(t, t_0)U(t, t_0)\, |\Psi(t_0)>$----------------------(60).
And $|\Psi_I(t)>=U_I(t,t_0)|\Psi_I(t_0)>$----------------------(61).

Using the fact that $U_0^+ =U^{-1}$,
then $U(t, t_0)=U_0(t, t_0)U_I(t,t_0)$,
which leads to $U(t, t_0) =U_0(t, t_0)[1-\frac{i}{\hbar}\int_{t_0}^{t} V_I(\tau)U_I(\tau, \tau_0)\, d\tau]$.

Now substitute for U_I and $|\Psi(t) > = U(t,t^-)\, |\Psi(t\text{-})>$ using the composition law (equation 50), then it is obtained that

$|\Psi(t) >= U(t, t^-)|\Psi(t^-)>=U(t, t\text{-})U(t^-, t^=)|\Psi(t^=>.$
Hence, $|\Psi(t) >= U(t, t^=)|\Psi(t^=)>$ ------------- (62).

Therefore, it is possible to write
$U(t^=, t^-)=U(t^=,t^-)\, U(t^-, t^=)=1$ (Show this as an exercise.)

Also, $U(t^=, t^-) =U^+(t^-, t^=)\Rightarrow U_0^+(t\text{-}, t_0)=U_0(t_0, t\text{-});$
therefore, $U_0(t, t_0)U^+(t^-, t_0)=U_0(t, t_0)U(t_0, t^-)=U_0(t, t^-).$

All these operational manipulations lead to

$$[U(t,t_0)=U_0(t,t_0)-\frac{i}{\hbar}\int_{t_0}^{t} U_0(t,t^-)\,V(t^-)U(t,t_0)\,dt^-]$$

as $t \to t^-$ $-\,-\,-\,-\,-\,-\,-\,-\,-\,-\,-\,-$ (63).

Now from equation 63, the following different orders can be obtained:,

1. Zeroth order: $U(t,t_0)=U_0(t,t_0)$----------------(64).
2. First order: $U(t,t_0)=$
   $U_0(t,t_0)-\frac{i}{\hbar}\int_{t_0}^{t} U_0(t,t^-)V(t^-)U_0(t^-,t_0)dt^-$------- (65).
3. Second order: $U(t,t_0)=U_0(t,t_0)-\frac{i}{\hbar}\int_{t_0}^{t} U_0(t,t^-)V(t^-)U_0(t^-,$
   $t_0)dt^-+(i/\hbar)^2\int_{t_0}^{t}\int_{t}^{t^-}[U_0(t,t^-)V(t^-)U_0(t^-,\qquad t^=)V(t^=)U_0(t^=,$
   $t_0)dt^-dt^=]$ --(66).
4. Third order: $U(t,t_0)=2^{\text{nd}}$
   order$+(\frac{i}{\hbar})^3\int_{t_0}^{t}\int_{t}^{t^-}\int_{t^-}^{t^=}\int_{t^=}^{t^\equiv} U_0(t,t^-)V(t^-)U(t^-,t^=)\times$
   $V(t^=)U_0(t^=,t^\equiv)V(t^\equiv)U_0(t^\equiv,t_0)dt^-dt^=dt^\equiv$ --------------(67).

In general, the nth order is written as

$U_n(t,t_0)=i\hbar^{-n}\int[\,dt_n^- \,dt_{n-1}^- ---dt_1^- \,U_0(t,t_n^-)V(t^-)U_0(t^-,t_{n-1}^-)V(t_{n-1}^-)-$
$----U_0(t_2^-,t_1^-)V(t_1^-)U_0(t^-,t_0),\; t>t_n^->t_{n-1}^->--->t_1^- >t_0$-----------(68).

By successive iterations, the following for $U_I{}^{(0)}(t,t_0)$ is obtained:

$$U_I{}^{(0)}(t,t_0)=1+\sum_{n=1}^{\infty} U_I^{(n)}(t,t_n)$$--------------------(69),

where $U_I^{(n)} \equiv U_I{}^{(n)}$ is

$U_I{}^{(n)} \equiv i\hbar^{-1}\int dt_n^- dt_{n-1}^- ----dt_1^- V_I(t_n^-)V_I(t_{n-1}^-)---V_I(t^-).$----(70).

Using this with the aid of $U_I(t, t_0) = U_0^+(t, t_0)U(t, t_0)$ and $V_I(t) \equiv U_0^+(t,t_0)V(t)U_0(t,t_0)$, as verified before, this is obtained:

$$U(t, t_0) = U_0(t, t_0) + \sum_{n=1}^{\infty} U_n(t, t_0), \text{------------------}(71)$$

[as an expansion of U]

[Note: $U_0 \equiv U^{(0)}$, $U_n \equiv U^{(n)}$]. Expansions of equations 69 and 71 are power series in $V(t)$ which converge the more rapidly, the closer $U_0(t,t_0)$ to $U(t,t_0)$. Therefore,

$U^{(0)}$ is the zeroth order approximation correction.
$U^{(1)}$ is the first order approximation correction.
$U^{(2)}$ is the second order approximation correction.
$U^{(3)}$ is the third order approximation correction.

$U^{(n)}$ is the nth order approximation correction.

So it is clear that as it goes higher in order, the problem gets more complicated. But with the availability of the high capacity and speed computers these days (2015), it becomes quite a workable problem.

8.3.1. Transition Probability (Fermi Golden Rule)

It is quite clear that if $H^{(0)}$ (the Hamiltonian) is time independent, the evolution operator $U^{(0)}(t, t^-)$ is simply given as $U^{(0)}(t, t^-) = e^{-iH^{(0)}(t-t^-)/\hbar}$ -------------------------------- (72).

Now the first requirement is to solve the eigenvalue problem for $H^{(0)}$. The spectrum will be assumed entirely discrete, unless otherwise assumed, for simplifying the writing. So let la>, lb>,---lk> be a complete set of eigenvectors of $H^{(0)}$, and $E_a^{(0)}, E_b^{(0)}$ --- $E_k^{(0)}$ are the corresponding eigenvalues. Take a case of transition from state of $E_k^{(0)}$ (energy) to state of $E_L^{(0)}$ (energy). This can be illustrated in the following diagram:

$$\begin{array}{l} \underline{\hspace{3cm}} \quad E_K^{(0)} \\ E_\gamma = \hbar\nu \\ \underline{\hspace{3cm}} \quad E_l^{(0)} \end{array}$$

As it is known from the beginning of the atomic era, the frequency of gamma radiation emitted from state $E_k^{(0)}$ to state $E_L^{(0)}$ is given by

$$\omega_{kl} = \frac{E_k^{(0)} - E_L^{(0)}}{\hbar} = \text{Bohr frequency,} \text{--------(73).}$$

And the matrix elements of the potential V(t) is given by:

$V_{KL} = <KlV(t)l\ L>$ ------------------------(74).

Let la> be the eigenstate of $H^{(0)}$ at the initial time t_0, and lb> is its eigenstate at later time t. Now the probability of the transition from the state Ia> to the state lb> is given by:

$W_{a \to b} = l<blU(t, t_0)la>l^2$ ---------------------------(75).

If the potentials were null, the vector representing the state of the system at time t would differ from the initial vector la>, by the time factor $e^{-i[E_a^{(0)}(t-t_0)]/\hbar}$, which leads to forbidden transition. So equation 75 will be zero because <bla>=0, orthogonal.

By using equation 71, the probability amplitude <blUla> can be expanded into power series in V, U(t, t_0) given by equation 71. Hence,

<blU(t, t_0)la>=$\sum_{n=1}^{\infty} < bl\ U^{(n)}\ la>$ ------------- (76).

Where $U^{(n)}$ is given by equation 68: $[U_n = U^{(n)}]$.

In the $H^{(0)} = H_0$ representation defined above, the contributions of the successive orders to the amplitude equation(76) are:

$$\underline{<b|U^{(1)}|a>} = -i\hbar^{-1} \int_{t_0}^{t} dt^- [e^{-iE_a^0(t-t_0)/\hbar} V_{ba}(t^-) e^{-iE_a^0(t-t_0)/\hbar}]$$
-----------------------(77). And

$$<b|U^{(2)}|a> =$$
$$(-i\hbar^{-1})^2 \sum_k \int_{t_0}^{t^-} dt^- \int_{t^-}^{t^=} dt^= [\quad e^{-iE_b^0(t-t^-)/\hbar} \quad V_{bk} \quad \exp(-E_k^0(t^- - t^=)/\hbar \times$$

$$V_{kL}(t^=) \exp(-E_L^0(t^= -e^=)/\hbar \; V_{La}(t^=) \exp(-iE_a^0(t^=-t_0)/\hbar \; ----- \; ----$$
---(78).

And $\quad \underline{<b|U^{(3)}|a>} = (-i\hbar^{-1})^3 [\sum_k \sum_L \int_{t_0}^{t} dt^- \int_{t_0}^{t^-} dt^= \int_{t_0}^{t^=} dt^{(=)} \exp.($
$-iE_b^0(t-t^-)/\hbar . V_{bk}(t^-) \; \exp.(-iE_k^0(t^- t^=)/\hbar \; V_{kL}(t^=) \; \exp.(-iE_L^0(t^= - t^=)/\hbar)$
$V_{La}(t^=) \exp.(-i E_a^0(t^= -t_0)/\hbar .$-------------------------(79).

It is to be noticed in all these expressions the summations are extended over the entire basis vectors of $H^{(0)}$. Equations 77, 78, and 79 can be schematically represented by the following diagrams from left to right, respectively:

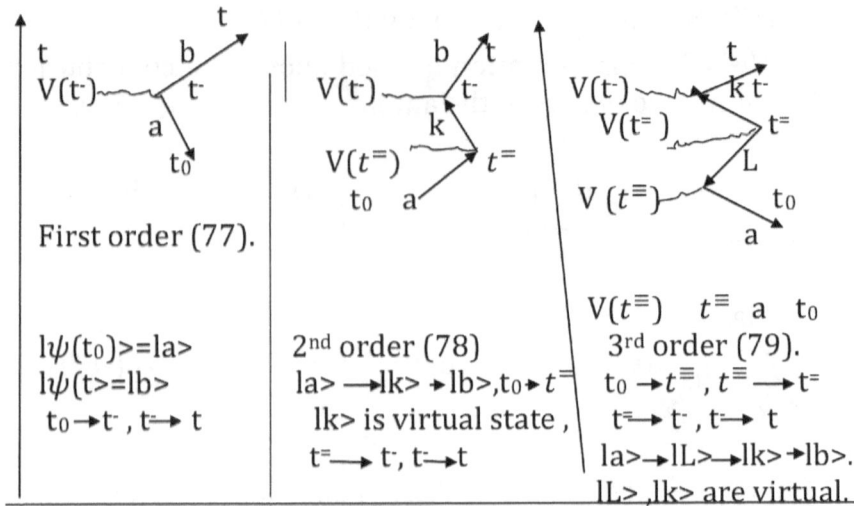

First order (77).

$l\psi(t_0)>=la>$	2nd order (78)	3rd order (79).
$l\psi(t>=lb>$	$la> \rightarrow lk> \rightarrow lb>, t_0 \rightarrow t^=$	$t_0 \rightarrow t^= , t^= \rightarrow t^=$
$t_0 \rightarrow t^- , t^= \rightarrow t$	$lk>$ is virtual state ,	$t^= \rightarrow t^- , t^- \rightarrow t$
	$t^= \rightarrow t^- , t^- \rightarrow t$	$la> \rightarrow lL> \rightarrow lk> \rightarrow lb>.$
		$lL> ,lk>$ are virtual.

It is important to notice that the number of the virtual states which take la> to lb> is (n-1), where n is the order rank. If n=1, no virtual state as in the first order; if n=2 the number of virtual states is one (k) as in the second order case, for n=3 there are two virtual states as in the third order case lk> and lL>.

The equation 75 for the probability of order n is written as

$\omega_{a \rightarrow b} \approx l<blU^{(1)}la>l^2 + l<blU^{(2)}la>l^2 + - - - - - - - - l<blU^{(n)}la>l^2$

Note: The expression obtained correctly represents the expansion of $\omega_{a \rightarrow b}$ in powers of V up to at least the order n+1. The precision obtained is of a higher order than n+1 if <blU^{(1)}la>=0.

The first order transition probability is given as

$\omega_{a \rightarrow b} \approx l<blU^{(1)}la>l^2 = \hbar^{-2}l\int_{t_0}^{t} e^{i\omega_{ba}t^-}V_{ba}(t^-)dt^-l^2$ - - - - - - - - - - - (80).

Where $\omega_{ba} = (E_b^0 - E_a^0)/\hbar$, Bohr frequency.

By approximation, it can be written that $\omega_{b \to a} \approx \omega_{a \to b}$, but for higher orders, this is not verified.

It should be noted clearly that the property of micro reversibility, $\omega_{kb} \to k_a = \omega_{a \to b}$ should not be confused with that, which is only satisfied if H is invariant under time reversal, where it is exactly true in all orders.

8.3.2. Some Examples for Brief Applications

a. Coulomb excitation of nuclei by charged particles

Consider beam of protons collide with a nucleus of charge Ze, as shown in the following schematic diagram.

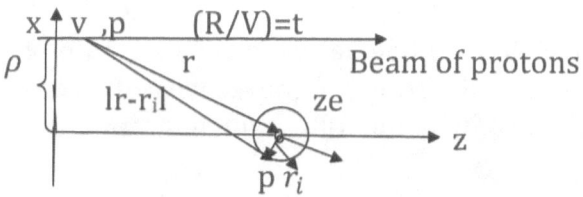

The force acting between the protons beam ($e^+ = e_p$) and the nucleus (Ze^+) is the known Coulomb force, given by $[F_c = (Ze.e_p)/lr-r_il^2]$. The Hamiltonian of the system is given by $H^{(0)} = H_{nuc} + P^2/2m + \sum_{i=1}^{Z} \frac{e_i^2}{|r-r_i|}$ the system is composed of a nucleus of positive charge (Ze) with radius R. Also with angular momenta J_α and J_β plus the energies E_α and E_β as shown in the following diagram,

Excited state β, E_β, J_β

Ground state α, E_α, J_α

This excitation of the nucleus is due to the collision between the protons and the nucleus. For the ground state (α), there are

$(2J_\alpha+1)$ linearly independent states which might be distinguished from one another by M_α of J_α along a given quantized axis. They are represented by the vectors $|\alpha J_\alpha M_\alpha\rangle$.

If H_n is the Hamiltonian of the nucleus, then the eigenvalue equation is $H_n|\alpha J_\alpha M_\alpha\rangle=E_n|\alpha J_\alpha M_\alpha\rangle$, $M_\alpha=-J_\alpha,-----,+J_\alpha$ for GS.

And similarly for the excited state:
$H_n|\beta J_\beta M_\beta\rangle=E_\beta|\beta J_\beta M_\beta\rangle$, $M_\beta=-J_\beta,---------,+J_\beta$.

These eigenvalue equations are quite familiar to physics students. Let $E_K=1/2\ Mv^2$ (kinetic energy) is the incident energy in the proton-nucleus center of mass system. The excitation energy of the nucleus is given by $E_{ex}=\Delta E=E_\beta-E_\alpha$. Now let Ω_α is the solid angle contained the direction of the incident proton, and Ω_β is that of the inelastically scattered proton, and Θ is the angle between the direction of the incident proton and the direction of the scattered proton. It can be represented simply by the following diagram:

From the scattering theory, the differential scattering crosses is given as $\frac{d\sigma}{d\Omega}$, which is the probability that the proton is in elastically scattered into the direction Ω_β and the nucleus is excited from the ground state (α) to the excited state (β). It can be easily calculated (refer to chapter 7). Then the total cross section of the scattering can be calculated, which represents the probability that the scattering operation has taken place.

This is the problem in its general scheme. The proton-nucleus interaction might be examined in more details as follows:

1. For large-distance r, from the incident proton to the nucleus center of mass ($r \gg R$), where R is the radius of the nucleus. The interaction in this case simply is the Coulomb interaction which is $\frac{Ze^2}{r}$; therefore, as r becomes smaller it increasingly deviates from this simple pure Coulomb force.

2. If $r > R$, then the deviation is purely of electromagnetic origin. It reduces essentially to the difference between the exact Coulomb interaction and the term $\frac{Ze^2}{r}$, namely the potential is given by

$$V(r) = e^2 \sum_{i=1}^{Z} \left(\frac{1}{|r-r_i|} - \frac{1}{r} \right)$$ ------------------------ (81).

Here, r_i is the position vector of the ith proton in the nucleus.

3. Once the proton is close to the nucleus, such as ($r < R$), nuclear interaction takes place which is prevailing largely over E$\propto$M interaction.

4. If the energy (E) is sufficiently small, then Coulomb repulsion potential (Ze^2/r) prevents the proton from approaching the nucleus, where (Ze^2/r) becomes the dominant interaction throughout the collision processes. The Hamiltonian will take the form,

$H^{(0)} = H_n + P^2/2m + Ze^2/r$, where $P^2/2m = E_K$ is the kinetic energy of the proton(p). This implies the motion of the proton separates completely from the nucleus motion, a case in which the nucleus stays in its ground state, but the proton p is scattered elastically with a $\frac{d\sigma(\theta)}{d\Omega}$ given by $\frac{d\sigma_R}{d\Omega} = [\frac{a^2}{4sin^4\theta/2}]$, Rutherford formula, where $a = 1/2$ (Ze^2/E). It is half the distance of the closest approach in classical mechanics.

This approximation is justified as long as $a \gg R$, $\Delta E/E \ll 1$, and $\gamma = \frac{a}{\lambda} = \frac{Ze^2}{\hbar v} \gg 1$. The potential time-dependent V(t) which is perturbing the dynamical variable of the nucleus is (see equation 81: $V(t)=e^2\sum_{i=1}^{Z}[\frac{1}{|r(t)-r_i|}-\frac{1}{r(t)}]$ ------------- (82).

where $r(t)$ is the trajectory of the proton.

The potential $V(t)$ can induce the transition from the ground state to the excited state, the transition probability $W_{a\to b}$ is being small $<<1$, so the first order approximation (equation 80) is:

$$W_{a\to b}=\hbar^{-1}\int_{-\infty}^{+\infty}e^{i\Delta Et/\hbar}<\beta J_\beta M_\beta \; |V(t)|\alpha J_\alpha \; M_\alpha> dt^2 \;\text{----}(83).$$

As had been shown before in the scattering theory, there is a relationship between the transition probability and the cross section of the interaction such as

$$\frac{d\sigma}{d\Omega}(\Omega_\alpha M_\alpha \to \Omega_\beta M_\beta)=[\frac{a^2}{4}\cdot\frac{1}{sin^4\frac{\theta}{2}}]\cdot W_{\alpha\to\beta,;}\;\text{therefore,}$$

$$d\sigma=[\frac{a}{4}\cdot\frac{1}{sin^4\frac{\theta}{2}}]\;W_{\alpha\to\beta}\cdot d\Omega \;\text{-------------------------}(84).$$

Hence, if $W_{\alpha\to\beta}$ is calculated; $d\sigma$ can be found. This problem starts by finding $V(t)$ explicitly first. To work for this, it is assumed the proton as a field which is a function of time interacts with the nucleus proton field (see the diagram in page 248) where $r(t)\equiv(x=\rho, y=0, z=vt)$. From equation 81 $(\frac{1}{r_i-r})$ can be expanded for $r>r_i$, to find that $\frac{1}{|r_i-r|}=\sum_{i=1}^{Z}e^2\sum_{\ell=1}^{\infty}\frac{r_i}{r(t)^{\ell+1}}\times$

$P_\ell(cos\theta)=\sum_{i=1}^{Z}\frac{e^2}{r(t)}+\sum_{i=1}^{Z}\frac{e^2 r_i}{r(t)^2}P_1(cos\theta)+\text{----, for }\ell=1.$

The term $\sum_{i=1}^{Z}\frac{e^2}{r(t)}$ contributes zero. (Why)?

The term $\sum_{i=1}^{Z}\frac{e^2 r_i}{(r(t))^2}P_1(cos\theta)=\sum_{i=1}^{Z}\frac{e\,r_i\;e\,r(t)}{(r(t))^3}P_1(cos\theta).$

$\sum_{i=1}^{Z}e\,r_i=q_i.r_i.$ It is the dipole moment of the nucleus.

V(t) can be written in general as:

$$V(t)=\sum_{\ell=1}^{\infty}\sum_{m=-\ell}^{\ell} (-)^m \, Q_{\ell}^m T_{\ell}^{-m} \quad \text{------------(85),}$$

$$\text{where } Q_{\ell}^m = \sum_{\ell=1}^{z} er_i^{\ell} \, Y_m^{\ell}(\Omega_{\ell}) \text{- -------------------------(86)}$$

$$\text{and } T_{\ell}^m(t)=\frac{4\pi e}{2\ell+1} \frac{Y_{\ell}^m(\Omega_{\ell})}{r^{i+\ell}}, \text{ for } r_i < r \text{ ----------(87).}$$

Q_{ℓ}^m and T_{ℓ}^m are well known in electrodynamics, and in studying the electromagnetic properties of the nucleus, see nuclear physics by the author or by de-Shalet. By substituting these physical dynamical quantities in equation 83, the transition probability takes the form

$$\mathbb{W}_{a \to b} = \left| \sum_{\ell m}(-)^{-m} \, S_{\ell}^{-m} <\beta J_{\beta} \, M_{\beta} \, | Q_{\ell}^m | \, \alpha J_{\alpha} M_{\alpha} > \right|^2 \text{----- (88).}$$

$$\text{Where } S_{\ell}^{-m} = \frac{1}{\hbar} \int_{-\infty}^{\infty} e^{-i(\Delta E)t/\hbar} \, T_{\ell}^m(t) \, dt \text{----------------- (89).}$$

S_{ℓ}^m depends only on the classical trajectory of the proton. It can be found by numerical integration. The $(2\ell + 1)$ operator Q_{ℓ}^m(m=-ℓ,---------.+ℓ) is the standard components of the 2^{ℓ}-dipole moment$Q^{(\ell)}$. The nonzero matrix elements of equation 88 are those satisfying the selection rules | J_{α} - J_{β}|≤ J ≤|J_{α} +J_{β} | and m =M_{β} -M_{α}, $\pi_{\alpha} \pi_{\beta}$=$(-1)^{\ell}$. J is the multipolarity order which is the angular momentum carried on by the transmitted or absorbed radiation between state α and state β, π_{α}, and π_{β} are the parities of the state α and the state β respectively.

Using Wigner-Eckert (WE) theorem, the following is obtained:

$$<\beta J_{\beta} \, M_{\beta} | \, Q_{\ell}^m | \alpha J_{\alpha} M_{\alpha}> =$$
$$\frac{1}{2J_{\beta}+1} <J_{\alpha} \ell M_{\alpha} m \, | J_{\beta} M_{\beta}> <\beta || \, Q_{\ell}^m || \, \alpha> \text{-------- (90).}$$

Owing to the selection rules, as mentioned above, the sum in equation 88 is limited to a finite number of ℓ values of well-defined parity and to a single value of m. Some rough calculation shows: all things are being equal; the $(2\ell + 1)$ contribution is about $(R/a)^2$, where R and a are defined earlier, times ℓ contribution. Therefore, it is quite justified in keeping only the term that corresponding to the smallest value of ℓ permitted by the selection rules, either $\ell=|J_\alpha - J_\beta|$ or $|J_\alpha - J_\beta|+1$ as might be the case. Calling ℓ as ℓ_0, the $\mathbb{W}_{\alpha \to \beta}$ can take the form

$$\mathbb{W}_{\alpha \to \beta} \cong \frac{<J_\alpha \ell_0 M_\alpha \, m J_\beta M_\beta>^2}{(2J_\beta+1)} |S_{\ell_0}^{-m}|^2 \, |<\beta|| \, Q^{\ell_0}||\alpha>|^2 \text{ --(91)}.$$

$$[m=M_\beta-M_\alpha]$$

Substituting this in equation 84, the differential scattering cross section will take the form

$$\frac{d\sigma}{d\Omega} = \frac{a^2}{4(2J_\alpha +1)(2\ell_0 +1)}|<\beta|| \, Q^{(\ell_0)}||\alpha>|^2 \, [sin^{-4}\theta/2 |\sum_m S_{\ell_0}^{-m}|^2 \, (92),$$

where $<\beta||Q^{(\ell_0)}||\alpha>$ is the reduced matrix elements; it is independent of M_α and M_β. Equation 92 is a theoretical expression for the cross section. It might be compared with the experimental data. In the experiment, the target nuclei are not oriented, so the polarization of the excited nuclei is not observed; therefore, the measured cross section is obtained as an average of the cross section that defined in equation 92 over the $(2J_\beta+1)$ possible values of M_α, and summing it over the $(2J_\beta +1)$ possible values of M_β, taking in consideration the orthogonality relation of Clebch-Gorgon (CG) coefficients.

Important Notes:

From equation 92, the following might be noted:

1. The angular dependence of the cross section is given by the expression in the bracket which must be numerically calculated.
2. The initial and the final states of the nuclear enter into the formula 92, only through their spin, their parity, and the square of the modulus of the 2^{ℓ}-pole transition moment, reduce matrix of $Q^{(\ell_0)}, <\beta \parallel Q^{(\ell_0)} \parallel \alpha>$.
3. This reduced matrix elements appear in equation 92 as a simple proportionality constant.
4. The comparison of equation 92 with experimental data for the cross section is, therefore, a very direct way to determine these characteristics quantities of nuclear structure.

Consider the case when the potential v(t) is time independent, conservation of unperturbed energy: when V is not explicitly time dependent, the integrals over time in equations 77, 78, 79 are easily affected, and certain simple properties appear. Consider, as example, the first order transition: take equation 78 at $t_0=0$, then it takes the form, $W_{a \to b} = |V_{ab}|^2 f(t, \omega_{ab})/\hbar^2$.----- -----------(93).

$$f(t, \omega) = |\int_0^t e^{i\omega t} dt|^2 = \frac{2(1-cos\omega t)}{\omega^2} ------(\text{verify it}).$$

Plotting $f(t, \omega)$ as a function of ω, the following diagram is obtained:

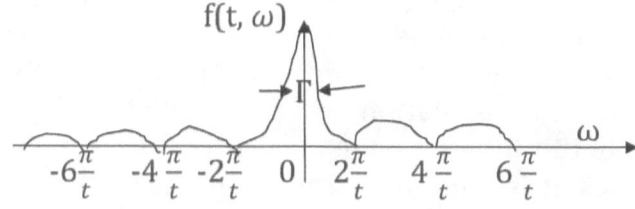

At $\omega \to 0$, there is a sharp peak; therefore,

$$W_{a \to b} \equiv |V_{ab}|^2\left(\frac{2(1-cos\omega t)}{\omega^2}\right)=2|V_{ab}|^2[\,(1-cos\omega t)/\omega^2]\text{---- (94)}.$$

Notice the sharp peak about $\omega = 0$; its width is simply $(2\pi/t)$. Also, it can be shown, $\int_{-\infty}^{+\infty} f(t,\omega)d\omega=2\pi t\text{-----(95)}.$

By using the residues method (see Boundary value problems by Churchill) the following result is obtained:

$$f(t,\omega)=\to \text{ (as } t\to \infty) \longrightarrow 2\pi t\delta(\omega)\text{ --------------(96)}.$$

Where $\delta(\omega)$ is the famous Dirac delta function (refer to appendix A). For a given value of t, $W_{a \to b}$ depends in a simple way on the final state (b); to within a constant, it is the squared modulus of the perturbation matrix elements V_{ab} weighted by the factor $f(t, \omega)$, which actually depends on Bohr frequency for the transition a → b.

It has a pronounced peak at $\omega_{ab}=0$, with width $\Gamma =2\pi/t$. The transition is preferentially goes toward the states whose energies are situated in a band of width $\delta E_0 \cong 2\pi\hbar/t$ about the energy of the initial state. In other words, the transition conserves the unperturbed energy to within $(2\pi\hbar/t)$. So it can be noticed that $\delta E_0 = 2\pi\hbar/t$ leads to $\delta E_0.t=h \Rightarrow \Delta E_0 .\Delta t \geq h$, it is analogous to the time-energy uncertainty relation.

But it must be noted here that the energy in this example is $H^{(0)}$, not the total energy of the system including the perturbation, and the time also is the time after which the measurement of $H^{(0)}$ is performed. It is not the time characterizing the evolution of the system. Due to the importance of this interaction in nuclear physics studies and in atomic studies, it is quite necessary to go farther to note and discuss the features of the transition probability $W_{a \to b}$.

Notes and Brief Discussion of $\mathbb{W}_{a \to b}$.

1. For a given state (b), $\mathbb{W}_{a \to b}$ is time dependent through the factor $f(t, \omega)$.
2. If the transition exactly conserves the unperturbed energy $[\omega_{ab}=(E_a^0 - E_b^0)]=0$, it implies it increases as t^2.
3. If it is not conserving the unperturbed energy, then the transition is an oscillating function oscillates between zero and $(4/\omega_{ab}^2)$ with period $(\frac{2\pi}{\omega_{ab}})$, $(\omega = 2\pi v, v = freq.)$.
4. So $\mathbb{W}_{a \to b}$ oscillates with the same period about the average value $\{2|V_{ab}|^2/[E_a^0-E_b^0]^2\}$.
5. Hence, $\mathbb{W}_{a \to b}$ has the (t^2) behavior only for t value, which is small compared with the period $(2\pi/\omega_{ab})$.
6. The transition might occur to group of states of neighboring energies rather than certain particular state.

Accordingly, it may conclude the following remarks:

(i) This type of transition always take place for transitions to states of continuous spectrum.
(ii) Under certain limiting conditions, to be specified, the transition is defined as the transition probability per unit time.
(iii) Therefore, consider a particular sequence of eigenvectors of $H^{(0)}$ belonging to the continuous spectrum where lb> is one of them, and E_b is the corresponding eigenvalue of $H^{(0)}$. Consider lb> as normalized eigenvector, i.e., $<b|b^-> = \delta(b - b^-)/n(b)$. [Remember for continuous spectrum $<\psi(\xi)| \psi (\xi^-) > = \delta(\xi-\xi^-)$], where n(b) is defined as some real positive function $(n>0)$. If the space domain is B, then the projector of the projection on to the states of B is given by $P_B = \int_{(B)}|b>n(b)db<b|$. If E(b) is taken as new variable and denoting B(E) as the corresponding

domain of integration, then $P_B = \int_{B(E)} |b> \rho_b (E) \, dE |<b|$ ------- (97).

Where $\rho_{b(E)} = n(b) \, db/dE$ is the density of states per energy interval unit. According to equation 97, the number of the vectors $|b>$ per unit of energy interval. In fact $\rho_b(E)$ depends on the normalized $|b>$. Then the transition $W_{a \to b}$, into one of the states of the domain (B) is given by $W_{a \to B} \equiv <a|U^+(t,t_0) \, P_B \, U(t,t_0)|a> = \int_{B(E)} W_{a \to b} \rho_b \, dE$ ----(98).

This equation is obtained by substituting for P_B from equation 97, and for $W_{a \to b} = |<b|U(t,t_0)|a>|^2$ (equation 75).

Now substituting equation 93 into in the right hand of equation 97 leads to the first order perturbation as follows:

$$W_{a \to b} \cong \frac{1}{\hbar} \int |V_{ab}|^2 \rho_b f(t, \omega_{ab}) \, dE \text{ ------------------(99).}$$

Where V_{ab} depends on E through the parameter b. As an example, let the transition to the levels b within the energy interval $[E-(1/2) \, \mathcal{E}$ and $E+(1/2)\mathcal{E}]$. Suppose the width $(\mathcal{E})$ is sufficiently small, which means V_{ab} and ρ_b can be taken out of the integral. Also, take the time t sufficiently large so that $\mathcal{E} \gg$ *the period of the oscillation of the function* f, i.e., $\mathcal{E} \gg (2\pi\hbar/t)$.

Then $W_{a \to b} = \frac{|V_{ba}|^2}{\hbar^2} \rho_b \int f(t, \omega_{ab}) \, dE$ ----- (100).

(As exercise, verify it.)

There are two cases out of this problem that must be noticed:

1. The central peak of the function (f) is outside the domain of integration, where the transition is not conserving the energy. Hence, f can be replaced by its

value averaged over several oscillations which leads to the following time-independent expression for the transition probability, $W_{a \to B} \cong (2\mathcal{E}\rho_b |V_{bal}|^2)/[E_b - E_a]^2$.

2. The central peak of the function (f) is inside of the domain of the integration—transition is conserving the energy—then the main contribution comes from this peak.

Only small contribution is involved in extending the limits of the integral to infinity, whence equation 96 takes the form:

$$W_{a \to B} \cong \frac{2\pi}{\hbar} |V_{bal}|^2 \rho_b (E_a) t \text{-------------------}(101).$$

Now due to $\mathcal{E} >> (2\pi\hbar)/t$, then $W_{a \to B}$ is bigger than the sum of all the others. Since $W_{a \to B}$ is the transition probability, then the transition probability per unit time is zero for the case 1 (time independent). But for time-dependent equation 101, the transition per unit time is given by:

$$\omega_{a \to B} = \frac{2\pi}{\hbar} |V_{bal}|^2 \rho_b (E_a), \text{ for case(2) -------------}(102).$$

So, in order that equation 102 validates, t must be sufficiently large to meet the condition $\mathcal{E} >> 2\pi\hbar/t$. At the same time, $(\omega_{a \to b})t$ is to be sufficiently small for the first order approximation is justified $(\omega_{a \to b})t << 1$. These demonstrations have the advantage of bringing the significance and the condition of the validity of the equation 102, which simply can be obtained by replacing the function $f(t, \omega_{ab})$ by its asymptotic form $(f \to 2\pi t \delta(\omega)$, as $t \to \infty,)$, (equation 96), in the right hand side of (equation 93) which gives:

$$W_{a \to b}(t \to \infty) \sim \frac{2\pi}{\hbar} |V_{bal}|^2 \delta(E_b - E_a) t \text{-------------}(103).$$

From equation 98, it can be found that:

$W_{a \to B} = \frac{2\pi}{\hbar} \int_{B(E)} |V_{bal}|^2 \rho_b(E) \; \delta(E_b - E_a) dE.t$, this implies that:

$\omega_{ab} = \frac{dW_{b \to a}}{dt} = \frac{2\pi}{\hbar} \int |V_{bal}|^2 \rho_b(E) \delta(E_b - E_a) dE$, using δ property (appendix A), this becomes

$\omega_{ab} = \frac{2\pi}{\hbar} |V_{bal}|^2 \rho_a(E)$ ------------- (104).

(It is the famous Fermi Second Golden Rule.)

8.3.3. *Some Applications of the Second Golden Rule*

Consider the following scattering problem:
An incident wave $\vec{k}$; its energy is $E_a = P^2/2m$. It is scattered by a target (G), as shown in the diagram below, where the scattered wave (outgoing wave $\vec{k^-}$) of energy $E_b = P^{-^2}/2m$. It is scattered within a solid angle $d\Omega^-$.

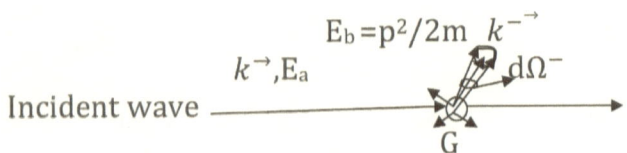

It is required to find the scattering cross section (σ).
Using Born approximation, first order in the interaction potential (V_I) between the incident wave and the target (G), this interaction potential is considered as a perturbation term. So the Hamiltonian of the system is written as:

$H = H^{(0)} + V_I$, where $H^{(0)} = P^2/2m = E_k =$ kinetic energy.

The eigenstate of $H^{(0)}$ is plane wave given by $e^{ik.r}$ which is also representing state of the momentum $P = \hbar k$, and it is normalized to unit density. So the ket vectors are $|k\rangle$; they obey the orthonormality and the closure relations, which can be shown as follows:

$<k|k^-> = (2\pi)^3 \delta(k-k^-)$, $\int |k> \frac{dk}{(2\pi)^3} <k| = 1$

$(P^2/2m)|k> \frac{\hbar^2 k^2}{2m}|k> = E_a^{(0)}|k>$

$\frac{P^{-2}}{2m}|k^-> = \frac{\hbar^2 k^{-2}}{2m}|k^-> = E_b^{(0)}|k^->$

The transition from k to $k^-(\omega_{k \to k^-})$ is given by:

$\omega_{k \to k^-} = \frac{2\pi}{\hbar} |<k^-|V(r)|k>|^2 \delta(E_k^- - E_k)$

Then the transition from k to k^- (ω_{k-k^-}) is given by:

$\omega_{k \to k^-} = \frac{2\pi}{\hbar} |<k^-|V(r)|k>|^2 \delta(E_k^- - E_K)$

Where, $E_K^- = \hbar^2 k^{-2}/2m$, $E_k = \hbar^2 k^2/2m$, kinetic energies.

Now, $<k^-|V|k> = \int \int <k^-|r><r|V(r)|r^-><r^-|k> dr\, dr^-$

$V(r) = e^2/r$, so this can be written as

$\int \int \psi_k^{+-}(r) \frac{e^2}{r} \delta(r - r^-)\psi_k(r^-)\, dr\, dr^-$, using Dirac delta function properties. $\int \psi_k(r^-)\delta(r-r^-)\, dr^- = \psi_k(r)$. So matrix elements is, $V_{K^- K} = \int \psi_K^-(r^-)e^2/r\, \psi_k(r^-)dr$

Therefore, $\omega_{k^- \to k} = \frac{2\pi}{\hbar} |V_{k^- k}|^2 \delta(E_{k^-} - E_k)$

Now take a normalization box such as:

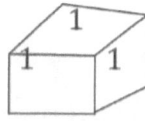 $\int |e^{-ik.r}|^2\, d\tau = 1$, normalized.

So $[<K^-| V(r)|k>|] = \int e^{-i(k^- - k).r}V(r)\, dr$

One only computes experimentally, $\omega_{k \to k^-}$ and to $k^- + dk^-$, so

$W_{k \rightarrow k^- \text{and } k^- + dk^-}$

$$=\frac{2\pi}{\hbar} \Sigma_{k^-}^{k^- + dk^-} |< k^- |V(r)| k >|^2 \delta (\frac{\hbar^2 k^{-2}}{2m} - \frac{\hbar^2 k^2}{2m}).$$

But $K \equiv (k_x = 2\pi n_x, \, k_y = 2\pi n_y, \, k_z = 2\pi n_z)$.

So $\Sigma_{n_{xn_yn_z}}^{dk^-} \int \frac{dk^-}{(2\pi)^3}$, which leads to,

$$W_{k \rightarrow k^- \text{and } k^- + dk^-} =\frac{2\pi}{\hbar} \int_{k^-}^{k^- + dk^-} |< k^- |V(r)| k |^2 \delta (F) \frac{dk^-}{(2\pi)^3}$$

Where $dk^- = \frac{1}{k^-} d\Omega_{k^-}$, if the change in the solid angle is small, the following is obtained:

$$\frac{dW_{k \rightarrow k^- \text{and } k^- + dk^-}}{d\Omega_{k^-}} =\frac{2\pi}{\hbar} |<k^-|V(r)|k>|^2 \frac{mk\hbar}{(2\pi\hbar)^3}$$

$$=\frac{mk}{(4\pi^2\hbar^3)}|<k^-|V(r)|k>|^2 \text{--------------------------------}(105).$$

This represents the probability of transition per unit time per unit solid angle. Consider the problem of finding the scattering cross section which differs only by the velocity for k^- and $k^- + dk^-$. Let it be illustrated in the following diagram:

Target
   Bean of particles

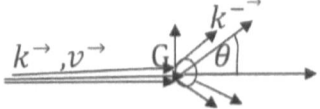

This means there are (v) particles; therefore,

$$d\sigma_{k \rightarrow k^-, k^- + dk^-} =\frac{1}{v} W_{k \rightarrow k^-, k^- + dk^-},$$

then, $\frac{d\sigma_{k \rightarrow k^-}}{d\Omega_{k^-}} =\frac{1}{v} \frac{dW_{k \rightarrow k^-}}{d\Omega_{k^-}} =\frac{2\pi}{\hbar v_k} |<k^-|V|k>|^2 \rho_{k^-} (E). \text{-----------}(106),$

where $\rho_{b=k^-}$ is the density of the final state (K$^-$) or (b).

It can be shown that $\rho_b =\frac{mk}{8\pi^3\hbar^2}$.$\text{--------------}(107)$

and $<K_b|V|k_a> \equiv \int e^{i(k_a - k_b).r} dr \text{------------} (108).$

Equations 105 and 108 represent the main core of the problem, where equation 108 gives the matrix elements of the potential of the interaction responsible for the transition.

Equation 106 is the differential scattering cross section for particle interaction with target. So having the potential [V(r)] explicitly and the eigenstates lk-> and lk>, σ_t easily can be found.

8.4. Variational Methods of Approximations

The variational methods are basically important to find out the ground state energy and its eigenfunction, where the eigenvalue problems are usually discrete. The Hamiltonian (H) is Hermitian operator; therefore, its eigenvalue is real. Usually the eigenvalue equation is written such as:

$$H_n^\wedge \, l\,\psi_n> = \mathcal{E}_n l\psi_n> \text{---} (109).$$

$H_n^\wedge$ is the Hamiltonian operator given as $[-i\hbar \frac{\partial^2}{\partial x^2} +V(r)=-i\hbar\nabla^2 +V(\,r)].l\psi_n>$ is the eigenfunction describing the system state with $H_n^\wedge$, and $\mathcal{E}_n$ is the eigenvalue. It is the measured value or the average value of (H_n). The problem is how to find $\mathcal{E}_n$. The variational principle is actually based on the proposal that: if it is given $\frac{<\psi \, l \, Hl \, \psi>}{<\psi \, l \, \psi>}$, which is the average value of H, it proposes that it is unchangeable, and H also is not change. So it is stationary, which means $l\psi>$ constant. Hence,

$$\delta_\psi[\frac{<\psi \, lHl \, \psi>}{<\psi l \, \psi>}]=0=\frac{\delta[<\psi l \, Hl\psi>]<\psi \, l \, \psi> -<\psi l \, Hl \, \psi> \delta(<\psi \, l \, \psi>}{(<\psi \, l \, \psi>)^2} =0$$

Completing the integration, the following is gotten:

$$\frac{<\delta\psi(H-\frac{<\psi|H|\psi>}{<\psi|\psi>})\psi>+<\psi \, (H-\frac{<\psi|H|\psi> \, \delta\psi>}{<\psi|\psi>})}{<\psi|\psi>^2}=0\text{----------------}(a)$$

259

If $|\psi>$ happens to be the eigenstate, then the eigenvalue is

$$\varepsilon_n = \frac{<\psi_n| H |\psi_n>}{<\psi_n|\psi_n>} = <\psi_n| H| \psi_n>, <\psi_n|\psi_n>=1, \text{ normalized.}$$

$H|\psi_n>=\varepsilon_n|\psi_n>$, so $<\psi_n| H|\psi_n>=\varepsilon_n$, these lead to

$H-\frac{<\psi| H| \psi>}{<\psi|\psi>}$ =0, this indicates that $|\delta\psi>$ and $<\delta\psi|$ are completely independent of each other because Hilbert space is an infinite dimension complex space. So it can be said $i|\delta^-\psi>\equiv|\delta^-\psi>$ then $-i<\delta\psi>=<\delta^-\psi|$. Therefore, not only (a) equals to zero, but its complex conjugate (multiply by $i=\sqrt{-1}$) also equals to zero. Also, the different terms should vanish which lead to ensure $|\delta\psi>$ and $<\delta\psi|$ are independent of each other. This fact ensures the difference terms in (a) are zeros.

This concludes the Hamiltonian is Hermitian, so its eigenvalue is real.

Theorem:
If ψ is an arbitrary ket and H is Hermitian, then

$\frac{<\psi |H| \psi>}{<\psi|\psi>} \geq E_0$ (ground state energy),the smallest eigenvalue.

The Proof: $H=\sum_n E_n P_n=\sum_n E_n|\psi_n><\psi_n|$, $P_n=|\psi_n><\psi_n|$ projection operator.

Hence, $\frac{<\psi_{vr}|H|\psi_{vr}> =<\psi|E_0|\psi>}{<\psi_{vr}|\psi_{vr}>}=\frac{<\psi|\sum_n(\varepsilon_n-E_0)P_n|\psi>}{<\psi|\psi>}\geq0$

$=\frac{\sum_n(E_n-E_0)<\psi| P_n|\psi>}{<\psi|\psi>}\geq0$. Now is $<\psi|P_n|\psi>\geq0$? $E_n-E_0>0$

To find the answer, the properties of the projector has to be reviewed. Hence, $<\psi|P_n|\psi>=(<\psi|P_n|)(P_n|\psi>=<\psi^-|\psi^->=1$ (norm). Therefore, the answer is yes, it is >0, and it implies that

$\frac{\sum_n(E_n-E_0)<\psi| P_n|\psi>}{<\psi|\psi>}>0$----------------------------(110).

On this basis, the variational method can be applied to show that. Suppose a trial wave function such as

$<r|\psi> \equiv \Psi_{\alpha\beta\gamma}(r) \equiv (r + ar^2 + b\ r^3)\ e^{-ar}$. Then take the derivative of $\frac{<\psi\mid H\mid\psi>}{<\psi\mid\psi>}$ $(\alpha\beta\gamma)$ with respect to each parameter of $(\alpha\beta\gamma)$ and equalizing to zero. This operational step is to find the parameters $(\alpha\beta\gamma)$, so first, one has to write the following:

(1.) $\frac{\partial}{\partial\alpha}[\frac{<\psi\mid HI\ \psi>}{<\psi|\ \psi>}\ (\alpha\beta\gamma)\ ]=0$

(2.) $\frac{\partial}{\partial\beta}[$ =   =   ]=0 $\Bigg\}$ --------------------(111).

(3.) $\frac{\partial}{\partial\gamma}[$ =   =   ]=0

The most important remarks to be acquainted with are

1. The correct choice to the trial function is quite necessary and so important.
2. $E_{vr.}$ (measured by variational method) $\geq E_0$ (ground state energy). Therefore, $E_{vr} = \frac{<\psi\mid H\mid\psi>}{<\psi\mid\psi>} \geq E_0$----------------(112).

8.4.1. Some Illustration Examples

Ground state of hydrogen atom $\frac{1}{1}H$:

Consider the ground state of the hydrogen atom. Let the trial function be $<r|\psi> = \psi(r) = \frac{u(\rho)}{\rho}$, where $u(\rho) = 0$ at $r \to 0$, and $u_\ell \propto e^{-cot\rho}$ as $r \to \infty$. For simplicity, let it be as $\rho e^{-b\rho}$. The aim of the solution is to find the parameter b (variational parameter). Using equation 112 with some mathematical manipulations, it can be found that (taking the minimum):

$\frac{<\psi| H|\psi>}{<\psi|\psi>}$ =E_0. So after substituting for $\psi(r)$ =$\rho e^{-b\rho}$ and carry on the required integrations, the following is obtained:

$\frac{<\psi| H |\psi>}{<\psi|\psi>}$ (-E_0)=-E_0 ($2b-b^2$). So the next step is to minimize this with respect to b such as, $\frac{\partial}{\partial b}[-2bE_0--b^2E_0]=-2E_0+2bE_0=0=2E_0[b-1]$.

But $2E_0 \neq 0$; hence, $b-1=0$, so $b=1$, $\Rightarrow \psi_{vr}(r)=\rho e^{-\rho}$, $E_{vr}=-E_0$. (As an exercise, verify this result.)

++*The ground state of helium atom* (^4_2He)

The Hamiltonian of this atom is given by:

$H=(\frac{P_1^2}{2m}+\frac{P_2^2}{2m}-\frac{ze^2}{r_1}-\frac{ze^2}{r_2})+\frac{e^2}{r_{12}}$ =$H_0+V(r)$, $V(r)=\frac{e^2}{r_{12}}$.

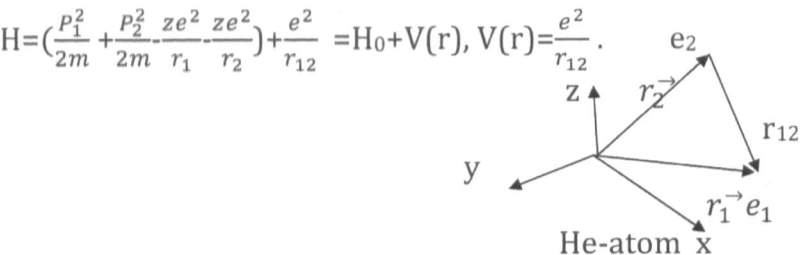

He-atom x

The trial wave function(close to reality) is given by, $<r_1 r_2|\psi>=\psi(r_1,r_2)=e^{-\alpha(r_1+r_2)}=\exp.[-z(\frac{r_1}{a_0}+\frac{r_2}{a_0})$. The required solution is to show that

$\frac{<\psi |H| \psi>}{<\psi|\psi>}=2E_0[z^{-2}-2(z-6/15)z^-]=K$. (Find it in details as an exercise.)

Taking the derivative, minimize it, i.e., $\frac{\partial}{\partial z^-}(K)=(2z-2)(z-5/16)z^-(2E_0)=0$, then $E_{vr}=-2E_0(z-5/16)$. Verify it.

Consider the excited states.

For this case, it is asked to find $\frac{<\psi I H I \psi>}{<\psi I \psi>}$ - E_1? Hence, it is necessary to evaluate

$$\frac{<\psi I H I \psi>-<\psi I E_1 I \psi>}{<\psi I \psi>}=\frac{<\psi I H-E_1 I\psi>}{<\psi I \psi>} \text{---------} (113).$$

Let the following theorem be used. It is: any Hermitian operator can be written as sum of eigenvalue times a projection operator. Therefore, Hamiltonian can be written as $H=\sum_n E_n P_n$. It implies that $\sum E_1 P_1 = E_1$, $P_1 = 1$. Therefore, $H-E_1 = \sum_n (E_n - E_1) P_n$, note that if $n=0$ then $E_n - E_0 = 0$. It is a negative result.

Whence, $H-E_1 = \sum_{n \neq 0}(E_n - E_1)P_n - (E_1-E_0)P_0$ ---------(114).

By substituting equation 114 into equation 113, the following is to be obtained:

$$\frac{<\psi|H-E_1|\psi>}{<\psi|\psi>}=\frac{<\psi|\sum_{n\neq0}(E_n-E_1)P_n|\psi>}{<\psi|\psi>}-\frac{(E_1-E_0)<\psi|P_0|\psi>}{<\psi|\psi>},$$

But it is known that $<\psi I P_0 I\psi>=<\psi I\psi_0><\psi_0 I\psi>=I<\psi I\psi_0>I^2=0$, orthogonal, $P_0 = I\psi_0><\psi_0 I$ - projection operator. Therefore, $\frac{<\psi|H \psi>-<\psi|E_1 \psi>}{<\psi \psi>} \geq 0 \Rightarrow \frac{<\psi I H I \psi>}{<\psi I \psi>}-E_1 \geq 0$, so $\frac{<\psi I H I \psi>}{<\psi I \psi>} \geq E_1$. ---------

---------(115).

It is the answer.

For the second excited state, it can be shown to be:

$\frac{<\psi \,|H|\, \psi>}{<\psi \,|\, \psi>} \geq E_2$. Show it as an exercise. Note: $<\psi \,|\, \psi_1> = 0$, orthogonal.

Important note: First, it is necessary to know the ground state eigenstate $|\psi_0>$ which can be estimated as

$|\psi_0> = |\phi_0> + |\Delta>$ (small) ----------------------------(116).

Now let $|\psi>$ be an arbitrary eigenstate which characterizes as $<\psi|\phi_0> = 0$, orthogonal. Hence $<\psi|\psi_0> = <\psi|\phi_0>$

$+<\psi|\Delta> = <\psi|\Delta>$ 0

How big the quantity $\frac{<\psi \,|\, \psi_0><\psi|\, \psi_0>}{<\psi|\, \psi>} = \frac{<\psi|\Delta><\Delta|\,\psi>}{<\psi \,|\, \psi>} \geq <$
$\Delta|\Delta>$ is? Take $<\psi_0 \,|\psi_0> = <\phi_0 \,|\phi_0> + 2\mathrm{Re}<\Delta \,|\phi_0> + <\Delta \,|\Delta>$

But $<\psi_0 \,|\psi_0> = <\phi_0|\phi_0> = 1$, normalized.
Hence, $<\Delta \,|\Delta> = 2\mathrm{Re}<\Delta \,|\, \phi_0>$---------------------------- (117).

Take $<\phi_0 \,|\psi_0> = <\phi_0|\phi_0> + <\phi_0 \,|\Delta> = 1 + <\phi_0|\Delta>$.
Then $|<\psi_0 \,|\phi_0>|^2 = 1 + 2\mathrm{Re} <\phi_0|\Delta> + |<\phi_0 \,|\Delta>|^2$.

So either one considers $|<\phi_0 \,|\Delta>|^2$ is very small (negligible) and of course $|<|\phi_0|\psi_0>|^2 = 0$ (orthogonal), then $2\mathrm{Re}<\phi_0|\Delta> = -1$.

Or the $|<\phi_0|\psi_0>|^2$ be left as it is for mathematical convenient.
Therefore, $-2\mathrm{Re}<\phi_0|\Delta> = 1 - |<\phi_0|\psi_0>|^2 = <\Delta \,|\Delta>$ (see 117).

Now go back to equation 115, $\frac{<\psi|H|\psi>}{<\psi|\psi>}-E_1 \geq E_1+E_0$. (Verify it.)

$\geq 2E_1+E_0$ which leads to $\frac{<\psi|H|\psi>}{<\psi|\psi>} \geq E_2$ --------------(118).

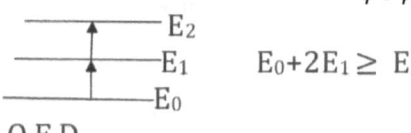

$E_0+2E_1 \geq E_2$

Q.E.D

Chapter Nine

Basic Elements of Relativistic Quantum Mechanics

9.1. *Introduction*

In 1905, Einstein announced his revolutionary concepts about the physical mechanics which was dominated by Newton and Kippler ideas. These concepts or ideas are dealing with the motion, the velocity, the frames of reference, the time concept, the simultaneity, and the speed of light. Albert Einstein, because of this revolutionary theory, was considered the most famous scientist in the scientific society, especially after his discovery of the general theory of relativity in 1915. So the special theory of relativity was discovered in 1905, which will be dealt with here, as basic elements which find its applications in the basic theory of quantum mechanics.

Hence, the relativity theory might consider two branches; the first is the special one concerned with the concepts mentioned above. Its applications take place when the physical dynamical variables in their motion approach close to the light velocity, which this theory proposed. It is absolute and constant about 2.9997×10^{10} cm.sec$^{-1} \approx 3 \times 10^8$ m.sec^{-1}. It is the highest velocity in the vacuum and cannot be reached by any massed particle in vacuum. The details of the special theory of relativity are well explained in many undergraduate physics books. Its famous proposals are very well treated in these books.

The second branch of the relativity theory is the general one which was developed by Einstein and announced in 1915. It is a generalization to the special relativity. It deals with the theory of the gravity and the universe structure. It explains the matter distribution among the space-time fiber structure of the universe; it shows the force of gravity is quite different from other known natural forces. Also, it shows the curvature of the space-time due to the matter distribution through it, is responsible for the gravity force, and then the movement of the universal constituents. It predicted many universe phenomena which discovered after fifty years later on, such as the gravity lenses, light deflection by stars passing by, and so on.

In this chapter, the only basic elements of the theory of the special relativity are needed. For the readers, be acquainted with the basic ideas of the subject. It is important to go through some main terminologies they might meet in the following paragraphs. These may be summarized as follows:

1. Energy-Momentum Four Vector. $P_\mu (\mu=1, 2, 3, 4)$. It consists of three spacial components vector $P_k (k=1, 2, 3)$ and fourth component $P_4 \equiv iE/c$, where E is the total energy of the system and c is the velocity of light about $3 \times 10^8 m.sec^{-1.}$ (Some scientific papers claim the photons individually can be slowed down by their velocity in vacuum, 2015,(an interesting claim need to be verified from others.) $i=\sqrt{-1}$, as it is known in complex variables field.

2. Coordinate-Four Vector. $X_\mu (\mu =1, 2, 3, 4)$ which consist of three spacial coordinates vector $X_k(k=1, 2, 3) \equiv \vec{r}$, and a fourth component $X_4=ict$. It implies that $\frac{\partial}{\partial x_4} = (i/c)\frac{\partial}{\partial t}$. It is the time component. So in the theory of special relativity the time is not absolute but a relative quantity and a part of the space in addition to the position space. So our space is a space-time fiber.

3. Gradient Four Vector. $\frac{\partial}{\partial X_\mu} (\mu=1, 2, 3, 4)$ consists of three spacial vector $\frac{\partial}{\partial X_k} (k=1, 2, 3) \equiv \nabla$, and a four-component (time component) $\frac{\partial}{\partial X_4} = \frac{1}{ic}\frac{\partial}{\partial t}$.

4. Four-Vector Potential. $A_\mu(\mu=1, 2, 3, 4)$ consists of a three spacial vectors: $A_k(k=1, 2, 3)=\mathbb{A}$, the vector potential, and $A_4 =i\phi$, where ϕ is the scalar potential.
5. The Invariant Sum. $\sum A_\mu B_\mu$, A_μ, and B_μ are four vectors. Sometimes it is written as AB.

9.2. Relativity Introduced in Quantum Mechanics

As it is historically known, the de- Broglie proposal of the duality behavior of any moving matter in 1924 and the quantization of energy before that in 1900 found by Plank initiated the thinking of finding mechanics to describe a system characterizes as wavelike behavior and obey the quantization property. Where Plank constant h plays the role of drawing the limit between quantum and classical physics, as $h \to o$ the classical physics there, and as $h =6.63\times 10^{-35}$ joule.sec. the quantum mechanics takes the role. Therefore, the more general theory is the quantum physics and the classical physics is the limit where $h \to 0$. So in 1926–1927, the quantum or the wave mechanics started to develop by Schrödinger and Heisenberg, almost independently, different approach but reaching the same result. Max Born and Dirac and others followed.

The important physical quantities that associated with any physical system under study are the Hamiltonian H, which is considered an operator in quantum mechanics and the wave function $\psi(x, y, z)$, the eigenfunction associated with H, describing the system and containing the whole information about it. It is of statistical characters. Hence, for nonrelativistic quantum mechanics, the eigenvalue equation is given by:

$H\psi =E\psi =i\hbar\frac{\partial}{\partial t}\psi$, E_{op} (operator)$=i\hbar\frac{\partial}{\partial t}$ (as shown before)

Also, $P_{op}=-i\hbar\nabla$, $E=P^2/2m+V(r)=$total energy=K.E+ potential.

So H=$-\frac{\hbar^2}{2m}\nabla^2$ +V(r), $\frac{h}{2\pi}$ =$\hbar$ reduced Plank constant

=1.054×10^{-34} J.S.

Hence, Schrödinger equation is ($\frac{-\hbar^2}{2m}\nabla^2$ +V(r))ψ =$i\hbar\frac{\partial}{\partial t}\psi$.
ψ is scalar, and the spin is not included.

Also, it is known that the probability of finding a particle somewhere is given by P(x, y, z)= $\psi^*\psi$. And the current density is given by j=$\frac{\hbar}{2mi}(\psi^*\nabla\psi -\psi\nabla\psi^*)$. Also, the continuity equation is $\frac{\partial\rho}{\partial t}+\nabla.j^\rightarrow$ =0.

These are the main basic concepts of nonrelativistic quantum mechanics. Now to introduce the relativity concepts to quantum mechanics, it is important to consider these concepts under the effect of the proposals of the theory of special relativity, to be familiar with the basic elements of the relativistic quantum mechanics for farther advanced studies in the same field.

Also, it is known the velocities of the electrons, the positrons, and the nuclear particles are close to the velocity of light; therefore, these basic principles of special relativity must be taken in consideration in treating the physical properties of these particles. As a starting step, consider the energy E which is given, relativistically, by the following formula:

E=$[(pc)^2+m^2c^4]^{1/2}$ --------------------------(1).
Or, $E^2-p^2c^2-(m\ c^2)^2$=0------------------------(2).
P is the momentum
m is the mass
c is the velocity of light,

Using the operational representation, the relativistic Schrödinger equation will be

$$[-\hbar^2 \frac{\partial^2}{\partial t^2} + c^2\hbar^2\nabla^2 - (mc^2)^2]\psi(r) = 0. \text{---------}(3).$$

Equation 3 is called Klein-Gordon equation in physics text books. It is a second order differential equation, which is not easy to solve. So it is not suitable for the physical required purposes. Also, it neglects the spin effect. But it might consider a good step toward the thinking of introducing the relativity theory into quantum mechanics. This concept evokes Dirac in 1927 to develop a mathematical and clever way to deal with such important mathematical physical problem. He proposed the particle exists in four states; he called them spinors, represented as:

1. $\psi = \begin{pmatrix} \psi_1 \\ \psi_2 \\ \psi_3 \\ \psi_4 \end{pmatrix}$, where, $\psi^* = [\psi_1^*, \quad \psi_2^*, \quad \psi_3^*, \psi_4^*]$ -------- (4).

The probability amplitude P(x, y, z) is given by $\psi^*\psi = \sum_v \psi_v^* \psi_v$.

2. To get rid of the second order in equation 3, Dirac proposed for this equation the form

$$\{\frac{1}{c}\frac{\partial}{\partial t} + \alpha_1 \frac{\partial}{\partial x_1} + \alpha_2 \frac{\partial}{\partial x_2} + \alpha_3 \frac{\partial}{\partial x_3} + \beta \frac{im_0 c}{\hbar}\}\psi = 0 \text{-------------}(5).$$

So the equation now is a linear, but Dirac assumed the relation between equation 5 and Klein-Gordon equation (3) is exactly as the relation between the Maxwell equations in electromagnetic theory and the wave equations, introducing that $\frac{\partial}{\partial t} \equiv \partial_t$ and $\frac{\partial}{\partial x_i} \equiv \partial_i$. Now multiply equation 5 by -1). This is obtained:

$\{\frac{1}{c}\partial_t -(\alpha_1\partial_1 +\alpha_2\partial_2 +\alpha_3\partial_3) +\beta \frac{im_0 c}{\hbar}\} \psi(r) =0$ -------------- (6).

Let the underlined(---------) be Π, then equation 6 takes the form $[\frac{1}{c^2}\partial_t^2 - (\Pi)^2]\psi(r) = 0$ ------------------(7).

Equation 7 can be written as:

$[\frac{1}{c^2}\partial_t^2 \quad -\sum_{k=1}^{3} \alpha_k^2 \partial_k^2 \quad -\sum_{k\, \ell\neq}(\alpha_\ell\alpha_k+\alpha_k\alpha_\ell)\partial_k\partial_\ell \quad +\beta^2(\frac{m_0 c}{\hbar})^2$
$+ (\frac{im_0 c}{\hbar}) \times \sum_k (\alpha_k\beta +\beta\alpha_k)\partial_k]\psi=0$, $([\partial_k, \partial_\ell]=0)$-----(8).

For solving equation 8, Dirac imposed the following relations conditions between the parameters (α^s and β):

$\alpha_k\alpha_\ell + \alpha_\ell\alpha_k =0, \alpha_k\alpha_\ell = -\alpha_\ell\alpha_k, \alpha_k^2 =1$ -----------(9).
$\beta\alpha_k+\alpha_k\beta =0, \beta^2 =1$----------------------------------(10).

These are similar to Pauli spin matrices (σ) which are represented as follows:

$\sigma_k\sigma_\ell + \sigma_k\sigma_\ell =0, \sigma_k\sigma_\ell =-\sigma_\ell\sigma_k, \sigma_k^2 =1,$

where: $\sigma_1 = \begin{bmatrix} 0 & 1 \\ 1 & 0 \end{bmatrix}, \sigma_2 = \begin{bmatrix} 0 & -i \\ i & 0 \end{bmatrix}, \sigma_3 = \begin{bmatrix} 1 & 0 \\ 0 & -1 \end{bmatrix}$ ----- (11).

α_k and β are related to Pauli matrices such as:

$\alpha_k= \begin{bmatrix} 0 & \sigma_k \\ \sigma_k & 0 \end{bmatrix}$, where $\alpha_1= \begin{bmatrix} 0 & \sigma_1 \\ \sigma_1 & 0 \end{bmatrix} = \begin{bmatrix} 0 & 0 & 0 & 1 \\ 0 & 0 & 1 & 0 \\ 0 & 1 & 0 & 0 \\ 1 & 0 & 0 & 0 \end{bmatrix}$ -- (12)

And $\beta = \begin{bmatrix} I & 0 \\ 0 & -I \end{bmatrix} = \begin{pmatrix} 1 & 0 & 0 \\ 0 & 1 & 0 \\ 0 & 0 & -1 \\ 0 & 0 & -1 \end{pmatrix}$ ------------------(13)

Therefore, equations 12 and 13 are called Dirac matrices.

Now define σ_k^4 as, $\sigma_k^4 = \begin{bmatrix} \sigma_k & 0 \\ 0 & \sigma_k \end{bmatrix}$, $\vec{\sigma} = (\sigma_1, \sigma_2, \sigma_3)$, then

$\alpha_k \sigma_k^{4\rightarrow} = \begin{bmatrix} 0 & I \\ I & 0 \end{bmatrix} \begin{bmatrix} \sigma_k & 0 \\ 0 & \sigma_k \end{bmatrix} = \begin{pmatrix} 0 & 0 & 1 & 0 \\ 0 & 0 & 0 & 1 \\ 1 & 0 & 0 & 0 \\ 0 & 1 & 0 & 0 \end{pmatrix}$ ---------- (14)

$\sigma_k^{4\rightarrow} = -\gamma_5 \sigma_k^{4\rightarrow}$

Where, $\gamma_5 = -\alpha_k$ (refer to equation 12); hence:

$\gamma_5 = \begin{pmatrix} 0 & -1 & 0 & 0 \\ 0 & 0 & 0 & -1 \\ -1 & 0 & 0 & 0 \\ 0 & -1 & 0 & 0 \end{pmatrix} = \begin{bmatrix} 0 & -I \\ -I & 0 \end{bmatrix} = -\alpha_K$, ----------(15).

γ_5 Matrix is very important to remember. Also, $\vec{\alpha}$ can be written as $\vec{\alpha} = -\gamma_5 \sigma_k^{(4)\rightarrow}$ ----------------------------------- (16).

Hence, Dirac matrices are, in fact, four-by-four matrices but can be written as two-by-two matrices:

$\rho_1 = \begin{bmatrix} 0 & I \\ I & 0 \end{bmatrix} = \begin{pmatrix} 0 & 0 & 1 & 0 \\ 0 & 0 & 0 & 1 \\ 1 & 0 & 0 & 1 \\ 0 & 1 & 0 & 0 \end{pmatrix}$ ----------------- (17).

$\rho_2 = \begin{bmatrix} 0 & -i \\ i & 0 \end{bmatrix}$, $\beta = \rho_3 = \begin{bmatrix} I & 0 \\ 0 & -I \end{bmatrix}$, $I = \begin{bmatrix} 1 & 0 \\ 0 & 1 \end{bmatrix}$ ----(18).

As for Pauli matrices, the relations between these are

$\rho_1\rho_2 + \rho_2\rho_1 = 0$, $\rho_1\rho_2 = i\rho_3$, $\rho_i^2 = 1$-------------(19).

(As an exercise, verify equation 19.). Now define as in equation 14.

$$\sigma^\rightarrow = \begin{bmatrix} \sigma & 0 \\ 0 & \sigma \end{bmatrix} \quad \text{------------------- (20).}$$

In terms of ρ_1 and σ, the following can be defined as

$$\alpha^\rightarrow = \rho_1, \sigma^\rightarrow = \begin{bmatrix} 0 & \sigma^\rightarrow \\ \sigma^\rightarrow & 0 \end{bmatrix}, \quad \alpha_4 = \beta \text{-------------- (21).}$$

$$\alpha_\mu\alpha_\nu + \alpha_\nu\alpha_\mu = 2\delta_{\mu\nu} \text{-------------------------- (22).}$$

Now let $\gamma_k = \rho_2\sigma^\rightarrow = \begin{bmatrix} 0 & -i\sigma \\ \sigma & 0 \end{bmatrix}$, $\gamma_4 = \alpha_4 = \beta$ ----- (23).

Then $\gamma_\mu\gamma_\nu + \gamma_\nu\gamma_\mu = 2\delta_{\mu\nu}$ ---------------------------- (24).

$\gamma_5 = \gamma_1\gamma_2\gamma_3\gamma_4 = -\rho_1$ -------------------------------- (25).

$\gamma_5\gamma_\mu + \gamma_\mu\gamma_5 = 0$ --(26).

$\sigma_{\mu\nu}^\rightarrow = \frac{1}{2i}(\gamma_\mu\gamma_\nu - \gamma_\nu\gamma_\mu) = \frac{1}{i}\gamma_\mu\gamma_\nu$, $\mu \neq \nu$ -----------(27).

So, $\sigma_{ij} = (1/i)\, \sigma_i\sigma_j = \sigma_k$, ijk cyclic,--------------------(28).

$\sigma_{k4}^\rightarrow = -\sigma_{4k}^\rightarrow = \alpha_k$ ------------------------------------- (29).

$\alpha_k = i\gamma_4\gamma_k$ --(30).

([As an exercise, verify all these important relations.)
In view of these information, the basic relativistic Schrödinger
equation is $[\frac{1}{c}\partial_t + \alpha^\rightarrow\nabla^\rightarrow + \frac{i\,cm_0}{\hbar}\beta]\psi$ ------- (31),
where m_0 is the rest mass ($m_0 = 0$ for photon) and $\alpha^\rightarrow\nabla^\rightarrow$ stands
for $\sum_k \alpha_k \partial_k$. The first equation was derived by Dirac in 1927. It
is given by $[(\alpha^\rightarrow.p^\rightarrow\, c + m_0\, c^2\beta)\psi = i\hbar\partial_t\psi]$ -- (32).

For canonical form, equation 32 can be written as

$$H_{\hat{R}}\psi = E\psi = i\hbar \frac{\partial}{\partial t}\psi, \; H_{\hat{R}} = \vec{\alpha}.\vec{p}\; c + m_0\, c^2 \beta + V(r) \; \text{(relativistic)}.$$

9.3. Covariant Form of Dirac Equation

As it is clear, the time in special relativity is considered a forth component of the spatial-time space. It is no longer an absolute physical quantity but a relative one as the motion and the space coordinate according to the frame reference considered. Therefore, it is a physical phenomenon which is dealing with a space consisting of spacial coordinates (x, y, z) and time (T=ict). So this space is of four coordinates (x_k, x_4), where k=1, 2, 3 and x_4=ict. Hence, their X_μ components, μ=1, 2, 3, 4.

The same category applies on the momentum space, where
$$P_\mu = (p_k, p_4), \; \mu = 1, 2, 3, 4., \; k = 1, 2, 3.$$

Therefore, $p_k = p_x, p_y, p_z$, and $p_4 = iE/c$.
Now take equation 31, multiply it from the left by $-i\beta$; it will take the form $-i\beta \frac{1}{c}\partial_t + \Sigma_k(-i\beta\,\alpha_k)\partial_k + \frac{m_0\,c}{\hbar}\beta^2)\psi = 0$.

Consider the relations between β and α_k, as shown before. This equation becomes $[\Sigma_\mu (\gamma_\mu\,\partial_\mu + \frac{m_0\,c}{\hbar})\psi = 0]$ ----- (33).

Equation 33 is the covariant form of Dirac equation.
Notice here that $-i\beta\,\alpha_k = -\gamma_k, \gamma_4 = \beta$. Or $[\alpha_k = i\beta\gamma_k = i\gamma_4\gamma_k]$.

9.4. Properties of Gamma (γ) Matrices

For the useful use of Dirac gamma matrices, it is important to be quite familiar with their properties. The understanding of these properties helps a lot to use them easily in dealing with a physical problem with the effect of the special relativity. In the following, a summary of these properties is given (verify them as exercise):

1. $\gamma_\mu \gamma_\nu = -\gamma_\nu \gamma_\mu, \mu \neq \nu$
2. $\gamma_\mu^2 = 1, [\gamma_k =\text{-}i\beta \alpha_k => (\text{-}i\beta \alpha_k)^2 = 1 = \alpha_k^2 \beta^2 => \gamma_\mu^2 = 1.$
3. $\gamma_\mu^* = \gamma_\mu$, so γ_μ is Hermitian.
4. $\gamma_k^* = (\text{-}i\beta \alpha_k)^* = i\ \alpha_k^* \beta^* = i\alpha\beta = \text{-}i\beta\alpha = \gamma_k$, so γ_k is Hermitian.
5. $\gamma_\mu^2 = \gamma_\mu^* \gamma_\mu = 1$; hence, γ_μ is a unitary.
6. $.\text{-}\gamma_5 \gamma_\mu + \gamma_\mu \gamma_5 = 0, \gamma_5^2 = 1,$
7. $\gamma_5 = \gamma_1 \gamma_2 \gamma_3 \gamma_4$

9.4.1. Dirac Adjoint: $\Psi^- = \gamma_4 \psi$

As an exercise: Show the covariant form of Dirac equation takes the following form in terms of Dirac adjoint:

$$\sum_\mu (\partial_\mu \Psi^- \gamma_\mu - \frac{m_0 c}{\hbar} \Psi^-) = 0 \text{--------------------} (34).$$

9.4.2. Four Vector (Probability Current)

Let $S_\mu^\rightarrow = \Psi^- \gamma_\mu \psi => S_4 = \psi^+ \gamma_4 \gamma_k \psi = \psi^+ \psi = P = $ probability density ρ.

$S_k = \psi^+ \gamma_4 \gamma_k \psi = \frac{1}{i} \psi^+ \alpha_k \psi = \frac{1}{ic} j_k^\rightarrow;$

therefore, $S_\mu^\rightarrow \equiv (S_K, S_4) = [(1/ic) j_\mu, \rho].$

Also, from the electrodynamics, it is known that
$S^{(4)\rightarrow} \equiv [(1/ic)j, \rho]$ and $F^{(4)\rightarrow}(A^\rightarrow, i\phi).$

9.5. Matrix Elements and Parity Operation in Dirac Theory

9.5.1. Matrix Elements:

It is a very established that classical physics in general and the classical mechanics in special are the limit of quantum physics as $h \rightarrow 0$, and $n \rightarrow \infty$, where h is Plank constant. Its value is about 6.623×10^{-34} joule.sec., and n is the principal quantum number which governs the spectrum of the microscopic system. Also, the condition to go from relativistic physics to nonrelativistic physics is $(v/c) \rightarrow 0$, where v is the velocity of the system, and c is the velocity of light. This condition implies, the energy of the system at rest is $m_0 c^2$. It is the ground state energy. So $E\psi = m_0 c^2 \psi = i\hbar \frac{\partial}{\partial t}$, which leads to the following: $\vec{\alpha} \rightarrow 0$, $\beta = 1$, and $\alpha_k = -\gamma_5 \sigma_k^4$. If $\alpha_k \rightarrow 0$, then either $\sigma_k^{\vec{4})} \rightarrow 0$ or $\gamma_5 \rightarrow 0$, but σ_k^4 is related to the intrinsic spin $(s = (1/2)\vec{\sigma})$ which cannot be zero; hence, $\gamma_5 \rightarrow 0$.

Therefore, as $v/c \rightarrow 0$, $\gamma_5 \rightarrow 0$ and $\alpha \rightarrow 0$. Going back to equations 31 and 32 under this condition, the relativistic equations go back to the nonrelativistic equations as it is the natural behavior of natural things. The matrix elements of an operator $\mathcal{A}$ in Dirac eigenfunction representation is written as

$$< f|\mathcal{A}|i > = (\Psi_f^- \mathcal{A} \psi_i) = \int \Psi_f^- \mathcal{A} \psi_i \, dv \text{-------------(35).}$$

Hence, knowing the type of interaction, then the operator, the wavefunction describing the system under investigation, the matrix elements, $\mathcal{A}_{fi}$, can easily be calculated because Dirac adjoint $\Psi^- = \psi \gamma_4$ will be known.

As it is mentioned earlier in the scattering theory, the matrix elements play a main role in calculating, for example, the rate of transition (ω_{12}), and consequently calculating the cross section of the interaction (scattering). In relativistic physics, they also play the same role, especially in beta-decay theory.

9.5.2. *Parity Operation in Dirac Theory*

The parity can be defined that

$$\Pi^{\wedge}\Psi(x_k, x_4) = \Psi(-x_k, ict) \text{-----------------------36).}$$

This is a kind of transformation such as $r \rightarrow -r$, $x_4 \rightarrow x_4$.
So in general, it can be written that $\Pi^{\wedge}\Psi = \Psi^\rho$, ρ indicates x, y, z→-x, -y, -z. It is inversion through the origin.

For Dirac relativistic theory, it can be started with

$$\sum_\mu (\gamma_\mu \partial_\mu + \frac{m_0 c}{\hbar})\Psi = 0,$$

where it can be written as

$$\sum(-\gamma_k \partial_k + \gamma_4 \partial_4 + \frac{m_0 c}{\hbar})\Psi^\rho = 0, k = 1, 2, 3. \text{------------- (37).}$$

Multiplying equation 37 from the left of Ψ by γ_4, it takes the form

$$\{\sum -\gamma_k \partial_k + \gamma_4 \partial_4 + \frac{m_0 c}{\hbar}\}\gamma_4 \Psi = 0 => \{\sum -\gamma_k \partial_k + \gamma_4 \partial_4 + \frac{m_0 c}{\hbar}\} \Psi^-.$$

Hence, $\Psi^\rho = +\Psi^+$ ----even $\gamma_4 \psi = +\Psi^-$
$$= -\Psi^- \text{----odd} = \gamma_4 \psi = -\Psi^-.$$

Chapter Ten
Introduction to Quantum Field Theory

10.1. Preface

After the Plank discovery of the discreteness of the energy of the black body radiation in 1900 and the de- Broglie proposal of the duality behavior of a moving system in 1924, the quantum theory had been greatly developed, plus the special and the general relativity theories in 1905 and in 1915 respectively by Albert Einstein. All these great scientific developments lead to a revolution in the theory of quantum theory of physics.

The black body radiation experiment by Plank has created great disturbance in physicists' thinking at that time, where Maxwell equations show the continuity of the energy, and the Lorentz classical equations were established, so they thought they can answer any natural phenomena might occur. Therefore, it was not easy for the physicists to accept the black body radiation first. But Plank's founding was experimentally confirmed, so the new concept were well established, where the energy of radiation is proved to be discrete as quanta each with energy E$\propto \nu$, where ν is the frequency of the radiation. The proportionality constant was measured to be about 6.623×10^{-34} joule .second (js). It is denoted by h; it is named as Plank constant. So the energy of each quanta is E=$h\nu$.

Hence, the energy is quantized, and the quantization is an intrinsic character in microphysical system, as seen in quantum physics. Hence, they finally believed in the new findings and developed the quantum physics greatly. As Dirac stated once, "The physicists have to replace prejudices by something more precise, leading to some entirely new conception of Nature."

In view of this, quantum theory started describing the microsystems and the submicrosystems, such as molecules, atoms, electrons, nuclei, nucleons, and other nuclear particles as particles, could be one particle, two particles, and three particles to many particles. The used technique is determined according the case under study. Thereafter, the force concept is replaced by the field concept. It is found more scientifically sound and fruitful. Therefore, the gravitational force is replaced by the gravitational field; its messenger particle carrying its effect between the interacting masses is the virtual "graviton" of spin two, zero rest mass, and speed of light not detected yet, long-range force.

Similarly, the electromagnetic force is replaced by the electromagnetic field $(A, i\phi)$; its messenger particle carrying its effect between interacting charged particles is the photon represented by γ, ($m_0=0$, spin=1, $c=3\times 10^8$ m/sec.). The nuclear force (strong) is replaced by the field whose messenger particles are the pions $(\pi^+\pi^0\pi^-)$ which carry the field effect between the interacting nucleons.

Also, the electroweak force (unifying E and M with nuclear weak force) is replaced by the field carrying its effect are the particles(w^+, z^0, w^-). On such basis the fields are considered instead of the forces. Also, the very strong nuclear force which combines the quarks inside the nucleons and other nuclear particles is replaced by field, its messenger particles proposed to be the gluons of eight favored colors, which is quite important in quantum chromodynamics (QCD). The use of the field is of very important reason; this can be recognized when the problem to be studied is of so many particles, where its solution using the force concept is very much complicated.

Just imagine a system of thousand particles, which needs thousands of equations to be solved, so it is clear that the field concept is too great. As it is known, the electromagnetic field is a result of unifying of the electric field (E) and the magnetic field (H) through the classical Maxwell equations. This was the first step of unification between two natural forces. This unification showed that the light is an electromagnetic wave with a speed about 3×10^8 m/second, where the light behaves as particles (photons) and as wave according to the physical phenomenon it subjected to. It has duality property according to the proposal of De Broglie in 1924. So now how the quantum theory can be applied on the field as it is applied on the particle systems individually such as the atoms, molecules, leptons, and baryons?

In the particles' systems, the requirement is the conjugate dynamical variables that usually are associated with the particles' motions ought to obey the Heisenberg uncertainty principle, which leads to the quantization concept to be applied between these canonical dynamical variables, such as $[p_i, q_i] = i\hbar$, $[p_i, p_j] = i\hbar\delta_{ij} = 0$, if i=j, also $[q_i, q_j] = i\hbar\delta_{ij}$. This is, in fact, the condition of quantization, which is sometimes called the first quantization.

In this chapter, both the particles quantization and the fields' quantization will considered. The fields' quantization is called the second quantization and generally is denoted as the quantum field theory.

The program of the quantization in this chapter will deal with the classical electromagnetic field ($\vec{A}, i\phi$) as well as with the so-called matter fields. The matter fields are governed by Klein-Gordon and Dirac equations (chapter nine), which have no classical counterpart.

In the case of electromagnetic fields, the second quantization is of importance and evidently of quite need. But for nonclassical fields (the so-called matter fields), such importance somehow is less because in the case of such fields, the quantization of Klein-Gordon and Dirac equations are already describing the behavior of the particles.

The complete set of solutions of these equations contain descriptive elements of particles in state of negative energy {Dirac Theory of spin 1/2 particles (electrons)}. This theory pointed out there are two solutions for the energy (relativistic solution): one of positive energy and the other is of negative energy, which considered the states in the negative energy that are in fact the anti-electrons particles called the positron(e^+). Such result was somehow complicating the physical interpretation concerning the possibility of spontaneous transition to these states of negative energy.

To get rid of this complexion, Dirac assumed there is sea of full states with electrons in this negative energy region. In such case, there is no place for additional electrons, which are fermions, due to Pauli exclusion principle. But Klein-Gordon equation actually describing particles with integer spin (Bosons) which do not obey Pauli principle. Therefore, if the presence of many particles is invoked in order to apply and interpret one particle equation, this might face a basic inconsistency. Hence, the satisfactory resolution of such inconsistency is embedded in the second quantization, the quantum field theory, the subject of this chapter. This is so because the structure and the formation of this elegant theory excuses the particle number (n) to change according to the process described above.

Therefore, the ability of this theory in describing a physical situation, such as the creation and annihilation of particles, makes it the real tool of dealing with the so-called nuclear particles in general and the fundamental particles in special. And it is the natural language of their properties.

This feature of the quantum field theory provides it with a wide chance of the applications in the contemporary fields of physics in addition to some other fields in science such as chemistry. So it is not beyond the reality that the development of quantum theory through about century and fifteen years

(1900–2015) has provided the physicists with a high level of conceptual thinking of nature understanding, and still more is coming.

But all that is a little of the creator science providing the human beings with to understand the universe in the limit of their benefit. No chance to fully be acquainted with the mind of the creator that is a red line.

In the following sections, the role of this elegant theory will be clarified with their principles and their applications.

10.2.a. *Field Quantization and Langrangian Formalism*

Introduction:

For the purpose of quantizing the fields, it is necessary, first of all, to review briefly the derivation of the field equations from the Langrangian formalism. For doing so, it will recall the Euler-Lagrange equation of motion. This equation is determined through the Hamiltonian principle known as the action principle, where the time integral of the Langrangian is taken as extremum. To start with, it is necessary to find Lagrangian from classical mechanics. Hence, in classical mechanics, the Lagrangian is denoted as $\mathcal{L}(x, x^{\cdot}, t)$, with appropriate boundary condition $x(t_0)$ and $x(t_1)$. The position $x(t)$ at time t can be determined by taking the extremum of the integral, $I = \int_{t_0}^{t} \mathcal{L}(x, x^{\cdot}\ t)\ dt$ ------------------------- (1),

where $x^{\cdot}=v=dx/dt$, the velocity. For more illustrations, let x be plotted versus t as in figure 1 below:

From the figure the variation is, $x(t) \rightarrow x(t) + \alpha\eta(t)$.

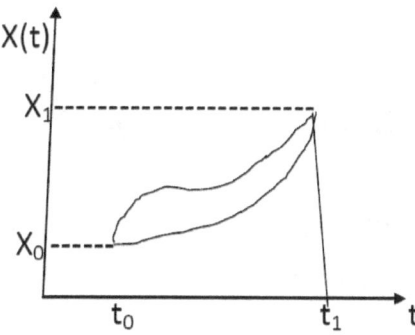

Since the end points are fixed the boundary condition is,
$\eta(t_0)=0$, since $x(t_0) \rightarrow x(t_0)+\alpha\eta(t_0)$,
   and $\eta(t_1)=0$, since,
      $X(t_1) \rightarrow x(t_1)+\alpha\eta(t_1)$.

Taking the above variation in x in consideration, equation 1 becomes $I(\alpha)=\int_{t_0}^{t_1} \mathcal{L}(x + \alpha\eta, x^{\cdot} + \alpha\eta^{\cdot},$ t), then   take   the

extremum of $I(\alpha)$, i.e., $\partial I(\alpha) = 0$ or equivalently $\frac{\partial I(\alpha)}{\partial \alpha} = 0$

Therefore, it is written as

$$\frac{\partial I(\alpha)}{\partial \alpha} = \int_{t_0}^{t_1} (\frac{\partial \mathcal{L}}{\partial x})(\frac{\partial x}{\partial \alpha})+(\frac{\partial \mathcal{L}}{\partial x^{\cdot}})(\frac{\partial x^{\cdot}}{\partial \alpha})\ dt = \int_{t_0}^{t_1} (\frac{\partial \mathcal{L}}{\partial x}\ \eta +(\frac{\partial \mathcal{L}}{\partial x^{\cdot}}\eta^{\cdot})dt$$

Hence, $\frac{\partial I(\alpha)}{\partial \alpha} = \int_{t_0}^{t_1} (\frac{\partial \mathcal{L}}{\partial x})\eta +(\frac{\partial \mathcal{L}}{\partial x^{\cdot}})\eta^{\cdot}\ dt$ --------------------(2).

Now integrating the second term in equation 2 by parts, using the boundary condition, the following is obtained:

$$\frac{\partial I(\alpha)}{\partial \alpha} = \int_{t_0}^{t_1} \eta (\frac{\partial \mathcal{L}}{\partial x})-\frac{d}{dt} (\frac{\partial \mathcal{L}}{\partial x^{\cdot}})dt=0, \text{ is the extremum condition.}$$

Since the region of the integration is arbitrary and the integral on the right hand side is zero, the integrand has to be vanished too. This leads to $(\frac{\partial \mathcal{L}}{\partial x} - \frac{d}{dt}(\frac{\partial \mathcal{L}}{\partial x \cdot}))=0$, so in general, for coordinates $x_1, x_2, x_3, \text{------}x_r, \dot{x}_1, \dot{x}_2, \dot{x}_3, \text{-------}\dot{x}_r, (\dot{x} \equiv x)$, it can be written as $\frac{\partial \mathcal{L}}{\partial x_r} = \frac{d}{dt}(\frac{\partial \mathcal{L}}{\partial \dot{x}_r})$, r=1, 2, 3, 4, ------------ (3).

Now consider the quantity: $[\sum_r \dot{x}_r \frac{\partial \mathcal{L}}{\partial \dot{x}_r} - \mathcal{L}]$, differentiate it with respect to time t. It gives the following:

$$\frac{d}{dt}[\sum_r \dot{x}_r \frac{\partial \mathcal{L}}{\partial \dot{x}_r} - \mathcal{L}] = \sum_r [\ddot{x}_r \frac{\partial \mathcal{L}}{\partial \dot{x}_r} + \dot{x}_r \frac{d}{dt}\frac{\partial \mathcal{L}}{\partial \dot{x}_r} - \frac{\partial \mathcal{L}}{\partial x_r}\dot{x}_r - \frac{\partial \mathcal{L}}{\partial \dot{x}_r}\ddot{x}_r] = 0$$

Since the time derivative of this quantity is zero (time independent), so it is constant of motion of the system.

Now define that $P_r = \frac{\partial \mathcal{L}}{\partial x \cdot}$, so $\frac{dP_r}{dt} = \frac{d}{dt}(\frac{\partial \mathcal{L}}{\partial x \cdot}) = \frac{\partial \mathcal{L}}{\partial x_r}$ (see equation 3); hence, $P_r = \frac{\partial \mathcal{L}}{\partial x_r}$ --- (4).

So the above quantity can be written as $\sum_r (\dot{x}_r p_r - \mathcal{L})$, which represents the Hamiltonian of the system H.

Therefore, $H(x_1, \text{------},P_r) \equiv \sum_r (\dot{x}_r p_r - \mathcal{L})$, so energy conservation is expressed as $\frac{dH}{dt} = 0$. It is valid for forces which is derivable from central potentials as such

$F = -\nabla V(r) (conservative)$.

As an example,
consider $\mathcal{L} = \frac{m}{2}(\dot{x}^2 + \dot{y}^2 + \dot{z}^2) - \phi(x, y, z) = T - V(r) = K.E - P.E.$ Then $p_x = m\dot{x}$, $\dot{x} = p_x/m$, and so on.

The Hamiltonian is written as $H = p_x \dot{x} + p_y \dot{y} + p_z \dot{z} - \mathcal{L}$.

Using the above information, H is written as

$$H=\frac{p^2}{2m}+\phi(x, y, z).=K.E.+ P.E.$$

Now it is important to investigate the properties of the Hamiltonian function H (p_i, x_i, t) which is should not be explicitly time dependent, where p and q are generalized canonical variables; hence,

$$\frac{\partial H}{\partial x_1}=\frac{\partial}{\partial x_1}\left(\sum_r p_r\, x_r -\mathcal{L}(x_r, x_r, t)\right)=-\frac{\partial \mathcal{L}}{\partial x_1} = -\,p_1,$$

in the same way, $\frac{\partial H}{\partial p_1}=x_1$. (As an exercise, verify these results.)

These relations are called the Hamiltonian equations of motion, summarized as $[x_r =\frac{\partial H}{\partial p_r}, p_r =-\frac{\partial H}{\partial x_r}]$.

Also, it can be written that if H is not explicitly time dependent, then $\frac{dH}{dt}$ =o. It means energy conservation. Therefore,

$$\frac{dH}{dt}=\frac{\partial H}{\partial t}+\frac{\partial H}{\partial x_r}\frac{\partial x_r}{\partial t}+\frac{\partial H}{\partial p_r}\frac{\partial p_r}{\partial t}=\frac{\partial H}{\partial t}+(-p_r x_r +x_r p_r)=\frac{\partial H}{\partial t}=0$$

So $\frac{dH}{dt}=0$, conservation of energy.

10.2.b. *Classical Field Theory (Functional Derivative)*

In classical field theory, a density of Lagrangian function can be introduced as L, when the Lagrangian is written as

$$\mathcal{L} =\int L\, d\tau,\ d\tau \equiv(dx_1, dx_2, dx_3),\ dX \equiv dx_1, dx_2, dx_3, dx_4.$$

Now let L be as $L=L(\phi,\frac{\partial \phi}{\partial x_1}, \frac{\partial \phi}{\partial x_2}, \frac{\partial \phi}{\partial x_3}, \frac{\partial \phi}{\partial x_4})\equiv L(\phi,\frac{\partial \phi}{\partial x_r}) \equiv L(\phi, \partial_r\, \phi)$. Where $\phi(\vec{x}, t)=\phi(x_1, x_2, x_3, x_4)= \phi(x)$.

Going back to the formula, $I=\int_0^t \mathcal{L}\, dt=\iint L(\phi, \partial_r\, \phi)d\tau dt$

$=\iint L(\phi, \partial_r\emptyset)\, dx$, $dx = d\tau\, dt$. Referring to figure 1, the following variation can be performed:

$\phi(x) \to \phi + \alpha\eta(x)$, which leads to:

$L \to L(\phi + \alpha\eta, \frac{\partial\phi}{\partial x_r} + \alpha\frac{\partial\eta}{\partial x_r})=L + \delta L$, δL is given by,

$\delta L = \frac{\partial L}{\partial\phi}\delta\phi + \frac{\partial L}{\partial(\frac{\partial\phi}{\partial x_r})}\delta(\frac{\partial\phi}{\partial x_r}) = \frac{\partial L}{\partial\phi}\alpha\eta + \frac{\partial L}{\partial(\frac{\partial\phi}{\partial x_r})}\alpha\frac{\partial\eta}{\partial x_r}$

Hence, $L \to L+\alpha + [\frac{\partial L}{\partial\phi}\eta + \frac{\partial L}{\partial(\frac{\partial\phi}{\partial x_r})}\frac{\partial\eta}{\partial x_r}]$.

Now, $\frac{\partial}{\partial x_r}(\frac{\partial L}{\partial(\frac{\partial\phi}{\partial x_r})}\eta)=\eta\frac{\partial}{\partial x_r}(\frac{\partial L}{\partial(\frac{\partial\phi}{\partial x_r})}) + \frac{\partial L}{\partial(\frac{\partial\phi}{\partial x_r})}\frac{\partial\eta}{\partial x_r}$, thus, I takes the form,

$I = \int_0^t \mathcal{L}\, dt=\iint L(\phi, \partial_r\, \phi)dx+\alpha \iint \eta(\frac{\partial L}{\partial\phi} - \frac{\partial}{\partial x_r}(\frac{\partial L}{\partial(\frac{\partial\phi}{\partial x_r})})\, dx$

$+ \alpha \iint \frac{\partial}{\partial x_r}(\frac{\partial L}{\partial(\frac{\partial\phi}{\partial x_r})}\eta)dx$. Now integrating the last term first, using the boundary condition that $\eta(t)=\eta(0)=0$, then,

$\alpha\int\frac{\partial}{\partial x_i}[(\frac{\partial L}{\partial(\frac{\partial\phi}{\partial x_r})})\eta]_0^t\, dt =0$, i=1, 2, 3. Therefore, $\delta I = \frac{\partial I}{\partial\alpha}$

$\alpha =0$

$=\iint \eta(\frac{\partial L}{\partial\phi} - \frac{\partial}{\partial x_r}(\frac{\partial L}{\partial(\frac{\partial\phi}{\partial x_r})})\, dx =0$; hence:

$\frac{\partial L}{\partial\phi} = \frac{\partial}{\partial x_r}(\frac{\partial L}{\partial(\frac{\partial\phi}{\partial x_r})}) = \frac{\partial}{\partial x_r}(\frac{\partial L}{\partial\partial_r\emptyset})$, (where summation convention understood).

This can be written such as

$$\{\frac{\partial L}{\partial \phi}-\frac{\partial}{\partial x_k}(\frac{\partial L}{\partial(\frac{\partial \phi}{\partial x_k})})\}-\{\frac{\partial}{\partial t}(\frac{\partial L}{\partial \phi \cdot})\}=0 => \frac{\partial \mathcal{L}}{\partial \phi}=\frac{d}{dt}\frac{\partial \mathcal{L}}{\partial \phi \cdot} \text{---------}(5).$$

Hence, it is found very important to study the functional concepts (function of a function) as concepts and properties because they are of great useful use in field theory in general.

For example, consider the following functional:

$I(u)=\int \int_{\Theta_I} F(x, y, u, u_x, u_y)dx\, dy,$

where $u = u(x, y)$, it is defined on a closed region Θ_I, such that

1. $u(x, y)$ is twice differentiable on Θ_I, and they are continuous.
2. u is defined on the boundary of Θ_I.

If now a subregion of Θ_I is considered, say Θ_{I^0}, then $\Theta_{I^0} \subset \Theta_I$

(see the diagram)

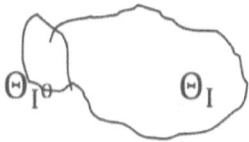

Now define a variation of u as δu, such that
$\delta u \in \Theta_{I^0}$ and $\delta u=0$ for $u \in$ boundary of Θ_I, then $(u + \delta u) \in \Theta_I$.
(ε means it belongs to)

Now taking a variation of u with these properties, then
$I(u + \delta u)=\int\int F(x, y, u + \delta u, u_x + \delta u_x, u_y + \delta u_y,)dx\, dy,$

where, $\delta u_x =\frac{\partial}{\partial x}(\delta \mu)=(\delta u)_x$ ----etc.

Then expanding F by Taylor series, it is obtained that

$F(x, y, u+\delta u, u_x+\delta u_x, u_y +\delta u_y)=F (x, y, u, u_x, u_y + Fu\delta u + F u_x\delta u_x +Fu_y\delta u_y +\sum Fuu(\delta u)^2+$ higher order terms.

Now integrating this expansion the following is obtained:

$I(u + \delta u)=\iint F(x,y,u,u_x,u_y)dx\ dy+\iint \sum F_u\ \delta u+F_{u_x}\delta u_x + F_{u_y}u_y +\iint \frac{1}{2}\sum F_{u\ u}(\delta u)^2 +$ ---higher order terms.

Or this might be written as follows:

$I (u + \delta u)= I (u) + I^{1}(u) + I^{2}(u) +$ -----

To the first order, however, $\delta I(u)$ can be written as

$\delta I (u)= I^{1}(u)=\iint [F_x\ \delta(u) + F_{u_x}\ \delta u_x + F_{u_y}\delta u_y\ ]dx$, it is a definition of $\delta I(u)$. An alternative and equivalent definition of $\delta I(u)$ is given by $\delta I(u) =\text{limit}(\epsilon \to 0)\frac{1}{\epsilon}[I (u + \epsilon\delta u)- I (\delta u)]$.

To show the equivalence of these two expressions, consider that

$\delta I =\text{limit} (\epsilon \to 0)\frac{1}{\epsilon}\ [\iint\{F(x,y,u, +\epsilon\ \delta u, u_x + \epsilon\delta u_x, u_y + \delta u_y - \iint F(x,y,u,u_x,u_y\}\ dx\ dy].$

Again, expanding F, *the following is gotten*:

$F(x, y, u + \epsilon\delta u, u_x + \epsilon\delta u_x, u_y +\epsilon\delta u_y)=F(x,y,u_x,u_y,u)+\epsilon [F_u\delta u+F_{u_x}\delta u_x +F_{u_y}\delta u_y]+\frac{1}{2}\epsilon^2 [F_{uu}(\delta u)^2+ $ ----]+ ----

Now $\delta I=\text{limit} (\epsilon \to 0)\{\frac{1}{\epsilon}\iint [\epsilon(F_u\ \delta u +F_{u_x}\delta u_x+F_{u_y}\delta u_y) +\frac{\epsilon^2}{2}[F_{uu}(\delta u)^2]\}dx\ dy$. However, since $\delta u =0$, at the boundary of Θ_I, *the integrals are identically zero, thus,*

$$\delta I(u) = \int\int_{\Theta_I} (F_u - \frac{\partial F_{ux}}{\partial x} - \frac{\partial F_{uy}}{\partial y}) \, \delta u \, dx \, dy.$$

Now define the integrand as K(x, y), then

$$\delta I(u) = \int\int_{\Theta_I} K(x,y) \, \delta u \, dx \, dy = L(\delta u) =>$$

$\frac{\delta I(u)}{\delta u} = K(x,y)$, thus, when $\delta I = 0$ for all $\delta u \in \Theta_{I^0}$, then

$\frac{\delta I(u)}{\delta u} = 0$, or K(x, y)=0, which implies that

$F_u - (\frac{\partial F_{ux}}{\partial x} + \frac{\partial F_{uy}}{\partial y}) = 0$ => the Euler-Lagrange equation which is

$\frac{\partial F}{\partial u} - \frac{\partial}{\partial x} \frac{\partial F}{\partial(\frac{\partial u}{\partial x})} - \frac{\partial}{\partial y} \frac{\partial F}{\partial(\frac{\partial u}{\partial x})} = 0$, or can be written as

$\partial_u F - \partial_x (\frac{\partial F}{\partial \partial_x u}) - \partial_y(\frac{\partial F}{\partial \partial_y u}) = 0.$

Also, the functional $L(\delta u)$ has the following linearity property: $L(a\delta u + b\delta u) = aL(\delta u) + b \, L(\delta(u))$. It is obvious from the above definition of $L(\delta u)$.

10.3. *Lagrange Formalism In Classical Field Theory*

To make things more clear about the physical ideas so far mentioned, it is necessary to show the analogies between the classical mechanics, dynamical variables, and their corresponding in classical field theory. This is shown in the following comparison:

Classical Mechanics	Classical Field Theory
q=generalized coordinate.	ψ =generalized field function.
i=coordinate subscript.	$\alpha, \vec{r}$ field subscript and variable.
$\sum_r$=summation over(x, y, z).	$\sum_\alpha \int d^3 r$ summation over field and integration over space.
F(q_i, t)=function	F($\psi_\alpha(\vec{r}$, t)=functional.

With this in mind, the functional can be written as:

$$F[\psi_\alpha(r^\rightarrow)] \equiv F[\psi_1(r_1), \psi_2(r_2), \psi_3(r_3)]$$

Now for convenience, let $\psi(r\rightarrow)$ as $\psi_{r^\rightarrow}$ and consider the following transformation:

$$\psi_{r^\rightarrow} \rightarrow \psi_{r^\rightarrow} + \delta\psi_{r^\rightarrow}, \; F \rightarrow F + \delta F \; \text{----------------------(6).}$$

A variational in the field functional $F(\psi_1(r_1), \psi_2(r_2), ---)$ due to the transformation given by equation 6. It will then be

$$\delta F = \delta F(\delta\psi_{r^\rightarrow}) = \text{limit}(\epsilon \rightarrow 0)\frac{1}{\epsilon}[F(\psi_{r^\rightarrow}) + \epsilon\delta\psi_{r^\rightarrow}) - F(\psi_{r^\rightarrow})].$$

Now consider the space being divided into small cells of volume $d\tau$, then a summation $\sum_\tau d\tau$ can be denoted, as $\sum_{r^\rightarrow} d^3r^\rightarrow$, the one going to be used. Thus, the above variation can be written as

$$\delta F(\delta\psi) = \sum_{r^\rightarrow} \frac{\partial F}{\partial\psi_{r^\rightarrow}} \delta\psi_{r^\rightarrow} \; \text{-----------(7).}$$

Multiply equation 7 from the right by $\frac{d^3r^\rightarrow}{d^3r^\rightarrow}$ (=1). It takes the form $\delta F(\delta\psi) = \sum_{r^\rightarrow} \frac{\partial F}{\partial\psi_{r^\rightarrow}} \partial\psi_{r^\rightarrow} \frac{d^3r^\rightarrow}{d^3r^\rightarrow} = \sum_{r^\rightarrow} \frac{\partial F}{\partial\psi_{r^\rightarrow}} d^3r^\rightarrow \frac{\delta\psi_{r^\rightarrow}}{d^3r^\rightarrow}$ -- (8).

If the space cells considered infinitesimally small such that $\frac{d\psi_{r^\rightarrow}}{d^3r^\rightarrow}$ is the same constant means value in all the sells, then

$$\delta F(\delta\psi_{r^\rightarrow}) = \frac{\delta\psi_{r^\rightarrow}}{d^3r^\rightarrow} \sum_{r^\rightarrow} \frac{\partial F}{\partial\psi_{r^\rightarrow}} d^3r^\rightarrow = \frac{\delta\psi_{r^\rightarrow}}{d^3r^\rightarrow} \cdot \frac{\partial F}{\partial\psi_{r^\rightarrow}} \; \text{-------------} \; (9).$$

By investigating equation 9, it can be seen that F in each term of the summation is describing a certain cell in space.

While in the right hand side term, F is describing the whole space. By rearranging equation 9, it gives:

$$\frac{\delta F(\delta\psi_{r\to})}{\delta\psi_{r\to}} = \frac{1}{d^3 r^\to}\frac{\partial F}{\partial\psi_{r\to}} \text{------------------------------- (10).}$$

Now multiply equation 10 by $\delta\psi_{r\to}$ and multiply by $\int d^3 r^\to$,. It will take the form

$$\int d^3 r^\to \delta\psi_{r\to}\frac{\delta F(\delta\psi_{r\to})}{\partial\psi_{r\to}} = \int \frac{d^3 r^\to}{d^3 r^\to}\cdot\frac{\partial F}{\partial\psi_{r\to}}\delta\psi_{r\to} \text{--------- (11).}$$

A comparing between the right side of equation 12 with equation 7, with the realization that in the limit as $d^3 r \rightarrow 0$, the summation over the cells of the space is equivalent to an integration, which can be seen as:

$$\int d^3 r^\to \delta\psi_{r\to}\frac{\delta F(\delta\psi_{r\to})}{\delta\psi_{r\to}} = \delta F(\delta\psi_{r\to}) \text{--------------------- (12).}$$

Let the Dirac three dimension delta function introduce as $\int \delta_{r\to,r^-\to} \, d^3 r^\to = 0$, if $r \neq r^-$, $=1$, if $r = r^-$,-------------(13).

Also, it has the property that $\delta_{r\to,r^-\to} \rightarrow = \delta_{r\to-r^-\to} = \delta_{r\to^--r,\to 0}$.

Now equation 13 for the special case of a variation that takes place only in a cell of the space, where $\delta\psi_r \to \delta_{r\to,r^-\to}$, takes the form: $\delta F(\delta_{r,\to r^-\to}) =$

$$\int d^3 r^\to \delta(r,\to r^{-\to})(\frac{\delta F}{\delta\psi_{r\to}}) \text{----------- (14).}$$

The integral in the right hand side of equation 14 is zero for every cell in the space, except where $r = r^-$ (property of delta function). Thus, integrating equation 14 over all space yields

$$\delta F(\delta_{r\to,r^-\to}) = \int d^3 r^\to(\delta_{r\to,r^-\to})(\frac{\delta F}{\delta\psi_{r\to}}) \text{--------(15).}$$

This implies that the variation of the functional is a variational of the delta function.

After the important and necessary physical concepts, concerning the Lagrangian formalisms, the Hamiltonian principle (least action), and the functional definition and properties, the next step is to deal with the classical field theory principles. To do so, the canonical conjugate to ϕ has to be defined. It is denoted here as $\Pi(x)$ related to $\phi((x)$ *through* the following relation:

$$\Pi(x) = \frac{\partial L}{\partial \phi^{\cdot}} = \frac{\delta \mathcal{L}}{\delta \phi^{\cdot}}, \text{ and } \Pi^{\cdot} = \frac{\delta \mathcal{L}}{\delta \phi} \text{ -------------------- (16).}$$

Where L is Lagrange density *and* $\mathcal{L}$ is Lagrange.

This, as it is known to physicists, is analogouse to the classical mechanics, where H=$\sum_r p_r q_r$-L, Hamiltonian density.

In the same *manner, where* $\phi^{\cdot} = \phi^{\cdot}(\Pi, \phi)$. H can be written as H=$\Pi\phi^{\cdot}$ -L$\equiv$ H($\Pi, \phi, \partial_k \phi$)------------------------(17).

So the total Hamiltonian $\mathcal{H}$ *is* given by $\mathcal{H} = \int H \, dx$ -----(18).

By analogy with the classical mechanics, it can be shown that

$$\frac{\delta \mathcal{H}}{\delta \Pi} = \frac{\partial H}{\partial \Pi} \text{ and } \frac{\delta H}{\delta \phi} = \frac{\partial H}{\delta \phi} - \sum_{k=1}^{3} \frac{\partial}{\partial x_k}\left(\frac{\partial H}{\partial\left(\frac{\partial \phi}{\partial x_k}\right)}\right) \text{--------(19).}$$

Then, the total variation in $\mathcal{H}$ will be given by

$\delta \mathcal{H} = \frac{\delta \mathcal{H}}{\delta \phi} \delta\phi + \frac{\delta \mathcal{H}}{\delta \Pi} \delta\Pi = \int [\frac{\delta H}{\delta \Pi} \delta\Pi + \frac{\delta H}{\delta \phi} \delta\phi] d\tau$, which from equations 17 and 19 it becomes as:

$\delta \mathcal{H} = \int \{\phi^{\cdot} \delta\Pi + \Pi\delta\phi^{\cdot} - \frac{\partial L}{\partial \phi} \partial\phi - \frac{\partial L}{\partial \phi^{\cdot}} \partial\phi^{\cdot}\} d\tau$. From classical mechanics, it can be generalized this by writing $\Pi = \frac{\partial L}{\partial \phi^{\cdot}}$ and $\Pi^{\cdot} = \frac{\partial L}{\partial \phi}$, so $\delta \mathcal{H}$ takes the form,

$$\delta\mathcal{H} = \int \{ \dot\phi \cdot \delta\Pi + \frac{\partial L}{\partial\dot\phi}\delta\dot\phi - \Pi\cdot\delta\dot\phi - \frac{\partial L}{\partial\dot\phi}\delta\dot\phi \} d\tau =$$
$$\int (\dot\phi\cdot\delta\Pi - \Pi\cdot\delta\phi)d\tau.$$

This yields to $\frac{\delta\mathcal{H}}{\partial\Pi} = \dot\phi$ and $\frac{\delta\mathcal{H}}{\partial\phi} = \dot\Pi$. ---------------------- (20).

This can be done another way: H=$\sum_\alpha \Pi_\alpha \dot\psi_\alpha$ –L -------- (21).

Then $\mathcal{H} = \int H d\tau$,

which gives $\delta\mathcal{H} = \int [\sum_\alpha \frac{\partial H}{\partial\Pi_\alpha}\delta\Pi_\alpha + \frac{\partial H}{\partial\psi_\alpha}\delta\psi_\alpha]d\tau$ -------------(22).

This, as mentioned before, gives the following form:

$$\delta\mathcal{H} = \int [\sum_\alpha \dot\psi_\alpha \delta\Pi_\alpha + \Pi_\alpha\delta\dot\psi_\alpha - \frac{\delta L}{\delta\psi_\alpha}\delta\psi_\alpha - \frac{\delta L}{\delta\psi_\alpha}\delta\dot\psi_\alpha]d\tau \text{ ----(23)}.$$

Using $\Pi_\alpha = \frac{\partial L}{\partial\dot\psi_\alpha}$, then equation 23 becomes

$$\delta\mathcal{H} = \int [\sum_\alpha \dot\psi_\alpha \delta\Pi_\alpha - \frac{\delta L}{\delta\psi_\alpha}\delta\psi_\alpha] \text{ -------------------(24)}.$$

Since $\frac{\delta\mathcal{H}}{\delta\Pi} = \frac{\partial H}{\partial\Pi}$, from equation 24, it can be shown that

$$\frac{\delta\mathcal{H}}{\delta\Pi} = \frac{\partial H}{\partial\Pi} = \dot\phi \text{ ----------------------------- (25)}.$$

Using the fact that $\frac{\delta L}{\delta\psi_\alpha} = \frac{\partial L}{\partial\psi_\alpha} - \frac{\partial}{\partial x_k}\frac{\partial L}{\partial\partial_k\psi_\alpha}$, --------(26).

Now, from equations 12 and 21, the following is obtained:

$$\delta\mathcal{H}(\delta\psi) = \int \frac{\partial H}{\partial\psi_\alpha}\delta\psi_\alpha d\tau = -\int \frac{\delta L}{\delta\psi_\alpha}\delta\psi_\alpha d\tau = -\int \{\frac{\delta L}{\delta\psi_\alpha}\delta\psi_\alpha d\tau =$$

$\int \{\frac{\partial L}{\partial\psi_k} - \frac{\partial}{\partial x_k}(\frac{\partial L}{\partial\partial_k\psi_\alpha})\}\delta\psi_\alpha d\tau$. Since $\frac{\partial L}{\partial\dot\psi} = \dot\Pi$,. This equation can be

written as $\int \frac{\partial H}{\partial\psi_\alpha}\delta\psi_\alpha d\tau = -\{\Pi_\alpha - \frac{\partial}{\partial x_k}(\frac{\partial L}{\partial\partial_k\psi_\alpha}) \delta\psi_\alpha d\tau =>$

$$\Pi_\alpha = -\frac{\partial H}{\partial\psi_\alpha} - \frac{\partial}{\partial x_k}(\frac{\partial L}{\partial\partial_k\psi_\alpha}) = -\frac{\partial H}{\partial\psi_\alpha} + \frac{\partial}{\partial x_k}(\frac{\partial H}{\partial\partial_k\psi_\alpha}) = -\frac{\partial\mathcal{H}}{\partial\psi_\alpha} \text{ ---(27)}.$$

As an example of this, consider the Klein-Gordon equation, working in the units that $\hbar = c = 1$, taking the special case m=0.

Take the following form of Lagrangian:

$$\mathcal{L} = \frac{1}{2}[\varphi^{\cdot 2} - (\frac{\partial \varphi}{\partial x_k})^2] \text{ ---------------------------- (28)}.$$

Using the Euler-Lagrange equation:

$$\frac{d}{dt}\frac{\partial \mathcal{L}}{\partial \varphi^{\cdot}} - \frac{\partial}{\partial x_k}\frac{\partial \mathcal{L}}{\partial \varphi_k} = 0, \text{ where } \partial \varphi_k = \frac{\partial \varphi}{\partial x_k}, k=1, 2, 3.$$

And the double summation subscript is understood; the above yields the Klein-Gordon equation as

$$\varphi^{\cdot\cdot} - \frac{\partial}{\partial x_k}(\frac{\partial \varphi}{\partial x_k}) = 0, \text{ or using Laplace operator, so}$$

$$\Delta \varphi - \sum_{k=1}^{3}\frac{\partial^2 \varphi}{\partial x_k^2} => [\varphi^{\cdot\cdot} - \Delta\varphi] = 0, \text{-------------(29)}.$$

From equation 28, $\frac{\partial \mathcal{L}}{\partial \varphi^{\cdot}} = \varphi^{\cdot}$. Also, it is known this equals Π,

$$=> \Pi = \varphi^{\cdot} \text{ ----------------------------(30)}.$$

Now Hamiltonian, as defined before, is

$\mathcal{H} = \Pi\varphi^{\cdot} - \mathcal{L}$. So substitute for Π and $\mathcal{L}$, the $\mathcal{H}$ takes the form. $\mathcal{H} = \frac{1}{2}\varphi^{\cdot 2} + \frac{1}{2}\sum_{k=1}^{3}(\frac{\partial \varphi}{\partial x_k})^2$, or $\mathcal{H} = \frac{1}{2}\Pi^2 + \frac{1}{2}(\nabla\varphi)^2$ --- (31).

This equation gives Hamiltonian $\mathcal{H}$ in terms of the canonical conjugates Π and φ. Hence, from equations 30 and 31,

$$\varphi^{\cdot} = \Pi = \frac{\partial \mathcal{H}}{\partial \Pi}.$$

Using equation 29, the following result is obtained: $\varphi^{\cdot} = \Delta\varphi$, and by the definition of Π, the following relation is true:

$$-\frac{\delta \mathcal{H}}{\partial \varphi} = +\Pi = \varphi'' = \Delta \varphi.$$

10.4. Conservation Laws (Noether Theorem)

The conservation of physical laws is very important to be understood. Therefore, it will be treated clearly in this section. So the results of the variational methods will be used to get the conservation of laws. It was found previously that

$$I(x) = \iint L(\varphi(x), \partial_v \varphi(x)) dx + \iint \eta [\frac{\partial L}{\partial \varphi} - \frac{\partial}{\partial x_v}(\frac{\partial L}{\partial x_v})] dx + \alpha \iint \frac{\partial}{\partial x_v}$$

$$(\frac{\partial L}{\partial v(\partial \varphi / \frac{\partial \varphi}{\partial x_v})}\ \eta)\ dx\text{--------------}(32).$$

And $\delta I = \iint \frac{\partial}{\partial x_v}(\frac{\partial L}{\partial(\frac{\partial \varphi}{\partial x_v}}\ \eta) dx$. Note: L is Lagrangian.

Since $L \equiv L(\varphi_1, \varphi_2)$, L is Lagrangian, then

$\frac{\partial L}{\partial x_\tau} = \frac{\partial L}{\partial \varphi}\frac{\partial \varphi}{\partial x_\tau} + \frac{\partial L}{\partial}\frac{\partial^2 \varphi}{\partial x_v \partial \varphi \partial x_\tau}$. This is a variation of the type

$L \longrightarrow L + \frac{\partial L}{\partial x_\tau}\alpha$. L is Lagrangian.

The second term in equation 32 corresponds to Euler-Lagrange equations; it is zero for a conservative system. The last term also has been shown zero; hence,

$I(x) = \iint L(\varphi, \varphi_v) dx., \varphi_v = \frac{\partial \varphi}{\partial x_v}$. And under variation

$I(x) = \iint L(\varphi, \varphi_v) dx + \alpha \iint \frac{\partial L}{\partial x_\tau} dx$ ----- (33).

Comparing equation 32 with equation 33, the following must be obtained: $[\frac{\partial L}{\partial x_\tau} = \frac{\partial}{\partial x_v}(\frac{\partial L}{(\partial \frac{\partial \varphi}{\partial x_v})}\cdot\frac{\partial \varphi}{\partial x_\tau})]$ ---------------------- (34).

This equation is a statement of the conservation laws.

For $\tau = 1, 2, 3$, it corresponds to conservation of momentum, where $\tau = 4$ is corresponding to conservation of energy.

To clear these points, consider $\tau = 4$.

1. *Conservation ofEenergy*, $x_4 = 4$

Start with the equation, $\frac{\partial L}{\partial t} = \frac{\partial}{\partial t}(\frac{\partial L}{\partial \varphi^{\cdot}}, \varphi^{\cdot}) + \sum_{k=1}^{3} \frac{\partial}{\partial x_k}(\frac{\partial L}{\partial(\frac{\partial \varphi}{\partial x_k})} \varphi^{\cdot})$.

or $\frac{\partial}{\partial t}[\frac{\partial L}{\partial \varphi^{\cdot}} \varphi^{\cdot} - L] + \sum_{k=1}^{3} \frac{\partial}{\partial x_k}(\frac{\partial L}{\partial(\frac{\partial \varphi}{\partial x_k})} \varphi^{\cdot}) = 0$.

From this and the fact that $\frac{\partial L}{\partial \varphi^{\cdot}} = \Pi$, $H = \pi \varphi^{\cdot} - L$, the continuity equation is gotten to be, $\frac{\partial H}{\partial t} + \nabla.S = 0$,-----------(35).

2. *Conservation of Momentum*

Let $\tau = 1$ in equation 34.
Then one obtains $\frac{\partial L}{\partial x_1} = \sum_{k=1}^{3} \frac{\partial}{\partial x_k}(\frac{\partial L}{\partial(\frac{\partial \varphi}{\partial x_k})} \frac{\partial \varphi}{\partial x_1}) + \frac{\partial}{\partial t}(\frac{\partial L}{\partial \varphi^{\cdot}} \frac{\partial \varphi}{\partial x_1})$ or $\frac{\partial}{\partial t}$

$(\frac{\partial L}{\partial \varphi^{\cdot}} \varphi_1) + \frac{\partial}{\partial x_1}(\frac{\partial L}{\partial \varphi_1} \varphi_1 - L) + \frac{\partial}{\partial x_2}(\frac{\partial L}{\partial \varphi_2} \varphi_1) + \frac{\partial}{\partial x_3}(\frac{\partial L}{\partial \varphi_3} \varphi_1) = 0$ --- (36).

Now define the momentum density as G, which is given by $P = \int G\, dx$, $G_k = \frac{\partial L}{\partial \varphi^{\cdot}} \varphi_k$, and P is the total momentum.

Also define the stress tensor as $T_{jk} = \frac{\partial L}{\partial \varphi_j} \varphi_k - L\, S_{ik}$, then equation 36 can be rewritten and generalized to $\tau = 1, 2, 3$ such as $\frac{\partial G_k}{\partial \tau} + \frac{\partial T_{jk}}{\partial x_j} = 0$ --------------------(37).

This equation represents the conservation of momentum.

Now equations 35 and 37 correspond to the conservation of the energy and the momentum respectively. They can be combined by expanding the stress tensor T_{jk} to 4×4 matrix $T_{\mu\nu}$ such as $T_{\mu\nu} = \frac{\partial L}{\partial \varphi_\mu} \varphi_\nu - L \delta_{\mu\nu}$, then

$$T_{44} = \frac{\partial L}{\partial \varphi_4} \varphi_4 - L = \frac{\partial L}{\partial \varphi^{\cdot}} \varphi^{\cdot} - L = \Pi\varphi^{\cdot} - L = H. \text{ Remember } \delta_{44} = 1.$$

And $T_{k4} = \frac{\partial L}{\partial \varphi_k} \varphi^{\cdot}$, $(k = 1, 2, 3.) = S_k$ (current density).

So $T_{k4} = S_k$, also $T_{4k} = \frac{\partial L}{\partial \varphi^{\cdot}} \varphi_k = G_k$ (momentum density).

Thus, $\frac{\partial H}{\partial \tau} = \frac{\partial T_{44}}{\partial \tau}$, and $\vec{j} \cdot \vec{S} = \frac{\partial T_{k4}}{\partial \tau}$.

Now, it can be claimed that $\sum_{\mu=1}^{4} \frac{\partial T_{\mu\nu}}{\partial x_\mu} = 0$, $\nu = 1, 2, 3, 4$.

It represent the conservation laws. To verify this claim, let $\nu = k$, then, $\sum_{\mu=1}^{4} \frac{\partial T_{\mu k}}{\partial x_\mu} = 0$. Or

$\sum_{j=1}^{3} \frac{\partial T_{jk}}{\partial x_j} + \frac{\partial T_{4k}}{\partial x_4} = \frac{\partial T_{jk}}{\partial x_j} + \frac{\partial G_k}{\partial \tau} = 0$. (Summation over j is understood; it is Einstein symbol for summation.) It is the conservation laws. Now equation 37 for $\nu = 4$, becomes

$\sum_{k=1}^{3} \frac{\partial T_{k4}}{\partial x_k} + \frac{\partial T_{44}}{\partial \tau} = \nabla \cdot S_k + \frac{\partial H}{\partial \tau} = 0$, which can be explicitly written for $T_{\mu\nu}$ such as

$T_{\mu\nu} = \begin{bmatrix} T_{jk} & G_i \\ S_k & H \end{bmatrix}$, in general $\frac{\partial T_{\mu\nu}}{\partial \nu}$ is written as $T_{\mu\nu/\nu} = \mathcal{K}_\mu = (\vec{k},$ $\ell) = (k_1, k_2, k_3, \ell)$. Where ℓ is the work density and

$\sum_\nu \frac{\partial T_{\mu\nu}}{\partial x_\mu} = 0$, $\nu = 1, 2, 3, 4$. For a general case becomes

$H^{\cdot} + \vec{\nabla} \cdot \vec{S} - \ell$, and $\vec{\nabla} \cdot \vec{T_m} + g^{\cdot} m = \mathcal{K}_m$, m=1, 2, 3.

However, for closed system, $T_{\mu\nu/\nu} = 0$, since $T_{\mu\nu}$ is symmetric.

Therefore, $[T_{\mu\nu}=T_{\nu\mu}$, and $S_k=G_k$]. It is the statement of the equivalence principle of Einstein. Define $T_{\mu\nu\rho}=x_\mu T_{\nu\rho}-x_{\mu\rho}=-T_{\nu\mu\rho}$. Then $[T\mu\nu\rho_{/\rho}=k_{\mu\nu}=x_\mu k_\nu -x_\nu k_\mu=k_{\nu\mu}]$. It is the statement of angular momentum conservation.

Noether Theorem: For a system of interacting fields, the work w is given by $w=1/z\int L(\psi_\alpha\partial_\mu\psi_\alpha)dx$-----------(38).

Now consider a transformation of the type $X_N \rightarrow X_{\bar{N}}$, $\psi_\alpha \rightarrow \psi_{\bar{\alpha}}$.

Assuming invariance of equation 38, so under this $=\int_R L(\psi_\alpha,\partial_N\psi_\alpha)dx$. If it is restricted to a continuous symmetry group (refer to any text book in group theory), there will be invariance with the full group if it is invariance under an infinitesimal transformation. Going further, the above transformation has to be defined as:

$X_N \rightarrow X_{\bar{N}} =X_N +\delta X_N$, $\psi_\alpha(x) \rightarrow \psi_{\bar{\alpha}}(x^-) = \psi_\alpha(x) +\delta\psi_\alpha(x)$.
Then $\delta w =\frac{1}{z^-}\int_{R^-} L(\psi_\alpha(x) +\delta\psi_\alpha(x),\partial_N\psi_\alpha+\partial_N\delta\psi_\alpha)dx-$

$\frac{1}{z}\int_R L(\psi_\alpha(x), \partial_N\psi_\alpha(x))dx$.

For the transformation from R to R^-, where $dx^- =\frac{dx^-}{dx}dx$ and the Jacobian for $\frac{dx^-}{dx}$ is given by

$$dx^-/dx \quad = \quad \begin{vmatrix} \frac{dx_1^-}{dx_1} & \frac{dx_1^-}{dx} -- & - \\ \frac{dx_2^-}{dx_1} & \frac{dx_2^-}{dx_2} -- & - \\ -- & - & - \\ ----------- & & \\ ----------------1+\frac{\partial\delta X_k}{\partial X_4} \end{vmatrix} \quad =N[1+\frac{\partial\delta X_N}{\partial X_N}] \, dx.$$

298

If only the first order terms are considered, then

$L(\psi_\alpha + \delta\psi_x, \partial_N\psi_X + \delta\partial_\mu\psi_x) =$

$\quad L(\psi_\alpha, \partial_n\psi_X) + \dfrac{\partial L}{\partial\psi_\alpha}\delta\psi_\alpha + \dfrac{\partial L}{\partial\,\partial_N\psi_\alpha}\delta_N\psi_\alpha$

If $L \neq L(x)$, L is not explicitly a function of x, then

$\partial_N((L\,\delta x_N) = L\dfrac{\partial\delta x_N}{\partial N}$; therefore:

$\int_R \sum \partial_N \{[(L \quad \partial_{N^2} \quad -\dfrac{\partial L}{\partial_N\,\partial_N\psi_\alpha}\partial_v\psi_\alpha)\delta x_v + \dfrac{\partial L}{\partial\,\partial_N\psi_\alpha}\delta\psi_\alpha] + [L]_\alpha(\delta\psi_\alpha -$

$\partial_N\psi_\alpha\delta x_v\} = 0$, where $[L]_\alpha = \dfrac{\delta L}{\delta\psi_\alpha} - \partial_N\dfrac{\partial L}{\partial\,\partial_N\psi_\alpha}$, and

$\partial_N L = \dfrac{\partial L}{\partial\psi_\alpha}\partial_N\psi_\alpha + \dfrac{\partial L}{\partial\,\partial_\rho\psi_\alpha}\partial_\rho\partial_N\psi_\alpha$. Now since $[L]_\alpha$ represents Euler-Lagrange equations, it follows that $[L]_\alpha = 0$; thus,

$\int_R \partial_N \{[L\delta_{Nv} - \dfrac{\partial L}{\partial_N\partial_\rho\psi_\alpha}\partial_v\psi_\alpha]\delta x_v + \dfrac{\partial L}{\partial\partial_N\psi_\alpha}\delta\psi_\alpha\} = 0$, let be F_N.

Then since the region R is arbitrary, it must lead to $\partial_N F_N = 0$.

It is the continuity equation on the four vector F_N and the differential form of a conservation law for the vector itself.

Suppose $\delta\psi_\alpha = 0$ (no change), and $\delta x_v \propto a$(certain parameter), then $F_N = L\delta_{Nv} - \dfrac{\partial L}{\partial\partial_N\psi_\alpha}\partial_v\psi_\alpha$, and $\partial_N F_N = 0$, gives,

$\dfrac{\partial L}{\partial x_v} = \dfrac{\partial}{\partial x_N}(\dfrac{\partial L}{\partial\,\partial_N\psi_\alpha}\delta_v\psi_\alpha)$, which is the result used to obtain the conservation laws and the energy-momentum tensor early.

10.5. Field Quantization in Analogy to That of Classical Mechanics

So far, the background in classical mechanics and the classical field theory in terms of Lagrangian formalism are briefly presented. Next is the field quantization to be treated thoroughly. To do this, the way should approach first the process of field quantization, noticing the existence analogies between the quantization of the classical mechanics and the classical field theory.

In classical mechanics, there are the canonical variables q_i and p_i and the dynamical equations of motion. After quantization, the solutions of the problem take on wave properties; while in the classical field theory, on the other hand, there are the canonical field functions $X_\alpha(r, t)$ and $\Pi(r, t)$. After quantization, the classical wave solution takes on the particle properties. In essence, q_i, p_i are classical point mechanics, while $\chi(r, t)$, $\Pi(r, t)$ are classical field theory.

In the classical point mechanics, the quantization conditions are $[p_{i(t)}, q_{j(t)}] = \hbar \delta_{ij} = \hbar$ if i=j and o if i≠j

And $[p_i, p_j] = [q_i, q_j] = 0$ $\qquad$ --------(39).

In similar way the quantization conditions for the classical fields are: $[\psi_\alpha(\vec{r}, t), \Pi_\alpha - (\vec{r}, t)]_\mp = z^- k \delta_{\alpha\alpha} - \delta(\vec{r} - \vec{r})$

And $[\psi_\alpha(\vec{r}, t), \psi_\alpha - (\vec{r}^-, t)]_\mp = [\Pi_\alpha(\vec{r}, t), \Pi_\alpha - (\vec{r}, t)]_\mp = 0$ --(40.)

When the commutator is indicated with (-) it is used for Bose fields (spins 1, 2, 3,----integer). And the anticommutator (+) for Fermi fields (half-integer fields, 1/2, 3/2 ----).

It is important to notice that the subscript α is referred to the field component, for example, the scalar field has only one component, while the vector field has three components, and the spinor field has two components, etc. Also, the scalar is a tensor of rank zero ($\lambda=0$); the vector is a tensor of rank one ($\lambda=1$), and the spinor is a tensor of rank 1/2($\lambda = 1/2$).

Hence, the components number is defined by ($2\lambda+1$). In addition to the commutation relation mentioned above, it is necessary to mention the commutation relation between the canonical field functions and the functions of these functions, the functional. Therefore, these commutation relations are:

300

Let I $=I(\chi_\alpha(r^\rightarrow), \Pi_\alpha(r^\rightarrow))$, then,

$$[p, \chi_\alpha] = -i\hbar \frac{\delta I}{\delta \Pi_\alpha}, \text{ and } [I, \Pi_\alpha] = -i\hbar \frac{\delta I}{\delta \chi_\alpha} \text{ --------------------(41)}.$$

This is analogous to what is actually done in the quantization of classical mechanics, for example, it is known that:

$[H, q_i] = -i\hbar \frac{\partial H(q_i, p_j)}{\partial p_j}$. However, because it is dealing with functional

in field theory, it is written that $\frac{\partial}{\partial p_j} \rightarrow \frac{\delta}{\delta \Pi_\alpha}$.

Now consider the commutator;

$[I\{\chi_\alpha(r^\rightarrow), \Pi_\alpha(r^\rightarrow)\}, \Pi_{\alpha^-}(r^-)] = i\hbar \frac{\delta I}{\delta \chi_{\alpha^-(r^{-\rightarrow})}}$, then the following

is found, $\frac{i}{\hbar}[I(\chi_\alpha(r^\rightarrow), \Pi_\alpha(r^\rightarrow)]\Pi_{\alpha^-}(r^{-\rightarrow}) + \delta_{\alpha\alpha^-} \frac{\delta I}{\delta \chi_{\alpha^-(r^{-\rightarrow})}} =$

$= \frac{i}{\hbar}[\Pi_{\alpha^-}(r^{-\rightarrow})] \times I[\chi_\alpha(r^\rightarrow), \Pi_\alpha(r^\rightarrow)]$.

Now the subscript α might be dropped out in the expression $\frac{\delta I[\chi_\alpha(r^\rightarrow), \Pi_\alpha(r^\rightarrow)]}{\delta \chi_\alpha(r^{-\rightarrow})}$, since the derivation is with respect to χ_α, then

$\frac{\delta I}{\delta \chi(r^{-\rightarrow})} = \delta I(\delta(r^\rightarrow - r^{-\rightarrow})$, as it has been shown before.

Thus, the commutator becomes as follows:

$\frac{i}{\hbar}[\Pi_{\alpha^-}(r^-\rightarrow), I(\chi_\alpha(r\rightarrow), \Pi_\alpha(r\rightarrow))] = I(\chi_\alpha, \Pi_{\alpha^-}) + \delta_{\alpha\alpha} - \delta I(\delta(r^\rightarrow - r^{-\rightarrow}))$, by using the commutation relations in equation 41. For the integral Hamiltonian $\mathcal{H}$, it is obtained that $[\mathcal{H}, \chi_\alpha] = -i\hbar \frac{\delta \mathcal{H}}{\delta \Pi_\alpha} = -i\hbar \dot\chi_\alpha$, and $[\mathcal{H}, \Pi_\alpha] =$

$i\hbar \frac{\delta \mathcal{H}}{\delta \chi_\alpha} = i\hbar \dot\Pi_\alpha \text{ --------------------(42)}.$

In general, for any operator H, which is a functional of χ_α and Π_α, their space derivatives are not explicitly depending on time. It can be written that $\frac{i}{\hbar}[\mathcal{H}, T] = \frac{dT}{dt}$, but in case it depends explicitly on time, then

$\frac{i}{\hbar}[\mathcal{H}, T] = \frac{dT}{dt} \cdot \frac{\partial T}{\partial t} => \frac{dT}{dt} = \frac{i}{\hbar}[\mathcal{H}, T] + \frac{\partial T}{\partial t}$. It is the dynamical equation of motion of the operator T. This equation is well known in classical mechanics in terms of Poison bracket { }, where in quantum mechanics the commutator [] is usually used. So if T is not explicitly time dependent, then $\frac{\partial T}{\partial t} = 0$; and if $[\mathcal{H}, T] = 0$, that means T commutes with the Hamiltonian $\mathcal{H}$, and $\frac{dT}{dt} = 0$. This implies that T is constant of the motion.

10.6. *Field Equation of Free Fields (Noncovariant Commutation Relations)*

In this section, the field equations for free fields will be discussed. Therefore, the equations of Schrödinger, Klein-Gordon, Dirac, and Maxwell to be discussed from the view point of the Lagrangian formalism are as follows:

(1.) Schrödinger Equation

Consider the scalar fields, where $\chi_\alpha \to \chi$, then

$i\chi\chi = -\frac{\hbar^2}{2m}D\chi + V(r) + \chi^*\chi$ ----------------- (43).

Now propose that the Lagrangian for this case is given by
$\mathcal{L} = i\hbar\chi^*\chi - (\nabla\chi^*, \nabla\chi) - V(r)\chi^*\chi$ ---------(44).

Here, $(\nabla\chi, \nabla\chi)$ is scalar or inner product.
In order to verify equation 44, the Euler-Lagrange relations,
$\delta\mathcal{L}/\delta\chi = \frac{d}{dt}\frac{\delta\mathcal{L}}{\delta\chi} $ or $\frac{\partial L}{\partial\chi} = \frac{d}{dt}\frac{\partial L}{\partial\chi} + \frac{\partial}{\partial x_\nu}\frac{\partial L}{\partial\chi_\nu}$, should be used.

This yields

$$-V(r)\chi^* - \{\frac{d}{dt}(i\hbar)\chi^* - \frac{\hbar^2}{2m}\frac{\partial}{\partial x_v}[\frac{\partial}{\partial \chi_v}(\chi_v^*, \chi)]\} = 0, \text{ or}$$

$$-V(r)\chi^* - i\hbar\dot\chi^* + \frac{\hbar^2}{2m}D\,\chi^* = 0.$$

Also, the following are true:

$$\frac{\partial L}{\partial \dot\chi} = \Pi = i\hbar\chi^*, \text{ then } \chi^* = \frac{\Pi}{i\hbar} \text{ and } \frac{i}{\hbar}[\chi(r,t), \Pi(r^-, t)] =$$

$$i\hbar\delta(r^{\rightarrow} - r^{-\rightarrow}), \text{ which becomes } [\chi(r,t), \chi^*(r^-, t)] = \delta(\text{r-r}^-).$$

Thus, taking the differential Hamiltonian (density), then

$$H = \Pi\dot\chi - L = i\hbar\chi^*\dot\chi - i\hbar\chi^*\dot\chi + \frac{i\hbar}{2m}(\nabla\Pi, \nabla\chi) - \frac{i}{\hbar}V\Pi\chi$$

$$= -\frac{\hbar^2}{2m}(\nabla\chi^*, \nabla\chi) + V\,\chi^*\chi,. \text{ So the integral Hamiltonian (total) is}$$

given by $\mathcal{H} = \int H d\tau = \int[-\frac{\hbar^2}{2m}(\nabla\chi^*, \nabla\chi) + V(r)\chi^*\chi]d\tau.$

In equation 44, L is not Hermitian due to the term $i\hbar\chi^*\dot\chi$.

But it can be made Hermitian by letting $i\hbar\chi^*\dot\chi \rightarrow \frac{i\hbar}{2}(\chi^*\dot\chi - \dot\chi\chi^*)$

Thus, $L = \frac{i\hbar}{2}(\chi^*\dot\chi - \dot\chi\chi^*)$. Then, $\frac{\hbar}{2i}\int_{\tau_1}^{\tau_2}\chi^*\dot\chi - \dot\chi\chi^* d\tau =$

$\frac{\hbar}{2i}\ [\int_{\tau_1}^{\tau_2}\chi^*\dot\chi d\tau\ -\int_{\tau_1}^{\tau_2}\dot\chi\chi^*\,d\tau\ ], \quad$ but $\quad \frac{\hbar}{2i}\int_{\tau_1}^{\tau_2}\dot\chi\chi^*d\tau = \chi\chi^*\ -$

$\int_{\tau_1}^{\tau_2}\chi\dot\chi^*d\tau$. Thus,

$\frac{\hbar}{2i}\int_{\tau_1}^{\tau_2}(\chi^*\dot\chi - \dot\chi\chi^*)d\tau = \frac{\hbar}{2i}\chi\chi^* + \frac{\hbar}{2i}\int_{\tau_1}^{\tau_2}(\chi^*\dot\chi + \chi\dot\chi^*)d\tau$.

(2.) Klein-Gordon Equation

Take the Klein-Gordon equation:

$$\Box\varphi(x) = k^2\varphi(x), \quad k = \frac{mc^2}{\hbar} \text{---------------- (45)}$$

And the Lagrangian is:

$$L = - (1/2) \partial_\mu \varphi \partial_\mu \varphi - (1/2) k^2 \varphi^2 \text{-- --------- (46)}$$

Also, it was found that $\Pi = \dfrac{\delta}{\delta \phi} = \dot{\phi}$ hence, the differential Hamiltonian H is given by:

$$H = \Pi \dot{\varphi} - L =$$
$$\dot{\varphi}^2 - \frac{1}{2} \partial_\mu \varphi \partial_\mu \varphi + \frac{1}{2} k^2 \varphi^2 = \dot{\varphi}^2 - \frac{1}{2} (\partial_\mu \varphi \partial_\mu \varphi + k^2 \varphi^2).$$

And $[\varphi, \Pi] = i\hbar \delta ([\vec{r} - \vec{r}^{\,-}) \rightarrow [\varphi(r,t), \frac{\partial \varphi}{\partial t}] = i\hbar \delta(\text{r-r}).$

(3.) Dirac Equation

First of all, let it be shown how the Klein-Gordon equation is arrived at. As pointed out in chapter nine, the relativistic energy equation is given by

$$E^2 = p^2 c^2 + m_0^2 c^4 \text{--------------------------- (47)}.$$

Where p is the momentum, c is the light velocity, and m_0 is the mass at rest. Also, the energy and the momentum operators are

$$E \rightarrow i\hbar \frac{\partial}{\partial t}, p \rightarrow -i\hbar \frac{\partial}{\partial x} \text{--------------- (48)}.$$

By substituting equation 48 into equation 47, the following are obtained:

$$-\hbar^2 \frac{\partial^2 \chi}{\partial t^2} = - c^2 \hbar^2 \nabla \chi + m^2 c^4 \equiv \Box \chi - k^2 \chi = 0 \text{------ (49)}.$$
$$\Box = \frac{1}{c^2} \frac{\partial^2}{\partial t^2} - \nabla. \text{ Let } m_0 \equiv m$$

It is equation 45, Klein-Gordon equation.

Alternatively, write the relativistic energy as

$$\left(\frac{E}{c}\right)^2 - p_x^2 - p_y^2 - p_z^2 - m^2 c^2 = 0, \text{------------------------ (50)}.$$

The wave equation (Schrödinger equation) is

$$\{ \left(\frac{H}{c}\right)^2 - p_x^2 - p_y^2 - p_z^2 - m^2 c^2 \}\chi = K\chi = 0 \text{----------------- (51)}.$$

This can be reduced to the form (chapter nine)

$$\{\frac{H}{c} - \alpha_1 p_x - \alpha_2 p_y - \alpha_3 p_z - \alpha_4 m^2 c^2 \} \chi = 0, \text{-------- (52)}.$$

With the requirement that $(\alpha\alpha\frac{H}{c} - \alpha_1 p_x - \alpha_2 p_y - \alpha_3 p_z - \alpha_4 m^2 c^2)(\frac{H}{c} + \alpha_1 p_x + \alpha_2 p_y + \alpha_3 p_z + \alpha_4 mc) = K$.

To be this true, the following relations between the α^s should be satisfied:

$$\left.\begin{array}{l} \alpha_1^2 = \alpha_2^2 = \alpha_3^2 = \alpha_4^2 \\ \alpha_i \alpha_j = -\alpha_j \alpha_i, \, i \neq j \\ \alpha_i \alpha_j + \alpha_j \alpha_i = 2\delta_{ij} \end{array}\right\} \text{---------------------------- (53)}.$$

These relations can be satisfied by choosing that:

$$\vec{\alpha} = \begin{bmatrix} \alpha_1 \\ \alpha_2 \\ \alpha_3 \end{bmatrix} = \begin{bmatrix} 0 & \vec{\sigma} \\ \vec{\sigma} & 0 \end{bmatrix} ; \quad \alpha_4 = \beta = \begin{pmatrix} 1 & 0 \\ 0 & -1 \end{pmatrix}$$

(See chapter nine) $\vec{\sigma}$ are Pauli matrices.

Now the solution to Dirac equation 52 is of the form:

$$\chi(x, y, z, t, s) = \mathcal{U}(s)\exp.\frac{i}{\hbar}[p_x x + p_y y + p_z z - Et] \text{------ (54)}.$$

Where $[\frac{E}{c} - \vec{\alpha}.\vec{p} - \beta mc]\mathcal{U}(s) = 0. \text{----------- (55)}$.

This equation has four solutions, so in unit of c =1, it reduces to
$[\vec{\alpha}.\vec{p}+ \beta m]\mathcal{U}(s)= E\mathcal{U}(s)$ --------------(56).

Let $\Pi = p_x +i\, p_y$, then it is possible to write

$$\vec{\alpha} \cdot \vec{p} = \begin{pmatrix} 0 & 0 & p_z & \Pi^* \\ 0 & 0 & \Pi & -p_z \\ p_z & \Pi^* & 0 & 0 \\ \Pi & -p_z & 0 & 0 \end{pmatrix} \quad \text{And } \mathcal{U}(s) = \begin{pmatrix} \mathcal{U}(1) \\ \mathcal{U}(2) \\ \mathcal{U}(3) \\ \mathcal{U}(4) \end{pmatrix}$$

Equation 56 may be written as $(E-\beta m)\ \mathcal{U}(s)=$
$(\vec{\alpha} \cdot \vec{p})\mathcal{U}(s).$Or

$$(E\text{-m}) \begin{pmatrix} 1 & 0 & 0 & 0 \\ 0 & 1 & 0 & 0 \\ 0 & 0 & -1 & 0 \\ 0 & 0 & 0 & -1 \end{pmatrix} =(\vec{\alpha}\vec{p})\mathcal{U}(s)$$

This yields the following four questions:

$(E\text{-m})\mathcal{U}(1) = p_z\mathcal{U}(3) +\Pi^*\mathcal{U}(4) \left.\vphantom{\begin{matrix}1\\1\end{matrix}}\right\}$
$(E\text{-m})\mathcal{U}(2)=\Pi\mathcal{U}(3) -p_z\mathcal{U}(4) \quad$ -------- (57a).

$(E\text{-m})\mathcal{U}(3)=p_z\mathcal{U}(1) +\Pi^*\mathcal{U}(2) \left.\vphantom{\begin{matrix}1\\1\end{matrix}}\right\}$
$(E +\text{m})\mathcal{U}(4)=\Pi\mathcal{U}(1) -p_z\mathcal{U}(2) \quad$ ----------- (57b).

The equations 57a correspond to the energy E<0 (negative).
But the equation 57b corresponds to positive energy E>0.
Therefore, there are four independent solutions. These can be
summarized in table 1 below

Table 1 Summary of Dirac equation solutions.

E>0, Positive   E<0, Negative

Solution	$\mathcal{U}^{(1)}(s)$	$\mathcal{U}^{(2)}(s)$	$\mathcal{U}^{(3)}(s)$	$\mathcal{U}^{(4)}(s)$	
1	1	0	$-p_z$	$\dfrac{\Pi^*}{\omega}$	where, ω
2	0	1	$-\dfrac{\Pi}{\omega}$	p_z/ω	= E/c+ mc.
3	p_z/ω	$\dfrac{\Pi^*}{\omega}$	1	0	and $\dfrac{1}{\sqrt{1+p^2c^2}}$ is
4	Π/ω	$-p_z/\omega$	0	1	normal. factor.

Now, $\chi' = \frac{i}{\hbar}[H, \chi] = c\alpha$, then $\frac{<\chi'>}{c} = <\chi|\alpha|\chi>$?

Take $\chi = \mathcal{U}^{(1)}(s) = \begin{pmatrix} 1 \\ 0 \\ \frac{p_z}{\omega} \\ \frac{\Pi}{\omega} \end{pmatrix}$ and $\alpha = \begin{pmatrix} 0 & 0 & 0 & 1 \\ 0 & 0 & 1 & 0 \\ 0 & 1 & 0 & 0 \\ 1 & 0 & 0 & 0 \end{pmatrix}$

Hence $\alpha\chi = \begin{pmatrix} \frac{\Pi}{\omega} \\ \frac{p_z}{\omega} \\ 0 \\ 1 \end{pmatrix}$, which yields $<\chi | \alpha_1 | \chi> =$

$(1\ 0\ p_z/\omega\ \frac{\Pi^*}{\omega})\ [\uparrow\ ] = \frac{\Pi}{\omega} + \frac{\Pi^*}{\omega} = \frac{\Pi + \Pi^*}{\omega} = \frac{p_x + ip_{y+}\ p_x - ip_y}{\omega} = \frac{2p_x}{\omega}.$

So, $<\chi | \alpha 1 | \chi> = \frac{2p_x}{\omega} => <\chi^{\cdot}> /c = 2p_x /\omega = \frac{2p_x}{\frac{E}{c} + mc} =>$

$<\chi^{\cdot}> = \frac{2p_x c^2}{E + mc^2} = \frac{c\,p_x}{\sqrt{p^2 + m^2\,c^2}} = \frac{mV_x^-}{mc} = \frac{V_x^-}{c}$, now $<\chi\ \alpha_1, \chi \geq 1 =>$

$V_x^{2-} = c^2 => V_x^- = \pm\ c$. It is a vibration motion of the electron.
Now consider different forms of Dirac equation, taking k=c=1 as follows:

$\alpha^{\rightarrow} \cdot\ p^{\rightarrow} + \beta m$. It is already known. This can be written as:

$(i\not{\nabla} - m)\ \chi =0, \nabla-$ 4-dimentional operator, --------- (58)
or $(\not{p} - m)\chi = 0$, p -4 dimensional momentum operator (59).
Sometimes, $(\gamma_N \partial_N +k)\chi = 0$ is used; it is a best form -- (60).

These equations are in general equivalent. For example, write, $i\not{p} = i\gamma_N\gamma_N = i\gamma_0\gamma_0 - i\gamma_k\partial_k$, where $\gamma_0 = \beta$ and $\gamma_k = \beta\alpha_k$. Then equation 58 becomes:

$(i\beta^2\alpha_0 - i\beta^2\alpha_k\partial_k - m\beta)\chi = 0,$ ------------------ (61).

But $i\alpha_0 = E$ and $i\partial_k = i\not{p}_k$; thus, equation 58 becomes

$(E - \alpha_k\not{p}_k - m\beta)\chi = 0$---------------------------- (62).

In a similar way, the other equations can be shown to be equivalent too. Next, the Lagrangian and the canonical variables will be considered for Dirac equation.

$(\gamma_N \partial_N + k)\chi = 0$, if $L = -\chi^-(\gamma_k \partial_k + k)\chi =$
$-\chi^-(\gamma_k \partial_k - \gamma_0 \alpha_0 + k)\chi$, then,

$\Pi = \frac{\partial L}{\partial \dot{\chi}} = \chi^- \gamma_0 = \chi^*$; $[\gamma_0 = \frac{\beta}{i} = -i\beta]$.

Thus, the Hamiltonian takes the following form:

$H = \chi^* \dot{\chi} + \chi^-(\gamma_k \partial_k - \gamma_0 \partial_0 + k)\chi$
$= \chi^- \gamma_0 \dot{\chi} + \chi^- \gamma_k \partial_k \chi - \chi^- \gamma_0 \dot{\chi} + \chi^- k\chi = \chi^-(\gamma_k \partial_k + k)\chi$, let $\chi = \psi$,
for convenience, then $H = \psi^-(\gamma_k \partial_k + k)\psi$ ---- (63).

The commutation relations are given by

$\left.\begin{array}{l} [\psi_\alpha, \Pi_\alpha -] = [\psi_\alpha(\vec{r},t), \psi_\alpha^{+-}(\vec{r}^{--},t)] = \delta_{\alpha\alpha} - \delta(\vec{r}^{-} - \vec{r}^{--}) \\ [\psi_\alpha, \psi_\alpha -] = [\psi_\alpha^*, \psi_\alpha^+ -] = 0 \text{-----------------------------} \end{array}\right\}$ ----(64).

Maxwell Equation:

Maxwell equations are considered the crown of the development of the electromagnetic theory during the second half of the nineteen century. They thought are the keys to answer any natural problems rise till 1900, where the quantum theory was discovered by the black body radiation experiment by Plank. These equations are:

$\left.\begin{array}{l} \vec{\nabla} \times \vec{H} = \frac{\partial \vec{E}}{\partial t}, \vec{\nabla} \cdot \vec{E} = 0 \\ \vec{\nabla} \times \vec{E} = -\frac{\partial \vec{H}}{\partial t}, \vec{\nabla} \cdot \vec{H} = 0 \end{array}\right\}$ ------ ----------- (66).

Their solutions are of the type

$\vec{E}(\vec{r},t) = \vec{E}(r)e^{-ikt}$, $(c = \hbar = 1)$ -------------------(67).

Where $\vec{\nabla} \times \vec{H} = -i\ k\ \vec{E}$; $\vec{\nabla} \times \vec{E} = i\ k\ \vec{H}$, $\vec{\nabla} \cdot \vec{E} = 0$, $\vec{\nabla} \cdot \vec{H} = 0$, $\vec{E}$ and $\vec{H}$ represent the electric and the magnetic fields' intensity respectively. Now for convenience to the quantization of these fields, let $H^{\rightarrow}$ be expressed in terms of a vector potential field $A^{\rightarrow}$, then $H^{\rightarrow}=\Delta^{\rightarrow} \times A^{\rightarrow}$,

$\Delta^{\rightarrow} \times E^{\rightarrow}=ik\nabla^{\rightarrow} \times A^{\rightarrow}$, and $\Delta^{\rightarrow} \times (E^{\rightarrow}- ikA^{\rightarrow})=0$, or finally,

$E^{\rightarrow} =ikA^{\rightarrow}+\nabla\phi, H^{\rightarrow}=\Delta^{\rightarrow} \times A^{\rightarrow}$. Where $A_N =(A^{\rightarrow},i\phi)$.

If the coulomb gauge is considered then $\phi =0$, but $A_N \neq 0$. So in applying this gauge, then $E^{\rightarrow} = ikA^{\rightarrow}$; $H =\Delta \times A^{\rightarrow}$.

So $\Delta \times(\Delta \times E^{\rightarrow})$

$=ik\Delta \times (\Delta \times A^{\rightarrow})=ik\Delta \times H^{\rightarrow}=ik\ E^{\rightarrow\cdot}=(ik)(-ik)E^{\rightarrow}=$

$k^2 E^{\rightarrow}+ \Delta E^{\rightarrow}$ or $(\Delta +k^2)E^{\rightarrow} =0$,it can be written in terms of the Dalembertian operator ($\square$ $^2 =\Delta +k^2$)as,

$\square^2 E =0$, $\square^2 H=0$, $\square^2 A^{\rightarrow}=0$, $\square^2\phi =0$.

Using $E^{\rightarrow}=ikA^{\rightarrow}$, then, $\Delta \cdot E^{\rightarrow}$ becomes $\nabla^{\rightarrow} \cdot E^{\rightarrow}=\nabla^{\rightarrow} \cdot A^{\rightarrow}=0$. Or $\partial_n A_N =0$.

The Lagrangian for Maxwell equations can be written in terms of the vector potential A_N, as, $L=-\frac{1}{2}\partial_N A_\nu \partial_N A_\nu$

$=-\frac{1}{4}F_{N\nu}F_{N\nu}$. Where $F_N\nu=\partial_\nu A_N - \partial_N A_\nu=F_{\nu N}$.

So, Maxwell equations can be written in terms of$F_{N\nu}$ as $\partial_\nu F_{N\nu}=0$, it represents $\nabla^{\rightarrow} \cdot E^{\rightarrow} =0$ for N=4, and $\nabla^{\rightarrow} \times H^{\rightarrow}= \frac{\partial E^{\rightarrow}}{\partial t}$, for N=1, 2, 3. Now it can be checked for N=4 to show that $\partial_\nu F_{4\nu}=0$. Using the whole previous information, this can be written such as $\partial_1 F_{41} + \partial_2 F_{42} + \partial_3 F_{43} +\partial_4 F_{44} =\partial_1[\partial_1 A_4- \partial_4 A_1]+ \partial_2[\partial_2 A_4 - \partial_4 A_2]+ \partial_3[\partial_3 A_4 - \partial_4 A_3]+$

$\partial_4\{ [\partial_4 A_4 - \partial_4 A_4]=0\}=\nabla^2 A_4 -\frac{\partial}{\partial}\nabla^{\rightarrow} \cdot A^{\rightarrow}=\nabla^2\rho - \nabla \cdot \frac{\partial A^{\rightarrow}}{\partial t} = 0 - \nabla \cdot E^{\rightarrow}=-\nabla E^{\rightarrow} = 0$. Thus, $\partial_\nu F_{4\nu} = 0$, as required.

The Maxwell equations $\nabla \cdot \vec{H}$, $\vec{\nabla} \times \vec{E} = -\frac{\partial \vec{H}}{\partial t} = -\vec{H}$ can be written in terms of $F_{N\nu}$ as, $\partial_\lambda F_{N\nu} + \partial_N F_{\nu\lambda} + \partial_\nu F_{\lambda N} = 0$

For example, if $\lambda = 1, N = 2, \nu = 3$, it will be

$\partial_1 F_{23} + \partial_2 F_{31} + \partial_3 F_{12} = 0$, or $\partial_1 [\partial_3 A_2 - \partial_2 A_3]$
$+\partial_2 [\partial_1 A_3 - \partial_3 A_1] + \partial_3 [\partial_2 A_1 - \partial_1 A_2] = -\vec{\nabla} \cdot \times \vec{\nabla} \vec{A} = \vec{\nabla} \cdot \vec{H}$
$= 0$, as required.

The Lagrangian can be written as:

$L = -\frac{1}{2}[\partial_k A_N \partial_k A_N - \partial_4 A_N \partial_4 A_N]$, where $\Pi_N = \frac{\partial L}{\partial A_N} = \partial_4 A_N = A_N$,
thus the Hamiltonian is given by,

$H = \Pi_\nu A_\nu - L = \frac{1}{2}\sum \Pi_\nu^2 + \frac{1}{2}\sum_k (\frac{\partial A_\nu}{\partial H_k})^2$, and the commutation relations are (t=t), $[\psi_\alpha (r, t), \Pi_{\alpha^-} (r^-, t)] = i\hbar(\delta_{\alpha\alpha^-}) \delta(r\text{-}r\text{-})$; and $[\psi_\alpha (r, t), \psi_{\alpha^-}(r^-, t)] = 0$.

Also, these commutation relations between $\vec{A}$ and Π are $[A_\nu$ (r, t), $\Pi_N (r^-, t)] = [A_\nu (r, t), \frac{\partial \Pi_\nu}{\partial t}] = i\hbar \delta_{N\nu} \delta(r\text{-}r\text{-})$, and $[A_N (r, t), A_\nu (r^-, t)] = [\Pi_N, \Pi_\nu] = 0$

Note: R and r⁻ are vectors.

10.7. *Interacting Fields; Field Equations*

The next step in the development of the noncovariant commutation relations is to consider the interacting microsystems, where the force usually is represented by its field. First, consider pion-nucleon interaction.

1) $\pi - N$ System,

The total Lagrangian of this system is given by:

$$L = L_{Dirac} + L_{Meson} + L_{Int} = [-\bar{\psi}(\gamma_N \partial_N + k)\psi - \frac{1}{2}\partial_N \varphi_i \partial_N \varphi_i]$$
$$-\frac{1}{2}k^2 \varphi_i \varphi_i + iG\bar{\psi}\gamma_5 \mathcal{T}^{\rightarrow}\psi\varphi.$$

Where G is the coupling constant, then

$$\frac{\partial L}{\partial \psi^-} - \frac{d}{dt}\frac{\partial L}{\partial \psi^{-\cdot}} \rightarrow -(\gamma_N \partial_N + k)\,\psi + iG\,\gamma_5 \mathcal{T}^{\rightarrow}\psi + \partial_N \varphi = 0$$

And $\frac{\delta L}{\delta \varphi} - \frac{d}{dt}\frac{\delta L}{\delta \psi^{-\cdot}} = 0$, which yields

$-K^2 \varphi + iG\bar{\psi}\gamma_5 \mathcal{T}^{\rightarrow}\psi + \partial_N \varphi_v = 0$, or,

$(\partial_N - k^2)\varphi = -iG\bar{\psi}\gamma_5 \mathcal{T}^{\rightarrow}\psi$. It is the field equation of the interacting pion-nucleon.

10.8. *The Noether Theorem:-*

It is well known that this theorem is concerned with the laws of conservation in physics. Its statement is if the physical system is independent of the change in its position coordinates, the momentum is conserved. And if it is independent of time, its energy is conserved. If it is independent of the rotational angle, its angular momentum is conserved. These facts are already mentioned in the previous section.

In this section, the derivation of this theorem will be verified. The derivation of this theorem will help in calculating the constants of the motion of the particular dynamical system that physicist are interested in. The equation will be performed using Minkousky space, with the use of Einstein convention. Let the system of interesting field be represented by integral action given by

$W = \frac{1}{i} \int_R L(\psi_\alpha, \partial_\mu \psi_\alpha) dx$ -------------------------(a).

Also, let a finite transformation of the following form be taking place, $\chi_\mu \rightarrow \chi_{\bar{\mu}}$, (-) prime notation.

$\psi_\alpha(x) \rightarrow \psi_{\bar{\alpha}}(x^-)$ (b).

Consider the system W remains invariant under this transformation, i.e., the field equation remains covariant under the transformation of equation b. Therefore, the condition W in equation a must stay unaltered with respect to equation b. Hence,

$\int_R - L(\psi_{\bar{\alpha}}(x^-), \partial_\mu \psi_{\bar{\alpha}}(x^-)) dx^- = \int_R L(\psi_\alpha(x) \partial_\mu \psi_\alpha(x)) dx$ ---(c).

The class of transformation (b) under it, W is invariant which leads to equation c. It is called the symmetry group of the system.

In the following, the treatment will be restricted to the continuous symmetry group. Invarience under the full group takes place if invariance under particular infinitesimal transformation that belongs to the group is determined. Now suppose the following transformation:

$\left.\begin{array}{l} x_\mu \longrightarrow x_{\bar{\mu}} = x_\mu + \delta x_\mu, \\ \psi_\alpha(x) \longrightarrow \psi_{\bar{\alpha}}(x^-) = \psi_\alpha(x) + \delta\psi_\alpha(x). \end{array}\right\}$ ------------ (d).

Taking c with the invariance under d, it implies that:

$\delta W = \frac{1}{i} \int_R - \{(\psi_\alpha(x) + \delta\psi_\alpha(x), \partial_\mu \psi_\alpha(x)) + \partial_\mu \delta\psi_\alpha(x)\} dx^- -$

$\frac{1}{i} \int_R L(\psi_\alpha(x), \partial_\mu \psi_\alpha(x) dx = 0$ ----------------------e).

Now the step is to transfer the integral over the region R^- into an integral over the region R. For this to be done, $dx^- = \frac{\partial x^-}{\partial x} dx$ has to be calculated, where $\frac{\partial x^-}{\partial x}$ is the Jacobian of the transformation.

$$\frac{\partial x^-}{\partial x} = \begin{vmatrix} \dfrac{\partial x_1^-}{\partial x_1} \dfrac{\partial x_2^-}{\partial x_2}, & \text{---} & \text{---} & \text{---} & \text{---} \\ \dfrac{\partial x_2^-}{\partial x_2} \text{------} & & & \text{---} \\ \text{--------} & & & \dfrac{\partial x_4^-}{\partial x_4} \end{vmatrix} = \begin{vmatrix} 1 + \dfrac{\partial \delta x_1}{\partial x_1} \dfrac{\partial \delta x_2}{\partial x_2} & \text{---} & \text{---} \\ \dfrac{\partial \delta_2}{\partial x_1} \text{---} & & \text{---} \\ \text{----} & & \dfrac{\partial \delta x_4^-}{\partial x_4} \end{vmatrix}$$

Now by expanding and only keeping the first order terms, it is obtained that $\frac{\partial x^{(-)}}{\partial x} = 1 + \frac{\partial \delta x_\mu}{\partial x_\mu}$ which yields that

$$dx^{(-)} = (1 + \frac{\partial \delta x_\mu}{\partial x_\mu})dx \text{ ----------------------------(f).}$$

Now the integrand of the first integral in equation e will be expanded in a Tylor series, keeping the first order terms only. The following is obtained:

$L(\psi_\alpha + \delta\psi_\alpha, \partial_\mu\psi_\alpha + \partial_\mu\delta\psi_\alpha) =$

$L(\psi_\alpha, \partial_\mu\psi_\alpha) + \frac{\partial L}{\partial \psi_\alpha} \delta\psi_\alpha + \frac{\partial L}{\partial\partial_\mu\psi_\alpha} \delta\partial_\mu\psi_\alpha \text{------------(g).}$

Next, by substituting equations f and g into equation e, with the use of $L\frac{\partial \delta x_\mu}{\partial x_\mu} = \frac{\partial}{\partial x_\mu}(L\delta x_\mu)$, L does not depend up onx_μ explicitly. So after integration by parts and reordering of terms, it is obtained:

$\int_R \{\partial_\mu[L\delta_{\mu\nu} - \frac{\partial L}{\partial\partial_\mu\psi_\alpha}]\partial_\nu\psi_\alpha \ \delta x_\nu + [\frac{\partial L}{\partial\partial_\mu\psi_\alpha} \delta\psi_\alpha] + [L]_\alpha(\delta\psi_\alpha - \partial_\nu\psi_\alpha\delta x_\nu)\}dx = 0 \text{ ---------------------------------(h).}$

Where $[L]_\alpha = \frac{\partial L}{\partial \psi_\alpha} - \partial_\mu \frac{\partial L}{\partial \partial_\mu \psi_\alpha}$, and $\partial_\mu L$ is as,

$\partial_\mu L \equiv \frac{\partial L}{\partial \psi_\alpha} \partial_\mu \psi_\alpha + \frac{\partial L}{\partial \partial_\rho \psi_\alpha} \partial_\rho \partial_\mu \psi_\alpha$, the field equation (Euler-Lagrange equations) are $[L]_\alpha = 0$. Then the region R in equation h is chosen. It follows that, $\partial_\mu \delta F_\mu = 0$ -------- (i),

where $[\delta F_\mu = (L\delta_{\mu v} - \frac{\partial L}{\partial \partial_\mu \psi_\alpha} \partial_v \psi_\alpha)\delta x_v + \frac{\partial L}{\partial \partial_\mu \psi_\alpha} \delta \psi_\alpha]$.

Equation i is, in fact, the continuity equation for the four vectors δF. After integrating equation i over infinitely large three-dimensional special region, using the Gauss theorem, and the fact that the field functions and their derivatives vanish at infinity, the conservation law then is given by

$$\frac{d}{dt}\frac{1}{i}\delta F_4 \equiv \frac{d}{dt}\frac{1}{i}\int [\,(L\delta_{4v} - \frac{\partial L}{\partial \partial_4 \psi_\alpha} \partial_v \psi_\alpha)\delta x_v + \frac{\partial L}{\partial \partial_4 \psi_\alpha} \delta \psi_\alpha]dx - (j).$$

But from equation i the integration equation j is a constant of the motion. It is the Noether theorem. So it is proved as required.

A particular continuous symmetry group induces then a conservation law for a definite physical quantity δF_4. So $\frac{1}{i}\delta F_4$ is real. From this theorem, the conservation laws of momentum, energy, angular momentum, charge and fermions, etc., are gotten. Take conservation of charge as an example. From the electromagnetism, the field equations are

$A_\mu = 0$, ---------------------------- (1).

$\partial_v A_v = 0$, --- ----------------------------- (2)

A is a potential vector field.

With the use of invariant under gauge transformation of the second kind, i.e., $A_\mu \rightarrow A_{\bar{\mu}} = A_\mu + \delta_\mu \mathcal{E}(x).\Big\}$ --------- (3).

$\Box\ \mathcal{E}(x) = 0$

Where $\mathcal{E}(x)$ is a particular scalar function with the condition that, $\Box(x) = 0$. It is a necessary condition for the Lorentz convention to remain invariant under the gauge transformation (3). This invariance should be retained even if A_μ interacts with another fields. Consider Dirac field Ψ is interacting with the potential vector field A_μ. So in Lagrangian, L will replace ∂_μ by, $\partial_\mu - ie\ A_\mu = D_\mu$ and $\partial_\mu + ieA_\mu = D_\mu^*$. Applying equation 3, the following are obtained: $D_\mu\Psi \rightarrow (\partial_\mu - ieA_\mu - ie\partial_\mu\mathcal{E})\Psi$ --------- ---------(4a)

$D_\mu^*\Psi^* \rightarrow (\partial_\mu + ie\ A_\mu + ie\partial_\mu\mathcal{E})\Psi^*$------- ---------- (4b)

The free electromagnetic Lagrangian density is given by

$L_{em} = -\frac{1}{2}\partial_\mu A_\nu \partial_\mu A_\nu = -\frac{1}{4}G_{\mu\nu}G_{\mu\nu}$ ----------------- (5).

(See Appendix C.) Where $G_{\mu\nu} = \partial_\mu A_\nu - \partial_\nu A_\mu$ remains invariant under equation 3. But the total L is given by

$L = -\frac{1}{4}G_{\mu\nu}G_{\mu\nu} - \Psi^*(\gamma_\mu\partial_\mu + m)\Psi + L_{int}$

$= -\ \frac{1}{4}G_{\mu\nu}G_{\mu\nu} - \Psi^*(\gamma_\mu D_\mu + m)\Psi$----------- (6).

It appears that the terms $(-ie\partial_\mu\mathcal{E})\Psi$ and $(+ie\partial_\mu\mathcal{E})\Psi^*$ are not wanted. To get rid of them, it is necessary to subject equations 1 and 2 to a simultaneously transformation

$\Psi_\alpha \rightarrow e^{ie\mathcal{E}(x)}\Psi$, and $\Psi_\alpha^* \rightarrow e^{-ie\mathcal{E}(x)}\Psi_\alpha^*$------------ (7).

It is called first kind of gauge transformation, and that of equation 3 is called the second gauge transformation.

Therefore, it can be written as

$$
\left.
\begin{aligned}
D_\mu \Psi &\rightarrow (\partial_\mu - ieA_\mu)\Psi e^{ie\mathcal{E}(x)} \\
D^{(*)}\Psi^* &\rightarrow (\partial_\mu + ieA_\mu)\Psi^* e^{-ie\mathcal{E}(x)}
\end{aligned}
\right\} \quad \text{-------------(8)}.
$$

Ψ^* and Ψ appear in the original free electromagnetic density only in bilinear forms, as the terms $e^{\pm ie\mathcal{E}(x)}$ in equation 6. The total L (see equation 6) of the coupling fields is invariant against gauge transformation of the first kind and the second kind. The infinitesimal transformation of this gauge group is given as

$$
\left.
\begin{aligned}
A_\mu &\rightarrow A_\mu^- = A_\mu + \partial_\mu \mathcal{E}(x) \\
\Psi_\alpha &\rightarrow \Psi_\alpha^- = \Psi_\alpha + ie\mathcal{E}(x)\Psi_\alpha \\
\Psi_\alpha^* &\rightarrow \Psi_\alpha^{-*} = \Psi_\alpha^* - ie\mathcal{E}(x)\Psi_\alpha^*
\end{aligned}
\right\} \quad \text{---------(9)}.
$$

Where $\mathcal{E}(x)$ is a particular infinitesimal function of x. So the general infinitesimal changes in the field functions are

$$
\left.
\begin{aligned}
\delta A_\mu &= \partial_\mu \mathcal{E}(x) \\
\delta \Psi_\alpha &= ie\mathcal{E}(x)\Psi_\alpha \\
\delta \Psi_\alpha^* &= ie\mathcal{E}(x)\Psi_\alpha^*, \text{ and } \delta x_\mu = 0
\end{aligned}
\right\} \quad \text{---------(10)}.
$$

The Noether theorem yields the following continuity equations:

$$
\partial_\mu ie \frac{\partial L}{\partial_\mu \Psi_\alpha} \Psi_\alpha \mathcal{E}(x) - ie \frac{\partial L}{\partial \partial_\alpha \Psi_\alpha^*} \Psi_\alpha^* \mathcal{E}(x) + \frac{\partial L}{\partial \partial_\mu A_\nu} \partial_\nu \mathcal{E}(x)
$$

$$
= \partial_\mu \delta F_\mu = 0 \text{------------------------} (11).
$$

Notice the field functions Ψ_α and Ψ_α^* are considered independent; $\partial_\mu A_\mu$ appears only in the free L$_{em}$ (see equation 6).

With $\frac{\partial L}{\partial \partial_\mu A_\mu} = - G_{\mu\nu}$, the following is obtained:

$$\partial_\mu [-G_{\mu\nu} \partial_\nu \mathcal{E}] = \partial_\mu [- \partial_\nu (G_{\mu\nu} \mathcal{E}) + \mathcal{E} \partial_\nu G_{\mu\nu}] \text{-------} (12).$$

But since $G_{\mu\nu} = -G_{\nu\mu}$; $\partial_\mu \partial_\nu (G_{\mu\nu} \mathcal{E}) = 0$, it can now define the current density through $\partial_\nu G_{\mu\nu} = j_\mu$ which is equivalent to

□ $A_\mu = -j_\mu$ with $\partial_\nu A_\nu = 0$. Therefore, $\partial_\mu [\mathcal{E} \partial_\nu G_{\mu\nu}] = j_\mu \partial_\mu \mathcal{E}$.

And $\partial_\mu \partial_\nu G_{\mu\nu} = 0$. Now going back to equation 11, the following is gotten:

$$j_\mu \partial_\mu \mathcal{E} + \partial_\mu [ie\mathcal{E} \frac{\partial L}{\partial \partial_\mu \Psi_\alpha} \Psi_\alpha - ie\mathcal{E} \frac{\partial L}{\partial \partial_\mu \Psi_\alpha^*} \Psi_\alpha^*] = 0. \text{----------} (13).$$

Or it can be written as:

$$[j_\mu -ie (\frac{\partial L}{\partial \partial_\mu \Psi_\alpha^*} \Psi_\alpha^* - \frac{\partial L}{\partial \partial_\mu \Psi_\alpha} \Psi_\alpha)] \partial_\mu \mathcal{E} - [\partial_\mu i \mathcal{E} (\frac{\partial L}{\partial \partial_\mu \Psi_\alpha^*} \Psi_\alpha^* - \frac{\partial L}{\partial \partial_\mu \Psi_\alpha} \Psi_\alpha] \mathcal{E} = 0.$$ As $\mathcal{E}$ and $\partial_\mu \mathcal{E}$ are chosen at a given point, then it can be written that:

$$j_\mu = ie(\frac{\partial L}{\partial \partial_\mu \Psi_\alpha^*} \Psi_\alpha^* - \frac{\partial L}{\partial \partial_\mu \Psi_\alpha} \Psi_\alpha) \text{-----------------------} (14)$$

and $\partial_\mu [ie(\frac{\partial L}{\partial \partial_\mu \Psi_\alpha^*} \Psi_\alpha^*) - \frac{\partial L}{\partial \partial_\mu \Psi_\alpha} \Psi_\alpha)] = \partial_\mu j_\mu = 0$ --------- (15).

Equation 14 shows how to calculate the current density from L. Equation 15 shows the current density is conserved; it fulfills the continuity equation. It is noted that for a Hermitian field, j_μ vanishes identically to □ $A_\mu = 0$; the Hermitian field has to be coupled to the electromagnetic field. So the gauge transformation of the first kind can only defined with complex fields. Therefore, only the complex fields can be connected with charge. The charge density of the field Ψ follows from the equation 14 as:

$\rho =-i j_4 = e(\frac{\partial L}{\partial \partial_4 \Psi_\alpha^*} \Psi_\alpha^* - \frac{\partial L}{\partial \partial_\mu \Psi_\alpha} \Psi_\alpha)$, and the total charge follows

as $Q = \int \rho\, dV = e \int (\frac{\partial L}{\partial \partial_4 \Psi_\alpha^*} \Psi_\alpha^* - \frac{\partial L}{\partial \partial_4 \Psi_\alpha} \Psi_\alpha) dV$ ----(16).

Equation 15 gives that $(dQ/dt)=0$; it implies that Q is a constant of the motion. Hence, the charge conservation is a consequence of the field's invariance under the combined gauge group.

10.9. *Charge Independence and Charge Conservation in Strong and Electromagnetic Interactions.-*

The interaction between $\pi - mesons$ and nucleons (nuclear strong force) with coupling constant G is of the form (see appendix C), $L = iG\psi^{(-)}\mathcal{T}^{\rightarrow}\gamma_5\psi\varphi$ --------------- (1).

With $\psi = \begin{bmatrix} \psi_p \\ \psi_n \end{bmatrix}$, and $\varphi^{\rightarrow} = \begin{bmatrix} \varphi_1 \\ \varphi_2 \\ \varphi_3 \end{bmatrix}$ ---------- (2).

Where $\varphi_{\pi(-)} \rightarrow \varphi = \frac{1}{\sqrt{2}}(\varphi_1 + i\varphi_2)$

$\varphi_{\pi(+)} \rightarrow \varphi^* = \frac{1}{\sqrt{2}}(\varphi_1 - i\varphi_2)$, $\left. \right\}$ -------------(3).

$\pi^0 \rightarrow \varphi_3 = \varphi_0$

(It is important to notice that the charged particles must be described by complex fields or the current density as obtained via the use of quantum electrodynamics will be zero.) Now equation 3 in terms of φ takes the form:

$\varphi_1 = \frac{1}{\sqrt{2}}(\varphi + \varphi^*)$, $\varphi_2 = \frac{1}{\sqrt{2}}(\varphi - \varphi^*)$, $\varphi_3 = \varphi_0$ --------(4).

These φ^s have the same properties as vectors or spinors in isospace. Thus, since they describe different fields corresponding to different charge states, their invariance, with respect to rotation, will ensure that any theoretical perturbation approach. Using these quantities and their transformation properties must be such that charge conservation and charge independence are satisfied. L is scalar quantity, thus in both isospin space and Minkousky space.

In Minkousky space, the quantity $\psi^{(-)}\gamma_5\psi$ transforms like a psuedoscalar. L is scalar; thus in both spaces, φ must be a vector (i.e., it must transform due to the fact that L is scalar, but it contains the isotopic spin $\mathcal{T}$.).

The electromagnetic gauge transformations of the first kind have the form $\psi = e^{i\epsilon}\psi$, where ϵ is phase factor --------- (5).

So the pions ($\pi^+\pi^-\pi^0$) fields take the forms

$$
\left.
\begin{aligned}
\varphi &\rightarrow e^{-i\epsilon}\varphi \text{ ----for } \pi^- \\
\varphi^* &\rightarrow e^{+i\epsilon}\varphi \text{----for } \pi^+ \\
\varphi_3 &\rightarrow \varphi_0 \text{ ---for } \pi^0
\end{aligned}
\right\} \quad \text{----------------- (6).}
$$

The invariance of this transformation utilizing standard group theoretical techniques is sufficient to state the charge is preserved. Thus, it is good to proceed farther to show this transformation is a little bit more than a rotation. Substituting these relations into equation 4 results the following:

$$
\varphi_1 = \frac{1}{\sqrt{2}}\sum(cos\epsilon - isin\epsilon)\varphi + (cos\epsilon + i\,sin\epsilon)\varphi^* =
$$

$$
\left.
\begin{aligned}
&= \varphi_1 cos\epsilon + \varphi_2 sin\epsilon \\
\varphi_2 &= -\varphi_1\,sin\epsilon + \varphi_2 cos\epsilon \\
\varphi_3 &= \varphi_3
\end{aligned}
\right\} \quad \text{------- (7).}
$$

The resulting form of this transformation is indicative of the fact that under gauge transformations of the first kind the

vector $\vec{\varphi} = \begin{bmatrix} \varphi_1 \\ \varphi_2 \\ \varphi_3 \end{bmatrix}$ transforms as would a three-dimensional vector rotated through an angle about the third axis. Since $\vec{\varphi}$ transforms like a vector in isospace, the fields $\varphi_1\varphi_2\varphi_3$ to which the charged and the uncharged particles belong is considered. Then correlate them with the components of a three-dimensional vector in isospace. As an example, show that $\Theta_i = \psi^{(-)}\tau_i\psi$ transforms like a vector in isospace.

Proof: ψ is a spinor in isospace, transforms as such, under any rotation, for example, as the transformation about the third axis, it will have the following forms:

$$\psi \to e^{i\frac{\epsilon\tau_3}{2}}\psi, \quad \psi^{(-)} \to \psi^{(-)} e^{-\frac{i\epsilon\tau_3}{2}} \text{-------------- (8)}.$$

$$\text{Thus, } \Theta_i = \psi^{(-)}e^{\frac{-i\epsilon\tau_3}{2}}\tau_i e^{i\frac{\epsilon\tau_3}{2}}\psi \text{ ----------------- (9)}.$$

Using the complex variables' properties, equation 9 can be written as

$$\Theta_i = \psi^{(-)}[(\cos\tfrac{\epsilon}{2} - i\tau_3 \sin\tfrac{\epsilon}{2})\,\tau_i(\cos\tfrac{\epsilon}{2} + i\tau_3 \sin\tfrac{\epsilon}{2})]\psi \text{ ------(10)}.$$

Find the product and remember that

$$\left.\begin{array}{l} \vec{\tau} \times \vec{\tau} = 2i\tau, \\ [\tau_i, \tau_k]_+ = 2\delta_{ik} \\ \tau_i\tau_j = \tau_k \end{array}\right\} \quad \begin{array}{l} \text{----------------- (11).} \\[4pt] i, j, k, = 1, 2, 3 \text{ (cyclic)} \end{array}$$

So the result will be

$$\left.\begin{array}{l} \Theta_1 = \psi^{(-)}(\tau_1 \cos\epsilon + \tau_2\sin\epsilon)\psi = \Theta_1\cos\epsilon + \Theta_2\cos\epsilon) \\ \Theta_2 = -\Theta_1 \sin\epsilon + \Theta_2 \cos\epsilon \\ \Theta_3 = \Theta_3 \end{array}\right\} \text{----- (12)}.$$

It is the required proof.

Now go back to L, $\psi^{(-)}\gamma_5\psi$ is a psuedoscalar, and thus $\psi^{(-)}\tau^-\gamma_5\psi$ is a vector since γ_5 is essentially a constant. As a result, it may be seen that since φ is a vector; L is scalar! Because L is scalar, so it is invariant under each rotation in isospace. Also, it is known from nuclear physics; and as shown before, the charge could be exchanged by the nucleons (protons, neutrons). That is to say

$$i\tau_2\psi_p = \psi_n \text{ and } -i\tau_{2=}[\cos\frac{\pi}{2}- i\sin\frac{\pi}{2}]\, \tau_2 = e^{\frac{-i\pi\tau_2}{2}} \text{ --------(13)}.$$

Therefore, $e^{\frac{-i\pi\tau_2}{2}}\psi_p = \psi_n$, (p=proton, n =neutron)-----(14),

where $e^{\frac{i\pi\tau_2}{2}}$ is the rotation operation by 180 degrees about the second axis of the isospace. Since the mutual exchange of the charge is invariant under each rotation, it is also invariant under this rotation corresponding to the transition p$\rightarrow$ n. The matrix elements for the three scattering processes (n-n, p-p, n-p) are invariant with respect to the exchange of charge states for one or both neucleons, whether the exchange occurs before or after the scattering. These processes have been observed experimentally, which have led to an interaction potential essentially the same for each process. *For this reason, it is stated that charged invariant states should exist.* Because the operators of the infinitesimal rotations in isospace are the components of the isospin, the invariance under isospin rotation usually includes a conservation of the isospin itself.

And because of the invariance of L, the isospin rotations about the three axes are constants of the motion. This is why the isospin is a good quantum number. Now it is important to investigate the conservation of charge of the $\gamma - p$ *system.*

It is an electromagnetic interaction. As it is known, the neutron is neutral (has no charge), so it has no interaction with the electromagnetic field. Therefore, the Lagrangian for this system is given by

$$L = e\psi^{(-)}[\tfrac{1}{2}(1+\tau_3)]\,\gamma_\mu\psi\,A_\mu \text{------------------ (15).}$$

Upon expansion of the term $\psi^{(-)}\,1/2(1+\tau_3)\psi$, it is found that only the proton (it is charged) interacts with the field. So

$$1/2\,\psi^{(-)}[1+\tau_3]\psi = \psi_p^{(-)}\psi_p \text{----------------------- (16).}$$

$\psi^{(-)}\psi$ is isoscalar, but $\psi^{(-)}\tau_3\psi$ is the third component of an isovector. Therefore, the total interaction is not scalar, so it cannot be invariant under general rotation in isospace. As a result of this, the isospin is not conserved in this interaction.

Charge conjugation of Dirac field coupled to the electromagnetic field. To do so, it is necessary to derive a matrix say C with appropriate properties to represent the charge conjugation. To start with let us consider a matrix B defined such as $\gamma^{\sim}_\mu = B^{-1}\gamma_\mu B$, where γ_μ is the well known Dirac matrices. Now to show is B antisymmetric? Taking the transpose of this relation it is obtained that, $\gamma_\mu = B^\sim B^{-1}\gamma_\mu\, BB^{-1}$, but $(B^\sim B^{-1})^\sim = BB^{\sim-1}$, then let $BB^{\sim-1} = k$, multiply by $B^\sim$ with some Algeria manipulation one gets that $k^2 = 1$. Therefore, $k = \pm 1$. Thus $B = \pm B^\sim$. Next step is to check which value +or −is the correct one. So, from this information the following relations can be obtained; $(B^{-1}1)^\sim = 1.\ B^{\sim-1} = k\,B^{-1}\,(1)$ ---------(17a), scalar, one component.

$(B^{-1}\gamma_\mu)^\sim = k\,B^{-1}\gamma_\mu$ -----------------(17b). vector, 4 components.
$(B^{-1}i\gamma_\mu\gamma_\nu)^\sim = k[i(\gamma_\mu\gamma_\nu)^\sim B^{-1}]$,----(18), Tensor, 6 components $\mu \neq \nu$.
By the same way the following might be obtained, $(B^{-1}i\gamma_\mu\gamma_\nu\gamma_\lambda)^\sim = -k\ B^{-1}i\gamma_\mu\gamma_\nu\gamma_\lambda$ -------------(19) pesudovector, 4 components.
$(B^{-1}i\gamma_1\gamma_2\gamma_3\gamma_4)^\sim = kB^{-1}\gamma_5$ --------------------(20), pesudoscalar, one component.

From 17a and 18 k=1, in this case we would have tenlinearly independent antisymmetric of 4×4 dimension. But the antisymmetric has only six independent elements, therefore k=1 is ruled out. Thus $B = -B^\sim$. -------------------------------- (21).

Now the question is, is B unitary or can be made unitary?

To do so, remember that γ_μ is Hermitian (chapter 8); therefore, one can write $(B^{-1}\gamma_\mu B)^+ = (B^{-1}\gamma_\mu B)$, $(\tilde{\gamma_\mu}^+ = \tilde{\gamma_\mu})$.

Hence, $(B^{-1}\gamma_\mu B)^+ = B^{-1}\gamma_\mu B$, multiply from the left by B and from the right by B^+; there will be $BB^+\gamma_\mu = \gamma_\mu BB^+$, so BB^+ must be at least proportional to the unit matrix $BB^+ = k.1$. The diagonal elements must be positive and real since $(BB^+)_{\alpha\alpha}$ $= \sum_\beta B_{\alpha\beta}B^+_{\beta\alpha} = \sum_\beta B_{\alpha\beta}B^*_{\alpha\beta} = \sum_\beta B_{\alpha\beta}{}^2$.

Now taking a matrix $\frac{1}{\sqrt{k}}$ in place of the original B—it is allowed since B is determined only up to a factor—it gives a new B which should be unitary, i.e., $B^+B=1$ or $B^{-1}=B^+$.

Using this B, it is possible to construct another matrix, say C, such that $[-\tilde{\gamma_\mu} = [C^{-1}\gamma_\mu C]$, which can be constructed as $C=-\gamma_5 B$, so that $C^{-1}\gamma_5 C = (-\gamma_5 B)^{-1}\gamma_\mu(-\gamma_5 B) = -\tilde{\gamma_\mu}$ (show it).

Then $C^\sim = (-\gamma_5 B)^\sim = -B^\sim\tilde{\gamma_5} = \gamma_5 B = -CC^\sim = -C\cdot \longrightarrow$ (proved).

For Dirac field coupled to the electromagnetic field illustrated by the following equations:

$[\gamma_\mu(\partial_\mu - \frac{ie}{\hbar c}A_\mu + k)]\psi = 0$----------------------------- (22).

$[(\partial_\mu + \frac{ie}{\hbar c}A_\mu)\gamma_\mu - k]\psi^- = 0$-----------------------------(23).

$[\square A_\mu = -ie\psi^-\gamma_\mu\psi] = 0$ ----------------------------- (24).

Since the spinor fields obey the anticommutation relations, so it is possible to make all the bilinear quantities antisymmetrical, so the quantity $\psi_A^-\psi_A$ can be changed to $\frac{1}{2}(\psi_A^-\psi_B - \psi_B\psi_A^-)$. This can be done since they only differ by an infinite C-number, or $\frac{1}{2}(\psi_A^-\psi_B - \psi_B\psi_A^-) = \frac{1}{2}\psi_A^-\psi_B +$

$\frac{1}{2}\psi_{\bar{A}}\psi_B - \frac{1}{2}(\psi_B\psi_{\bar{A}} + \psi_{\bar{A}}\psi_B)=\psi_{\bar{A}}\psi_B$ +a.   Now   to   make

antisymmetrization, $\psi^-\gamma_\mu\psi$ is written as

$\frac{1}{2}(\psi^-\gamma_\mu\psi - \psi\gamma_\mu\psi^-)$, so equation 24 will take the form.

$\Box A_\mu =-\frac{1}{2}\, ie\, (\psi^-\gamma_\mu\psi - \psi\gamma_\mu\psi^-)=-j_\mu$------------(25)

(Current density)

To put equation 23 into more suitable form, the transpose of the equation is taken such as $[\gamma_\mu^{\sim}(\partial_\mu+\frac{i}{\hbar c}A_\mu)-k]\psi^{-\sim}=0$

Multiply from the left by C defined as $-\gamma_5=C^{-1}\gamma_\mu C$; hence,

$C\,[\gamma_\mu(\partial_\mu +\frac{ie}{\hbar c}A_\mu)-k]\,C\psi^{-\sim}=0$, since $-C\gamma_\mu=\gamma_\mu C$, then it becomes

$[\gamma_\mu(\partial_\mu+\frac{i}{\hbar c}A_\mu)+k]C\psi^{-\sim}=0$ ---------------(26).

From equations 22, 24, and 17, the field equations take the following forms:

$[\gamma_\mu(\partial_\mu - \frac{ie}{\hbar c}A_\mu) +k]\psi=0,$ ----------------------- (27).

$[\gamma_\mu(\partial_\mu+\frac{ie}{\hbar c}A_\mu) +k]C\psi^{-\sim}=0,$ -------------------- (28).

$\Box = -\frac{1}{2}(\psi^-\gamma_\mu\psi - \psi\gamma_\mu\psi^-)=-j_\mu,$----------(29).

Now consider the system of the following transformations:

$\psi(x) \rightarrow \psi^-(x) =C\psi^{-\sim}(x)$ --------------------(30).
$\psi^-(x) \rightarrow \psi^{(-)-}(x)=-\psi^\sim C^{-1}$------------------ (31).
$A_\mu(x) \rightarrow A_\mu^-(x)= -A_\mu(x)$ ------------------------(32).

From equation21, it can be seen that equation 28 goes into equation29. Also, substituting equation 32 into equation27, the result is obtained: $C\,\psi^{-\sim} \rightarrow C(-\psi^\sim(x)C^{-1})^\sim=-C\,C^{\sim-1}\psi(x)$.

But $C^\sim = -C$, then, $C\psi^{-\sim} \rightarrow \psi(x)$.

Equation 32 is compatible with equation 31 since taking the Hermitian adjoint of equation 31 gives:

$$\psi^+ \rightarrow (C\psi^{-\sim})^+ = \psi^\sim \gamma_4^\sim C^{-1}, \text{ (show that)} \text{ ------------ (33)}.$$

Thus, $\psi^- = \psi^+\gamma_4 \rightarrow \psi^\sim \gamma_4^\sim C^{-1}\gamma_4 = -\psi^\sim C^{-1}$ (show this).

Now to see, is equation 13 invariant under these transformations? Take the transformations in component form such as $\psi_\beta \rightarrow C_{\beta\rho}\psi_\rho^-$, $\psi_\alpha^- \rightarrow -\psi_{\sigma\alpha}C_{\sigma\alpha}^{-1}$-------- (34).

Consider the right side of equation 17:

$\psi^-\gamma_\mu\psi$ $-\psi\gamma_\mu\psi^-$ $=\psi_\alpha^-\gamma_{\alpha\beta}^\mu\psi_\beta$ $-\psi_\beta\gamma_{\alpha\beta}^\mu\psi_\alpha^-$. Substitute for ψ_β and ψ_α. Using equation 22 with some manipulations, one finds it as $=\psi_\sigma\gamma_{\rho\sigma}^\mu\psi_\rho^-$ - $\psi_\rho^-\gamma_{\rho\sigma}^\mu\psi_\sigma$. It is the negative of the current density expression before the transformation. Thus, the system of equations28, 26, and 30 is invariant under charged conjugation defined by $\psi^c = c\psi^{-\sim}$, $\psi^{-c} = -\psi^-c^{-1}$.

These transformed fields are called charge conjugate fields.

The antisymmetrical current density can be written in an equivalent form using the charged conjugate fields, as follows:

$j_\mu = \frac{1}{2}ie$ $(\psi_\alpha^-\gamma_{\alpha\beta}^\mu\psi_\beta$ - $\psi_\beta\gamma_{\alpha\beta}^\mu\psi_\alpha^-)$, also $\psi^c =c$ $\psi^{-\sim}$, which can be written as $c^{-1}\psi^c = \psi^{-\sim}$ --- (35)

and $\psi^{(-)c} = -\psi^\sim c^{-1}$ or $-\psi^{(-)c}$ $c =\psi^\sim$ ---------------------- (36).

Substituting equations into the current density with some manipulation gives

j_μ $=(1/2)ie(\psi_\alpha^-\gamma_{\alpha\beta}^\mu\psi_\beta$ $-\psi_\sigma^{-c}\gamma_{\sigma\rho}^\mu\psi_\rho^c)=\frac{1}{2}(j_\mu$ $-j_\mu^c)$ [in matrix notation]. So in the current density, it was shown that the Dirac covariant vector quantity goes into its negative undercharged conjugation. Similarly, also this effect can be shown on the other Dirac covariants. As an example, consider the scalar,

$S = \frac{1}{2}(\psi_\alpha^- \psi_\alpha - \psi_\alpha \psi_\alpha^-)$ ------------- (36).

In the component form, the charged conjugation has the form $\psi_\alpha^- = -\psi_\sigma c_{\sigma\alpha}^{-1}$ and $\psi_\alpha = c_{\alpha\rho}\psi_\rho^-$. So under this transformation, equation 24 takes the following steps to reach the same form:

$S \to S^c = \frac{1}{2}(-\psi_\sigma c_{\sigma\rho}^{-1} c_{\alpha\rho}\psi_\rho^- + c_{\alpha\rho}\psi_\rho^- \psi_\sigma c_{\sigma\alpha}^{-1}) =$

$\frac{1}{2}(-\psi_\sigma \delta_{\sigma\alpha}\psi_\rho^- + \delta_{\sigma\rho}\psi_\rho^- \psi_\sigma] = \frac{1}{2}(-\psi_\alpha \psi_\alpha^- + \psi_\alpha^- \psi_\alpha)$ ⟶

$S \longrightarrow S^c = \frac{1}{2}(\psi_\alpha^- \psi_\alpha - \psi_\alpha \psi_\alpha^-)$ which is as equation 36. Hence,

S is invariant under charged conjugation.

By the same way, it can be shown that:

$A^c = A$, ---------pesudovector.

$T^c = -T$,-------tensor.

$P^c = P$, ----pesudoscalar.

Consider the partial number representation of Dirac field:

$\psi\psi = \frac{1}{\sqrt{N}} \Sigma (a_{ks} u(\vec{k},s)e^{i\vec{k}\cdot\vec{r}} + (b_{ks}^* v^*(\vec{k},s)e^{-i\vec{k}\cdot\vec{r}})$ (37).

$\psi^- = \frac{1}{\sqrt{N}} \Sigma (a_{ks}^* u^-(\vec{k}\cdot s) \ e^{-i\vec{k}\cdot\vec{r}} + b_{ks}v^-(\vec{k}\cdot s)e^{i\vec{k}\cdot\vec{r}})$ (38),

where u and v are defined as

$Hu(\vec{k}) = \text{I}E\text{I}\, u(\vec{k})$ ----------------------------------- (39).

$Hv(\vec{k}) = -\text{I}E\text{I} v(\vec{k})$ --------------------------------- (40).

And $H = c\gamma_4\gamma_i\, k_i + m_0 c^2\gamma_4$ --------------------------- (41).

Now, it is to show that $u^c \equiv cu^{-\tilde{}} = v$ ----------------(42).

And $v^c \equiv cv^{-\tilde{}} = u$ ----------------------------------- (43).

Take the adjoint of equation39, noticing equation 41 for H, then $(c\gamma_4\gamma_i k_i + m_0 c^2\gamma_4)\, u(\vec{k}) = \text{I}E\text{I}\, u(\vec{k})$.

H is Hermitian, so $u(\vec{k})(c\gamma_4\gamma_i k_i + m_0 c^2 \gamma_4) = 1$ E l $u^-(\vec{k})$
Take the transpose of this eigenvalue equation; one gets
$(i\widetilde{\gamma_4}\widetilde{\gamma_i} k_i + m_0 c^2 \widetilde{\gamma_4})u^{-\widetilde{}} = 1$ E l$u^{-\widetilde{}}$.

Now multiply by c, remembering that $c\widetilde{\gamma_i} = -\gamma_i c$. The following is obtained: $(i\gamma_4\gamma_i k_i - m_0 c^2 \gamma_4)cu^{-\widetilde{}} = lEl\ cu^{-\widetilde{}}(-\vec{k})$ Or H c $u^{-\widetilde{}}(-k) = $- lEl $cu^{-\widetilde{}}(-k)$.

Thus, it can be noticed that $cu^{-\widetilde{}}(-k)$ satisfies the same equation as $v(-k)$. Consequently, using equation 26, one gets the following:

$$\psi^c \equiv c\,\psi^{-\widetilde{}} = a^+_{ks}\, c\, u^{-\widetilde{}}e^{-i\vec{k}\cdot\vec{r}} + b_{ks}c\, v^{-\widetilde{}}e^{i\vec{k}\cdot\vec{r}}$$
$$= a^+_{ks}v(k,s)e^{-i\vec{k}\cdot\vec{r}} + b_{ks}u(k,s)e^{i\vec{k}\cdot\vec{r}}$$

Comparing this with equation 25, it is clear that
$a_{ks} \rightarrow b_{ks}$, $b^+_{ks} \rightarrow a^+_{ks}$, and similarly, $a^+_{ks} \rightarrow b^+_{ks}$, $b_{ks} \rightarrow a_{ks}$.

Clearly and explicitly, this showed that the charge conjugation takes the particle into its antiparticle.

10.10. *Solution of Dirac Equaton with the Sources Present*

10.10.1. *Introduction*

In the previous section, the Dirac field equations, for the case of its interaction with the electromagnetic field, were derived. In this section, it is decided to use the technique of Green function and an iterative type of calculation to solve these equations. In the course of the problem, a physical insight into the origin and the meaning of the scattering matrix will be considered. In order to demonstrate the physics of the problem, the scattering of the electrons by the protons will be studied using the solution of Dirac equations that were derived for the interaction with the electromagnetic fields.

10.10.2. *Interaction field Equation and Green Function*

The Dirac equation in the presence of the electromagnetic field which was derived in the previous section is given by

$(i\not{\partial} - e\not{A} - m)\psi(x) = 0$ ------------------------- (1).

This can be written as $(i\not{\partial} - m)\psi(x) = e\not{A}(x)\psi$ ------------ (2).

The complete solution of equation 2 is the sum of the solution of the humongous equation where A=0 and the particular solution (referred to any textbook in differential equation).

The homogenous solution is just the field solution of free Dirac field ψ (no external force). Therefore, equation 2 can be written as $(i\not{\partial} - m)\psi(x) = \int e\not{A}(r)\psi(r)S^4(x-y)\, d^4r$ (3).

Now Green function is defined as $S_f(x-y) = \frac{S^4(x-y)}{(i\not{\partial} - m)}$ ---- (4).

Then equation 3 becomes $\psi(x) = \int e\not{A}(r)\psi(r)S^4(x-y)d^4r$ (5).

Thus, the complete solution of equation 1 is

$\psi = \psi(x) + \int d^4r\, e\not{A}(r)\psi(r)S_f(x-y).$ ------------------(6).

In order for the solution to be a complete meaning, the Green function has to be determined. So this important step must be done. By taking the Fourier transformation of equation 4, the following relations are obtained:

$i\not{\partial}_x \xrightarrow{\text{F.T}} \not{p}.$

$m \longrightarrow m.$

$S_f(x-y) \longrightarrow S_f(p).$

$S(x-y) \longrightarrow 1.$

$\left. \right\}$ -------------------- (7a).

With the inverse relations given by

$$S_P(x-y) = \frac{1}{(2\pi)^4} \int d^4p \; S_f(p) \, e^{-ip(x-y)}$$
$$\delta_4 = \frac{1}{(2\pi)^4} \int d^4p \, e^{-ip(x-y)}$$

-------------- (7b).

Equation 4 takes the form $(\not{p}-m)S_f(p)=1$,------(8).

Multiplying *equation 8 by* $(\eth+m)$, it gives the form

$$((\eth+m)(\eth-m)S_f(p) = (\eth+m)$$ ----------------------(9).

Or $S_f(p) = \frac{\eth+m)}{\eth^2-m^2}$ ------------------------------------(10).

Where $\eth=\gamma_\mu p_\mu$, so $\eth^2=\gamma_\mu p_\mu \gamma_\mu p_\mu = p_0^2 - p_k^2 = p_0^2 = \mathrm{lp}^{\to}\mathrm{l}^2$.
Also, the relativistic energy is $E_p^2 = \mathrm{lp}^{\to}\mathrm{l}^2 + m^2$ [in unit of c and $\hbar$, i.e., let $\hbar=c=1$].

So it leads to $\eth^2 = p_0^2 - E_p^2 + m^2$ and then $S_f(p) = \frac{(\eth+m)}{p_0^2-E_p^2}$ ------(11).

If this problem is limited to an electron, then the following three-field solutions are to be noticed:

(1.) For positive energy $E_p = +\sqrt{p^2+m^2}$, the solutions are

$$\psi^+(x) = e^{-ipx}u(p,s)\sqrt{\frac{m}{EV}}$$
$$(\not{p}-m)u(p,s)=0$$

--------------- (12a).

(2.) For negative energy $E_p = -\sqrt{p^2+m^2}$ the solutions are

$$\psi' = e^{ipx}v(p,s)\sqrt{\frac{m}{EV}}; \; (p+m)v(p,s)$$--------------- (12b).

These are mentioned now because in order to determine the inverse transform of $S_f(p)$, it must appeal to the physics of the problem. Since it is dealt here with the electrons scattering, so the positive energy states will be considered. It is propagating forward in time, while those electrons with negative energy states are propagating backward in time, i.e., into the past. With this idea in mind, from equations 7b and 11, it is possible to get

$$S_f(x-y) = \frac{1}{(2\pi)^4} \int d^4 p \, S_f(p) \, e^{-ip(x-y)} \rightarrow$$

$$S_f(x-y) = \frac{1}{(2\pi)^3} \int d^3 p \, e^{-iP(x-y)} \int_{-\infty}^{\infty} \frac{dp_0}{2\pi} S_f(p) \, e^{-ip_0(x_0-y_0)}, - \text{(13a)}.$$

And

$$S_f(x-y) = \frac{1}{(2\pi)^3} \int d^3 p \, e^{i\vec{p}.(\vec{x}-\vec{y})} \int_{-\infty}^{\infty} \frac{dp_0}{2\pi} \frac{(\eth+m)e^{-ip_0(x_0-y_0)}}{(p_0-E_p)(p_0+E_p)}, \text{-----------}$$

--------(13b).

Now the attention will be restricted on the integral (I), i.e.,

$$I = \int_{-\infty}^{\infty} \frac{dp_0}{2\pi} \frac{(\eth+m))e^{-ip_0(x_0-y_0)}}{(p_0-E_p)(p_0+E_p)} \text{---------------------------} \text{(14)}.$$

This equation, as it is clear, has poles at $p_0 = \pm E_p$. If p_0 is considered a complex variable, then it is possible to evaluate the integral I by complex integration methods. In the complex integration, the path of integration depends on the boundary conditions of the problem. For the electrons scattering, as mentioned before, there are two cases:

Case 1: $(x_0 - y_0) > 0$, positive time intervals. For this case, in order to propagate only positive energy solutions into the future, only the positive energy pole will be included.

Case 2: $(x_0 - y_0) < 0$, negative time intervals. For this case, it is wanted to propagate the negative energy solutions into the past; hence, only the negative energy pole is included.

These two cases will be treated in details as follows:

For case 1, choose a path of integration as shown in the contour figure below.

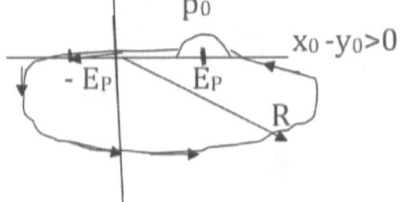

Using Cauchy Residue theorem (see complex variables by Churchill) to evaluate the integral I, where

$$\int_R^{-R} + \int_{|R|} 2\pi i \text{ Residue of } p_0 = E_p = 2\pi i R_{E_p}; \text{ hence, } I = 2\pi i R_{E_p}.$$

In the lower complex plane, since $(x_0-y_0)>0$, the term in equation 14, $| e^{-i p_0 (x_0-y_0)} | \leq 1$, thus, $limit_{R\to\infty} \int_{p_0=|R|} = 0$, and

$$limit_{R\to\infty} \int_R^{-R} = -\ I = 2\pi i R_{E_p}, \text{ so } I = -2\pi i R_{E_p} \Rightarrow$$

$$R_{E_p} = \frac{1}{2\pi i} I = \frac{1}{2\pi} e^{-i E_p (x_0-y_0)} \left(\frac{E_p \gamma_0 - P^{\to} r^{\to} + m)}{2E_p} \right)$$

Hence, $I = -i e^{i E_p (x_0-y_0)} \left(\frac{(E_p \gamma_0 - P^{\to} r^{\to} + m}{2E_p} \right)$

Substituting this in equation 13b, then $S_f(x-y)$ for $(x_0 -y_0)>0$ case is given by, $S_f(x - y) =$

$$= -i \int \frac{d^3 p}{(2\pi)^3} \cdot \frac{e^{i p^{\to} (x^{\to}-y^{\to})} e^{-i E_p (x_0 -y_0)} (E_p \gamma_0 - p^{\to} r^{\to} + m)}{2E_p} \text{------ (15).}$$

Next is to evaluate $S_f(x - y)$ for case 2, where $(x_0-y_0) < 0$.

The same way is to be followed, so the path of integration to be drawn over a semicircle as $R\to \infty$, follows below:

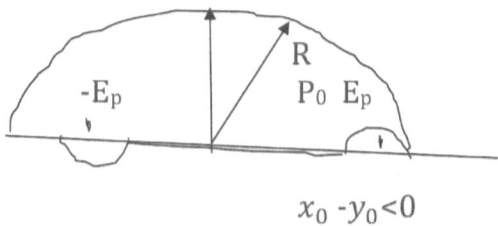

$$x_0 - y_0 < 0$$

Thus, for this case, $S_f(x - y)$ is given by,

$$S_f(x - y) = \int \frac{d^3p}{(2\pi)^3} e^{i\vec{p}\cdot(\vec{x} - \vec{y})} 2\pi i R_{E_p}, \text{ and}$$

$$R_{-E_p} = \frac{1}{2\pi} e^{iE_p(x_0 - y_0)} \cdot \frac{(-E_p\gamma_0 + \vec{p}\cdot\vec{r} + m)}{-2E_p}$$

Or $S_f = -i \int \frac{d^3p}{(2\pi)^3} \times e^{i\vec{p}\cdot(\vec{x} - \vec{y})} e^{iE_p(x_0 - y_0)} \frac{(-E_p\gamma_0 - \vec{p}\cdot\vec{r} + m)}{2E_p}$

Now if the variable of the integration is changed from $\vec{p}$ to $-\vec{p}$, then S_f will take the following form:

$$\left[S_f = -i \int \frac{d^3p}{(2\pi)^3} e^{-i\vec{p}\cdot(\vec{x} - \vec{y})} \cdot e^{iE_p(x_0 - y_0)} \frac{(-E_p\gamma_0 - \vec{p}\cdot\vec{r} + m)}{2E_p} \cdots \right.$$

$$\left. \text{--- (16).} \right]$$

Equations 15 and 16 are Green functions for the cases $(x_0 - y_0) > 0$ and $(x_0 - y_0) < 0$, respectively.

10.10.3. *Projection Operators:*

For further evaluation to the Green function S_f, the projection operators might be recalled. These operators are defined as $\Lambda_+(p) = \frac{\eth + m}{2m}$ and $\Lambda_- = -\frac{-\eth + m}{2m}$, with the following properties:

(1.) $\Lambda_+(p)u(p,s) = u(p,s)$,
(2.) $\Lambda_+(p,s)\, v(p,s) =0$,
(3.) $\Lambda_-(p)\, u(p,s)=0$,
(4.) $\Lambda_-(p)\, v(p,s) =v(p,s)$ ---------------- (17).
(5.) $\Lambda_+^2(p) =\Lambda_+(p)$,
(6.) $\Lambda_-^2(p) =\Lambda_-(p)$,
(7.) $\Lambda_-\Lambda_+=0$.

Now $\dfrac{E_p\gamma_0 - \vec{p}\,\vec{r}}{2E_p} = \dfrac{\not{p}+m}{2E_p} = \dfrac{\not{\delta}+m}{2m}\cdot\dfrac{2m}{E_p} = -\Lambda_+(\not{p})\dfrac{m}{E_p}$, ----- (18).

And $\dfrac{-E_p\gamma_0 - \vec{P}\,\vec{r}+m}{2E_p} = \dfrac{-\not{p}+m}{2m}\cdot\dfrac{m}{E_p} = \Lambda_-(\not{p})\dfrac{m}{E_p}$ --------(19).

Hence equations (15 and 16) can be written as

$S_f(x-y)=-i\int \dfrac{d^3p}{(2\pi)^3}\, e^{-ip(x-y)}\dfrac{m}{E_p}\Lambda_+(\not{p})$, -- for $(x_0-y_0)>0$—(20).

$S_f(x-y)= -i\int \dfrac{d^3p}{(2\pi)^3}\, e^{ip(x-y)}\dfrac{m}{E_p}\Lambda_-(\not{p})$,--for$(x_0-y_0)<0$,----(21).

10.11. *Evaluation of the Field Function ψ*

So far, the tools for the evaluation of the field function which is given by equation 6 are prepared. To do this, the following iterative calculation of successive approximation is to be used:

(1.) Use $\psi_0=\psi$ in the right hand side of equation 6 to calculate ψ_1; (2.) Again use this in the right hand side of equation 6 to find ψ_2; (3.) and by the same way, using ψ_2 to find ψ_3, and so on. Therefore, the field functions ψ is given as:

$\psi(x) =\psi(x)+\int S_f(x-y)e\not{A}(r)\psi(r)d^4r$
$+\int S_{r\,(x-y)}\, e\not{A}(r)[\int S_f(x-y)e\not{A}(z)\psi(z)d^4z]d^4y+$----(22).

Now the completeness relations might be used for the spinors, such as:

$$\sum_{\uparrow\downarrow,r=1,2} u_\alpha^r (p,s) u_\beta^{r^-}(p,s) - v_\alpha^{r+2}(p,s)\, v_\beta^{r+2^-}(p,s) = \delta_{\alpha\beta} \text{---(23)}.$$

If it is multiplied from both side by the operator $[\Lambda_+(p)]_{\tau\alpha}$, summing over α will give (remember that $\Lambda_+ v=0$):

$$\sum_{\uparrow\downarrow,r=1,2} u_\tau^r(p,s)\, u_\beta^{r^-}(p,s) = [\,\Lambda_+(p)]_{\tau\beta}, \text{-------------(24)}.$$

If it is multiplied by the operator $[\Lambda_-(p)]_{\tau\alpha}$, and sum over α, it gives:

$$\sum_{\uparrow\downarrow,r=1,2} v_\tau^{r+2}(p,s)\, v_\beta^{r+2^-}(p,s) = -[\Lambda_-(p)]_{\tau\beta}, \text{--------} (25).$$

Therefore, equations 12a and 12b take the following forms:

$$\psi_p^r(x)=e^{-ipx} u^r(p)\sqrt{\frac{m}{E_p V}}\,; r=1,2\,.\, E_p >0, \text{ for 12a,}$$

$$\psi_p^{r^-}(p)=e^{ipx} v^r(p)\sqrt{\frac{m}{E_p V}}\,; r=1,2.\, E_p <0, \text{ for 12b.}$$

$$\left.\begin{array}{c} \\ \\ \end{array}\right\} \text{----- (26).}$$

Combining equations 24, 25, and 26 with 20 and 21, it is found for the case (x-y) >0. Green function is given by

$$S_f(x-y)=-i \int \frac{d^3 p}{(2\pi)^3} v \sum_{r=1,2} \psi_p^r(x)\, \psi_p^{r^-}(x), \text{----------------(27)}.$$

And for(x-y) <0 is given by

$$S_f(x-y)=-i \int \frac{d^3 r}{(2\pi)^3} v \sum_{r=3,4} \psi_p^r(x)\psi_p^{r^-}(x), \text{--------------(28)}.$$

Now if $\mathbb{Q}(x)$ is a function defined as
$\mathbb{Q}(x-y)=1, for\ (x-y)>0$ and=0 for (x-y)<0, then it is possible to combine equations 27 and 28 in one equation given by

$$S_f(x\text{-}y) = -i \int \frac{d^3p}{(2\pi)^3} \, v \, [\Sigma_{r=1,2} \, \psi_p^r(x)\psi_p^{r-}(p)\mathbb{Q}(x-y) +$$

$$\psi_p^{r+2}(x)\psi_p^{(r+2)^-}(x)\mathbb{Q}(x-y)]\text{-----------------------}(29).$$

So it is clear if $\mathbb{Q}$ (x-y)=0, the only equation 27 remains.

10.12. *Scattering Amplitude*

So far, all required formalisms and tools to cope with describing the scattering events, which are types of interactions between the projectiles and the targets are delt with. The important factor to be calculated is the scattering amplitude.

To start toward the calculation of this amplitude,

let $\psi_{sc.} = \not\psi - \psi$.

It represents the scattering solution in the limiting cases. So the field functions ψ is given as

$$\not\psi = \psi(x) + \int S_f(x\text{-}y) \, e \, \not A(r)\psi(r) \, d^4(r).\text{-------------}(30a).$$

And in the limit x→ ∞, (positive time intervals), S_f is given as S_f

$$(x\text{-}y) \rightarrow^{x\to\infty} - i \int \frac{d^3}{(2\pi)^3} v \, \Sigma_{r=1,2} \, \psi_p^r(x)\psi_p^{(r)^-}(r), \text{ and}$$

$$\psi_{sc}\underrightarrow{x \to \infty} - ie\int \frac{d^3p}{(2\pi)^3} v \, \Sigma_{r=1,2.} \, \psi_p^r(x)\int \psi_p^{(r)^-}(r)\not A(r)\psi(r)d^4r.$$

--------- (30b).

This corresponds to electron scattered via an external potential, a positive-energy solution propagated into future.

On the other hand, for $x \to -\infty$, $S_f(x-y)$ is given by

$$S_f(x-y) \longrightarrow (x \to -\infty) \to -i \int \frac{d^3p}{(2\pi)^3} \quad v \quad \Sigma_{r=3,4.} \psi_p^r(x) \quad \psi_p^{(r)^-}(x),$$

which yields

$$\psi_{sc} \text{ (as } x \to -\infty) \to -i \int \frac{d^3p}{(2\pi)^3} v \times$$
$$\Sigma_{r=3,4.} \psi_p^r(x)[-i \int d^4 r \psi_p^{r-}(r) A(r) \psi(r)] \text{------------(31a)}.$$

This describes a scattered positron through an external potential and propagated into the past.

For electron scattering (positive time intervals), going to equation 22, ψ can be written as

$$\psi = \psi(x) + \int \frac{d^3p}{(2\pi)^3} v \times$$
$$\Sigma_{r=1,2.} \psi_p^r(x)[-ie \int d^4r \ (\psi_p^{r-}(r) A(r) \psi_p^r(x)) + \text{----}] \text{------(31b)}.$$

The term between the bracket[ ] is denoted as S_1. It is the scattering amplitude to a first approximation. Since $A(r) = A_N \gamma_n$, the product in S_1 is relativistically invariant, i.e.,

$$(\psi_p^{(r)^-}(r) A(r) \psi_p^r(r)) = (\psi_p^{(r)^-}(r) \gamma_N \psi_p^r(x) A_N(r)).$$

10.13. *Electrons Scattering by the Protons*

In the previous section, the electron scattering due to external potential was briefly treated. S in this section, the electrons-protons scattering will primarily be studied. If the electron is scattered by the proton as a target, then A is given by the equation, $\square A_N = -J_N$, where J_N is the proton current. The solution of this equation is similar to the solution of Dirac equation.

One gets $D_f(x-y) = \int \frac{d^4q}{(2\pi)^4} e^{-iq(x-y)} D_f (q^2)$ ------------- (34).

Thus, equation 32 becomes $q^2D_f(q^2) = +1$

And $\qquad\qquad\qquad\qquad\qquad D_f(q^2) = +\dfrac{1}{q^2}$ ----- (35).

Or making D_f analytic, then $D_f(q^2) = limit_{\epsilon\to 0}\ \dfrac{+1}{q+\epsilon}$ -----(36).

Hence, equation 33 takes the form

$A_N(x) = \dfrac{+1}{(2\pi)^4}\int J_N(r)[\dfrac{d^4q}{q^2+i\epsilon}e^{-iq(x-y)}]d^4r$ ------------(37).

For proton has state ψ_1 before scattering and ψ_2 after scattering, the proton current is given by

$J_N(r) = e_p\psi_{p_2}^-(r)\ \gamma_n\psi_{p_1}(r),$ -------------------------(38).

Thus, equation 31b becomes:

$S_{21}^1 = -ie\int d^4r\ \psi_2^{e-}(r)\ \gamma_N\psi_1^e(r)\ A_N(r),$

Hence, $S_{21}^1 = -ie\int d^4r\ \psi_2^{e-}(r)\ \gamma_N\psi_1^e(r)\int d^4z\ D_f(r-z)J_N(z).$
Now substitute for $J_N(z)$ along z-axis. S_{21}^1 becomes as

$[S_{21}^1 = -ie_p\int d^4r\ \psi_2^{e-}(r)\ \gamma_N\psi_1^e(r)\int d^4z D_f(r-z)e_p\psi_2^{p-}(z)$
$\times\ \gamma_N\psi_1^p(z).$ ---(39)]

Where $\psi^p(r) = \sqrt{\dfrac{m_p}{E_pV}}u(p,s)e^{-ip\cdot r}$, for the proton.

And $\psi^e(r) = \sqrt{\dfrac{m_e}{E_eV}}u(p,s)e^{-ip\cdot r}$, for the electron.

Now use these information and the relation in equation 39.

The following are obtained:

$$S_{21}^1 = ie^2 \sqrt{\frac{m_e m_p}{E_p^1 E_p^2 E_e^1 E_e^2}} \frac{1}{V^2} \int d^4 z (e^{i(p_2 - p_1 + p_2 - p_1)z})$$

$\int d^4(r-z) \, e^{i(p_2 - p_1)(r-z)} \, D_f(r-z)$. With the use of the delta function definition and properties, it yields

$$S_{21}^1 = [-ie^2 \sqrt{\frac{m_e m_p}{m_p^1 m_p^2 \, m_e^1 m_e^2}} \frac{(2\pi)^4}{V^2} \delta^4(p_2 - p_1 + p_2 - p_1) u^-(p_2, s_2) \gamma_N \times$$

$$u(p_1, s_1) \frac{1}{q^2 + i\epsilon} u^-(p_2, s_2) \gamma_N u(p_1, s_1) \text{------------------ (40).}$$

Thus, $\psi = \psi_1^e(x) + \int \frac{d^3 p}{(2\pi)^3} V \psi_1^e(x) S_{21}^1 + -- = \psi_1^e(x) + \psi_{sc}.(41).$

So $\psi_{sc}(x) = \int \frac{d^3 p \, V}{(2\pi)^3} \psi_1^e(x) S_{21}^1 + --- = \psi - \psi_1^e(x) \text{ --------(42)}.$

If the energy is conserved, then $\delta^4 \rightarrow \delta^3(p_2 - p_1 + p_2 - p_1)$. Therefore,

$$\psi_{sc} = -ie^2 \sqrt{\frac{m_e^2 m_p^2}{E_{e1} E_{e2} E_{p1} E_{P1}}} \frac{(2\pi)^4}{V} \int \psi_1^e(x) \delta^3(p_2 - p_1 + p_2 - p_1) \text{- (43)}.$$

Or

$$\psi_{sc} = -$$

$$ie^2 \sqrt{\frac{m_e^2 \, m_p^2}{m_{e1} m_{e2} m_{p1} p_2}} \frac{(2\pi)^4}{V} \psi_1^e(x) u^-(p_2, s_2) \gamma_N u(p_1, s_1) (\frac{1}{q^2 + i\epsilon}) \times$$
$$u^-(p_2, p_2) \, \gamma_N u(p_1, s_1)$$

Where $\vec{p_2} - \vec{p_1} + \vec{p_2} - \vec{p_1} = 0. \rightarrow \delta^3 = 1$, then the matrix is given by
$M_{21}^1 = <\psi_1^e(x) \mid \psi_{sc}(x) > = $
$-ie^2 \sqrt{K^2} u^-(p_2, s_2) \, \gamma_N u(p_1, s_1) (\frac{1}{q^2 + i\epsilon}) u^-(p_2, s_2) \gamma_N u(p_1, s_1).$

$\cong j_N^e \frac{1}{q^2} j_N^p = H_{int.}$ (Hamiltonian). ------------------ (44).

j_N^e is the electron current density, and

j_N^p is the proton current density. $q^2 = \dfrac{1}{D_f(q^2)} = \dfrac{1}{D_f(x-y)}$.

Where D(x-y) was already defined as

$$\Box \, xD_f(x\text{-}y) = \delta^4(x-y).$$

10.14. *Some Important Applications*

10.14.1 *Introduction to Isospace and Isospin*

The concepts of isospin and isospace are of great importance in nuclear physics and the so-called physics of elementary particles. Therefore, it is quite important and very useful to summarize them in this section, pointing out their roles in these two fields of physics. As it is well known to physicists in these fields that the underlying symmetry transformations belong, roughly speaking, to two classes:

(1.) Transformations which are affecting space-time.
(2.) Transformations which are independent of space and time behavior of the fields, such as the charge conjugation and electromagnetic gauge transformation.

It had been found that the neutron (n) and the proton (p) have the same properties apart from the charged and slight difference in the mass which might be explained as due to self-energy effect. In fact, not only the properties of the members of the charged multiplet (n.p), ($\pi,^+\pi^0$, π^-)are alike but also behave similarly in their interaction. Hence, this suggests that there is a deeper relationship among them; it also indicates that there must be some additional degree of freedom as intrinsic property of the field, and there exists some transformation groups related to these new variables.

To show how to deal with these problems, it is useful to introduce φ as the pion field which is a complex field. Its exchange particles are (π^+, π^0), which are quanta called pions.

Also, there is the neutral field φ_0 which its exchange particle (quanta) is the neutral pion (π^0). The electromagnetic gauge transformation for these fields $(\varphi, \varphi^*, \varphi^0)$ is

$$\varphi \to e^{-i\epsilon} \; ; \varphi^* \to e^{+i\epsilon} \; ; \varphi_0 \to \varphi_0. \text{------------------------} (1).$$

Where φ is a complex field; it can be split into two real (Hermitian) parts:

$$\varphi = \frac{1}{\sqrt{2}}(\varphi_1 + i\varphi_2); \; \varphi^* = \frac{1}{\sqrt{2}}(\varphi_1 - i\varphi_2); \; \varphi_0 = \varphi_0 \text{--------------}(2).$$

By rearranging terms in equation 2 such that it can be rewritten as $\varphi_1 = \frac{1}{\sqrt{2}}(\varphi + \varphi^*); \; \varphi_2 = \frac{1}{i\sqrt{2}}(\varphi - \varphi^*) \; ; \; \varphi_3 = \varphi_0$ (3).

From equations 1 and 3, φ_1, φ_2, and φ_3 take the following forms:

$$\varphi_1 \to \frac{1}{\sqrt{2}}[(\cos\epsilon - i\sin\epsilon)\varphi) + (\cos\epsilon + i\sin\epsilon)\varphi^*]$$

$$= \cos\epsilon \cdot \varphi_1 + \sin\epsilon\varphi_2.$$

Similarly, $\varphi_2 \to -\sin\epsilon \cdot \varphi_1 + \cos\epsilon \cdot \varphi_2 \; ; \varphi_3 \to \varphi_3$

This can be written as:

$$\begin{bmatrix} \varphi_1 \\ \varphi_2 \\ \varphi_3 \end{bmatrix} \to \begin{bmatrix} \cos\epsilon & \sin\epsilon & 0 \\ -\sin\epsilon & \cos\epsilon & 0 \\ 0 & 0 & 1 \end{bmatrix} \begin{bmatrix} \varphi_1 \\ \varphi_2 \\ \varphi_3 \end{bmatrix} \text{--------------} (4).$$

Hence, in general, $\varphi^{\rightarrow} \rightarrow A\varphi^{\rightarrow}$ or $\varphi^- = A\varphi$; therefore:

$$A = \begin{bmatrix} cos\epsilon & sin\epsilon & 0 \\ -sin\epsilon & cos\epsilon & 0 \\ 0 & 0 & 1 \end{bmatrix}$$

This shows that $\varphi^{\rightarrow}$ transforms under gauge transformation with the phase angle(ϵ), precisely in the same manner as a three-dimensional vector transformation under rotation about the third coordinate axis by an angle ϵ. Therfore, this leads to consider these fields which are referred to the charged and neutral pions, as the components of three-vector in some abstract space (which is completely distinct from the ordinary space-time). It is called isospace or isotopic space, and some books called it as charge space.

Now what is the isobaric space? Or what is the physical concept of the isobaric spin? The isobaric spin concept is physically based on the fact that the proton and the neutron are strikingly similar in their properties, as mentioned before, which had led the nuclear physicists to suggest their common representation to a particle called nucleon. So the proton and the neutron are different energy states for this nucleon as pointed out before.

The state function of the nucleon is given as $\psi(x, s, \tau)$, where x is the ordinary space coordinates and s is the spin coordinate and τ is the isospin coordinate. It distinguishes between the two possible charged states, so τ has the values (± 1). Hence, $\psi_p(x, s, +1)$, p-state, $\tau = +1$.

And $\psi_n(x, s, -1)$, n-state, $\tau = -1$.

Now instead of using τ) as internal coordinate, the nucleon state function might be written as $\psi = \begin{bmatrix} u_1(x, s) \\ u_1(x, s) \end{bmatrix}$ ----------- (5).

Let τ_3 be introduced as 2×2 matrix operator associated with the isotopic spin τ, given as $\tau_3 = \begin{bmatrix} 1 & 0 \\ 0 & -1 \end{bmatrix}$ -------- (6),

where $\left.\begin{array}{l} \tau_3\psi_p = +1\ \psi_p \\ \tau_3\psi_n = -1\psi_n \end{array}\right\}$ ---------- (7).

As it is known, ψ_p *corresponds to the eigenvalue* (+1) and ψ_n correspondds to the eigenvalue (-1).

From equation 5, ψ can be written as

$$\psi = \mu_1 \begin{bmatrix} 1 \\ 0 \end{bmatrix} + \mu_2 \begin{bmatrix} 0 \\ 1 \end{bmatrix} = \begin{bmatrix} \mu_1 \\ 0 \end{bmatrix} + \begin{bmatrix} 0 \\ \mu_2 \end{bmatrix} = \begin{bmatrix} \mu_1 \\ \mu_2 \end{bmatrix}$$

$$\psi_p = \mu_1 \begin{bmatrix} 1 \\ 0 \end{bmatrix} \ ; \psi_n = \mu_2 \begin{bmatrix} 0 \\ -1 \end{bmatrix} \ ; \rightarrow \psi_p = \eta_p\mu_1, \ \psi_n = \eta_n\mu_2 \ .$$

Where $\eta_p = \begin{bmatrix} 1 \\ 0 \end{bmatrix}$, $\eta_n = \begin{bmatrix} 0 \\ -1 \end{bmatrix} \longrightarrow \psi = \eta_p\psi_p + \eta_n\psi_n$ ------- (8).

μ_1 *and* μ_2 are ordinary four-component Dirac spinors (chapter 8), in the usual Minkawiski space-time. So altogether, ψ has eight components. Now let $\vec{T}$ be defined as the isotopic spin which is given as $\vec{T} = \frac{1}{2}\vec{\tau}$, where $\vec{\tau} = (\tau_1, \tau_2, \tau_3)$, and $\vec{T} = (T_1, T_2, T_3).\tau_1, \tau_2, \tau_3$, are the familiar Pauli matrices (in special representation) which satisfy the following known relations:

$$\tau_i\tau_j = i\tau_k, \tau_i\tau_k + \tau_k\tau_i = 2\delta_{ik}, \text{---------------------- (9)}.$$

τ_1, τ_2, τ_3 are 2×2 matrices, where

$$\tau_1 = \begin{bmatrix} 0 & 1 \\ 1 & 0 \end{bmatrix} \ ; \tau_2 = \begin{bmatrix} 0 & -i \\ i & 0 \end{bmatrix} \ ; \tau_3 = \begin{bmatrix} 1 & 0 \\ 0 & -1 \end{bmatrix} \text{--------- (10)}.$$

τ_i Operates on ψ_p and ψ_n as follows:

$$\tau_1\psi_p = \psi_n \; ; \tau_1\psi_n = \psi_p \; ; \tau_2\psi_p = i\psi_n \; ;$$
$$\tau_2\psi_n = -i\psi_p \; ; \tau_3\psi_p = \psi_p \; ; \tau_3\psi_n = -\psi_n \; ;\text{----------}(11).$$

(As an exercise, verify equation 11.
Now with the use of the knowledge of the angular momenta, concerning the concept of the rising and the lowering operators, analogically, it is possible to construct τ_+ as rising operator and τ_- as lowering operator, which are given by

$$\tau_+ = \frac{1}{2}(\tau_1 + i\tau_2), \text{ and } \tau_- = \frac{1}{2}(\tau_1 - i\tau_2) \text{ ------------- } (12).$$

Where $\tau_+\psi_p = 0; \; \tau_+\psi_n = \psi_p \; ; \tau_-\psi_p = \psi_n \; ; \tau_-\psi_n = 0.$

From equations 10 and 12, it can be shown that:

$$\tau_+ = \begin{bmatrix} 0 & 1 \\ 0 & 0 \end{bmatrix} ; \tau_- = \begin{bmatrix} 0 & 0 \\ 1 & 0 \end{bmatrix} \text{ ----------------- } (13).$$

Now define $\vec{T_i} = \frac{1}{2}\vec{\tau_i}$, where $\vec{T_i}$=isobaric spin, and$\vec{\tau_i}$ is its associated operator. Therefore, $T_1 = \frac{1}{2}\tau_1 \; ; T_2 = \frac{1}{2}\tau_2; T_3 = \frac{1}{2}\tau_3;$ hence,

$$T_1 = \frac{1}{2}\begin{bmatrix} 0 & 1 \\ 1 & 0 \end{bmatrix}, \; T_2 = \frac{1}{2}\begin{bmatrix} 0 & -i \\ i & 0 \end{bmatrix}, \; T_3 = \frac{1}{2}\begin{bmatrix} 1 & 0 \\ 0 & -1 \end{bmatrix} \rightarrow$$

$$T_1 = \begin{bmatrix} 0 & \frac{1}{2} \\ \frac{1}{2} & 0 \end{bmatrix}, \; T_2 = \begin{bmatrix} 0 & \frac{-i}{2} \\ \frac{1}{2} & 0 \end{bmatrix}, \; T_3 = \begin{bmatrix} \frac{1}{2} & 0 \\ 0 & \frac{-1}{2} \end{bmatrix} \text{ --------- } (14).$$

Now it is important to construct the isospace coordinates as

$\vec{\xi} \rightarrow (\xi_1, \xi_2, \xi_3).$

$\psi^- = \psi^- (x, s, \xi_1 + \epsilon\xi_2, \xi_2 - \epsilon\xi_1, \xi_3).$

$= \psi + \epsilon\xi_1 \dfrac{\partial\psi}{\partial\xi_2} \cdot \epsilon \dfrac{\partial\psi}{\partial\xi_1} + \cdots \xi_2\xi_1 \cong$

$1 + \epsilon(\xi_1 \dfrac{\partial}{\partial\xi_2} \cdot \xi_2 \dfrac{\partial}{\partial\xi_1})\psi = (1 + i\epsilon T_3) = e^{\frac{i\epsilon\tau_3}{2}}\psi.$

Hence, $\psi^- = e^{\frac{i\epsilon\tau_3}{2}}\psi$, this indicates that τ_3 is the generator of the transformation in isospace about ξ_3. Similarly the transformation about the coordinates ξ_1 and ξ_2 is given by

$\psi^- = e^{\frac{i\epsilon\tau_1}{2}}\psi$ and $\psi^- = e^{\frac{i\epsilon\tau_2}{2}}\psi$. So $\psi^-(x, s, \xi^-) = A\psi(x, s, \xi),$

where $A = e^{\frac{i\epsilon\tau_1}{2}} e^{\frac{i\epsilon\tau_2}{2}} e^{\frac{i\epsilon\tau_3}{2}} = e^{\frac{i\epsilon(\tau_1 + \tau_2 + \tau_3)}{2}}.$

So $A\,A^+ = 1$, and also its det. $= 1$; therefore, A is a unitary operator. So far, it has been shown that ψ is a vector in isospace; and also, $\varphi = \begin{bmatrix} \varphi_1 \\ \varphi_2 \\ \varphi_3 \end{bmatrix}$ is a vector in isospace too. In the previous treatment, a fair detail was given to the isospace and the isobaric spin with their properties and formalism. It might be important, for more clarification, to the underline subject to consider the pion-nucleon interaction $(\pi - N)$ as another application. This might check $(\psi^- \gamma_5 \psi)$ as pseudoscalar.

To find the Lagrange of the interaction L$_{int.}$ and L$_{EM.}$, also to show that L is scalar in isospace and in the Minkawiski space too. These are in the following section.

10.15. $\pi - N$ Interaction

The $\pi - N$ interaction is quite well represented by the well-famous Feynman diagrams which are as follows:

(i.) $L_P = iG\bar{\psi}_p\gamma_5\psi_p\varphi_0$.

(ii.) $L_n = iG\bar{\psi}_n\gamma_5\psi_n\varphi_0$

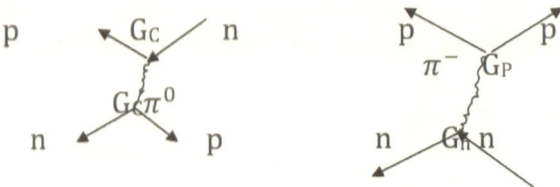

(iii.) $L_{np} = iG\bar{\psi}_p\gamma_5\psi_n\varphi^* + iG_c\bar{\psi}_n\gamma_5\psi_p\varphi$

Hence, $L_{int.} = L_p + L_n + L_{np}(L_{np(1)} + L_{np(2)}) =$
$iG_p\bar{\psi}_p\gamma_5\psi_p\varphi_0 + iG_n\bar{\psi}_n\gamma_5\psi_n\varphi_0$
$+iG_c\bar{\psi}_p\gamma_5\psi_n\varphi^* + iG_c\bar{\psi}_n\gamma_5\psi_p\varphi$. ($G_c$ = coupling constant to n-p coupling.)

Since all these diagrams give the same contribution to the scattering, the following relations should hold:

$G_P^2 = G_n^2 = G_pG_n + G_C^2 \rightarrow G_p = G_n = G_c = 0$ is excluded.

Or, $G_p = -G_n = \frac{1}{\sqrt{2}}G_C = \frac{1}{\sqrt{2}}G$, -- G is the coupling strength.

So Lagrangian of interaction shown by these diagrams is:

$L_{int.} = iG[\bar{\psi}_p\gamma_5\psi_n\varphi^* + \bar{\psi}_n\gamma_5\psi_p\varphi + \frac{1}{\sqrt{2}}(\bar{\psi}_p\gamma_5\psi_p\varphi^* - \bar{\psi}_n\gamma_5\psi_n)\varphi_0]$ --------(1).

Now $\psi = \begin{bmatrix} \psi_p \\ \psi_n \end{bmatrix}$; $\quad \psi^- = [\, \psi_p, \ \psi_n]$

Therefore $\psi^- \tau_1 \psi = \psi_p^- \psi_n + \psi_n^- \psi_p$

And $\quad \psi^- \tau_2 \psi = -i\psi_p^- \psi_n + i\,\psi_n^- \psi_n$

These implies the following:

$$\left. \begin{aligned} &\psi_p^- \psi_n = \tfrac{1}{2} [\psi^- \tau_1 \psi + i\,\psi^- \tau_2 \psi\,]. \\ &\psi_n^- \psi_p = \tfrac{1}{2} [\psi^- \tau_1 \psi - i\psi^- \tau_2 \psi\,]. \\ &\text{Also,} \\ &\psi^- \tau_3 \psi = \psi_p^- \psi_p - \psi_n^- \psi_n, \ \varphi_0 = \varphi_3, \\ &\varphi = \tfrac{1}{\sqrt{2}} (\varphi_1 + i\varphi_2), \ \varphi^* = \tfrac{1}{\sqrt{2}}(\varphi_1 - i\varphi_2). \end{aligned} \right\} \text{--------------(2).}$$

Now substituting equation 2 into equation 1 and letting 1/2 G≡G, then L$_{int}$ is given by L$_{int}$ = $iG\psi^- \tau_i \gamma_5 \psi \varphi$ ------------(3).

So after finding the interaction Lagrangian, the next is to show that $\psi^- \gamma_5 \psi$ is pseudoscalar. From chapter eight, it is known that $\gamma_5 = \gamma_1 \gamma_2 \gamma_3 \gamma_4$, where this is given by

$\gamma_5 = \begin{bmatrix} 0 & 0 & -1 & 0 \\ 0 & 0 & 0 & -1 \\ -1 & 0 & 0 & 0 \\ 0 & -1 & 0 & 0 \end{bmatrix}$, where $\gamma_1 \gamma_2 \gamma_3 = i\alpha_1 \alpha_2 \alpha_3 \beta^4 =$

$= i\alpha_1 \alpha_2 \alpha_3, \ \beta^4 = \gamma_1^4 = \gamma_0^4 = 1.$

So $\alpha_k = \begin{bmatrix} 0 & \sigma_k \\ \sigma_k & 0 \end{bmatrix}$, for k =1, 2, 3.

$$\alpha_1 \alpha_2 \alpha_3 = \begin{pmatrix} 0 & 0 & 0 & 1 & 0 \\ 0 & 0 & 1 & 0 & 0 \\ 0 & 1 & 0 & 0 & 0 \\ 1 & 0 & 0 & 0 \end{pmatrix} \begin{pmatrix} 0 & 0 & 0 & -i \\ 0 & 0 & i & 0 \\ 0 & -i & 0 & 0 \\ i & 0 & 0 & 0 \end{pmatrix} \begin{pmatrix} 0 & 0 & 1 & 0 \\ 0 & 0 & 0 & -1 \\ 1 & 0 & 0 & 0 \\ 0 & -1 & 0 & 0 \end{pmatrix}$$

$$= \begin{pmatrix} 0 & 0 & i & 0 \\ 0 & o & 0 & i \\ i & 0 & 0 & 0 \\ 0 & i & 0 & 0 \end{pmatrix}, \text{ so } \gamma_1\gamma_2\gamma_3\gamma_4 = i\alpha_1\alpha_2\alpha_3 = \begin{pmatrix} 0 & 0 & -1 & 0 \\ 0 & 0 & 0 & -1 \\ -1 & 0 & 0 & 0 \\ 0 & -1 & 0 & 0 \end{pmatrix}$$

$$= \gamma_5 .$$

Now if $\psi^- \to -\psi$ under inversion and $\to \psi$ under rotation, it means ψ is pseudoscalar; hence, $\psi^-\gamma_5\psi = \psi^+\gamma_0\gamma_5\psi \to$
$\psi^- = A\psi$, $\psi^+ = A^+\psi^+ \to \psi^-\psi^+ = A A^+\psi^+\psi = \psi^+\psi$.

Hence, $\psi^-\gamma_5\psi \to \psi^-\gamma_5\psi$, under rotation, so $\psi^-\gamma_5\psi = (\psi^-\gamma_0\gamma_5\psi)$, but $\Pi(\psi^-\gamma_5\psi) =- (\psi^-\gamma_5\psi)$. Π is parity operator, inversion operation. Since $\Pi\gamma_5 = \gamma_0^{-1}\gamma_5\gamma_0 = -\gamma_0^{-1}\gamma_0\gamma_5 = -\gamma_5 \Rightarrow \psi^-\gamma_5\psi^-\psi\gamma_5\psi = -\psi^-\gamma_5\psi$, which implies that ($\psi^-\gamma_5\psi$) is pseudoscalar.

Now $L = iG\psi^-\tau_i\gamma_5\psi\varphi$ is scalar in isospace. To show this, first remember that γ_5 is a constant in isospace, so take

$\psi^-\tau_i\psi = R_i$. Also, it was shown that $\psi \to e^{\frac{i\epsilon\tau_3}{2}}\psi$, and
$\psi^- \to e^{\frac{-i\tau_3}{2}}\psi^-$. Therefore, $\psi^-\tau_i\psi = \psi^- e^{\frac{-i\epsilon\tau_3}{2}}\tau_i e^{\frac{i\epsilon\tau_3}{2}}\psi =$
$\psi^-(\cos\frac{\epsilon}{2} - i\tau_3 \sin\frac{\epsilon}{2})\tau_i (\cos\frac{\epsilon}{2} + i \tau_3\sin\frac{\epsilon}{2})\psi.$

Now complete the multiplication, remembering the mentioned tau (τ_i) relations property. The following are obtained:

$R_1 = \psi^-(\tau_1 \cos\epsilon + \tau_2\sin\epsilon)\psi = R_1 \cos\epsilon + R_2 \sin\epsilon$,
$R_2 = -R_1 \sin\epsilon + R_2 \cos\epsilon, R_3 = R_3.$

Therefore, $\psi^- \tau_i \, \varphi$ transforms like a vector in isospace. It was early proven that φ transforms in isospace as a vector. Hence, L=$iG\psi^-\gamma_5\tau\psi\varphi$) is scalar.

Now it is important to show that the Lagrangian of the electromagnetic field is given by L$_{em}$ =$\psi^-Q\gamma_\mu\psi A_\mu$. Where Q is the charge given by Q =(e/2) (1+τ_3). Therefore, L$_{em}$=$e\psi^-(\frac{1}{2})[(\tau_3 +1)]\gamma_\mu\psi A_\mu$, where $\psi^-(1/2)(\tau_3 +1)\psi=$

$[\psi_p^-\psi_n\,]\,1/2[\begin{bmatrix} 1 & 0 \\ 0 & -1 \end{bmatrix} + 1]\begin{bmatrix} \psi_p \\ \psi_n \end{bmatrix}$ $=\psi_p^-\psi_p$.

Hence, [L$_{em}$=$e\psi_p^-\psi_p\gamma_\mu A_\mu$]-----------------------(4).

10.16. *More Illustration Examples*

In this section, calculations and derivations will concern the following important physical dynamical variables:

   (1.) Lagrangian for the electromagnetic field, L$_{em}$.
   (2.) $\partial A_\mu=0$
   (3.) L$_{el.}$ =$-\frac{1}{2}\partial_\mu A_\nu\partial_\mu A_\mu$= $-\frac{1}{4}F_{\mu\nu}F_{\mu\nu}$
   (4.) $\square$ $A_\mu=0$.
   (5.) $T_{\mu\nu}$(momentum-energy tensor, $T_{\mu\nu}/\nu=0$.
   (6.) S$_k$, the pointing vector, $G_k^{\rightarrow}$ the momentum field density($E^{\rightarrow} \times H^{\rightarrow}$), and T$_{ik}$ the energy momentum tensor.

To work out all these, the basic formalisms of the following have to be prepared:

   (1.) Maxwell equations, electromagnetic tensor $F_{\mu\nu}$.
   (2.) Gauge transformation, Lorentz condition.
   (3.) $\frac{\partial T_{\mu\nu}}{\partial x_\nu}$ =$T_{\mu\nu}/\nu$ =0, in general.

(4.) Maxwell equations, div. $D^{\rightarrow}=0$, Curl $H^{\rightarrow}=\dfrac{\partial D^{\rightarrow}}{\partial t}$.

Using the variational principle to obtain,

(5.) $\square$ $A_\mu =0$, to show $L_{el.}=-\dfrac{1}{4}F_{\mu\nu}F_{\mu\nu}$.

(6.) *To obtain $T_{\mu\nu}$ in symmetrical form from Lel.*

(7.) Then obtain S_k, G_K, and T_{ij}.

In the following sections, the answers for these to be worked out:

(1.) Maxwell Equations and Electromagnetic Tensor $F_{\mu\nu}$

From the classical electrodynamics, $E^{\rightarrow}$ *and* $H^{\rightarrow}$ form the antisymmetrical electromagnetic field tensor which is denoted as $F_{\mu\nu}$, is given by:

$$F_{\mu\nu} = \begin{pmatrix} 0 & H_z & -H_y & -iE_x \\ -H_z & 0 & H_x & -iE_y \\ H_y & -H_x & 0 & -iE_z \\ iE_x & iE_y & iE_z & 0 \end{pmatrix} \qquad \text{-----------------(1).}$$

Where $F_{14}=iE_x$, $F_{24}=-iE_Y$, $F_{34}=-iE_z$, ----------- (2).

$F_{12}=H_z$, $F_{31}=H_y$, $F_{23}=H_x$,------------------------ (3).

And $F_{\mu\nu}=-F_{\nu\mu}$, $F_{ii}=-F_{ii}=0$, diagonal.----------------(4)

The free field Maxwell equations are given by:

Div· $B^{\rightarrow}=0$, Curl $E^{\rightarrow}=0$.}, $\dfrac{\partial F_{\mu\nu}}{\partial x_\rho}+\dfrac{\partial F_{\nu\mu}}{\partial x_\mu}+\dfrac{\partial F_{\rho\mu}}{\partial \nu}=0$, -------- (5).

And Div· $D^{\rightarrow}=0$, Curl$H^{\rightarrow}=\dfrac{\partial D^{\rightarrow}}{\partial t}$}, $\dfrac{\partial F_{\mu\nu}}{\partial x_\nu}=0$, -----------(6).

The tensor components are defined in terms of the four-vector potential $A_\mu(A^{\rightarrow}, iA_4)$ such as:

$$F_{\mu\nu}=\dfrac{\partial A_\nu}{\partial x_\mu}-\dfrac{\partial A_\mu}{\partial x_\nu}=\partial_\mu A_\nu - \partial_\nu A_\mu \text{----------------------- (7).}$$

By taking the cyclic permutation of the derivative $\frac{\partial F_{\mu\nu}}{\partial x_\rho}$, equation 5 can be obtained from equation 7.

(2.) Gauge Transformation, Lorentz Condition

It is known that A_μ is not uniquely determined by the tensor $F_{\mu\nu}$, since the addition to A_μ of the gradient of any scalar quantity does not alter the field. It is due to the fact that the Maxwell equations are invariant under the transformation

$A_\mu \rightarrow A_\mu^- +\frac{\partial \eta}{\partial x_\mu}$; this is known as gauge or gradient transformation, where η is an arbitrary scalar, which might be φ in some text books. Since the potential A_μ is not uniquely defined, a subsidiary condition may be imposed such as

$$\frac{\partial A_\mu}{\partial x_\mu} = 0, \text{--- (8)}.$$

This imposed condition is called Lorentz condition.
Equation 8 restricts gauge transformations to those functions of (η) for which $\square \eta = 0$. This implies that to satisfy equation 8, it is sufficient to require that at t=0, $\chi = \frac{\partial A_\mu}{\partial x_\mu}$ and $\frac{\partial \chi}{\partial t}$ should vanish (=0).

(3). To show that $\frac{\partial T_{\mu\nu}}{\partial x_\nu} = T_{\mu\nu/\nu} = 0$. Consider $\frac{\partial L}{\partial x_\nu}$, where L=

=L $(A_P, \frac{\partial A_P}{\partial x_\sigma})$, and use the known Euler-Lagrange equation, as mentioned before. So

$\frac{\partial L}{\partial x_\nu} = \frac{\partial L}{\partial \left(\frac{\partial A_\rho}{\partial x_\sigma}\right)} \frac{\partial \left(\frac{\partial A_\rho}{\partial x_\sigma}\right)}{\partial x_\nu} + \frac{\partial L}{\partial A_\rho}\frac{\partial A_\rho}{\partial x_\nu}$. Now from Euler-Lagrange

equation, substitute for $\frac{\partial L}{\partial A_\rho}$, i.e., $\frac{\partial L}{\partial A_\rho} = \frac{\partial}{\partial x_\sigma}\frac{\partial L}{\partial \left(\frac{\partial A_\rho}{\partial x_\sigma}\right)}$. Therefore,

$$\frac{\partial L}{\partial X_\nu} = \frac{\partial}{\partial X_\nu}\left(\frac{\partial A_\rho}{\partial X_\sigma}\right)\frac{\partial L}{\partial\left(\frac{\partial A_\rho}{\partial X_\sigma}\right)} + \frac{\partial A_\rho}{\partial X_\nu}\frac{\partial}{\partial X_\sigma}\left(\frac{\partial L}{\partial\left(\frac{\partial A_\rho}{\partial X_\sigma}\right)}\right),$$ which might be

combined into, $\dfrac{\partial L}{\partial X_\nu} = \dfrac{\partial}{\partial X_\sigma}\left(\dfrac{\partial A_\rho}{\partial X_\sigma} \cdot \dfrac{\partial L}{\partial\left(\frac{\partial A_\rho}{\partial X_\sigma}\right)}\right),$--------------(9).

Since the second derivations are continuous, let $\mu = \sigma$, then equation 9 takes the form:

$$\frac{\partial}{\partial X_\mu}\left(L\delta_{\mu\nu} - \frac{\partial L}{\partial\left(\frac{\partial A_\rho}{\partial X_\mu}\right)}\frac{\partial A_\rho}{\partial X_\nu}\right)=0, ------------- (10).$$

Now by rearranging indices, the term between the parentheses in equation 10 is the energy-momentum tensor, so

$$T_{\mu\nu} = \left[L\,\delta_{\mu\nu} - \frac{\partial L}{\partial\left(\frac{\partial A_\rho}{\partial X_\nu}\right)}\frac{\partial A_\rho}{\partial X_\mu}\right]----------- (11).$$

Therefore, from equations 9 and 10, it can be found that:

$$\frac{\partial T_{\mu\nu}}{\partial X_\nu} = T_{\mu\nu/\nu} = 0. \text{ (As it is required to be.)}--------- (12).$$

Now it is required that $\dfrac{\partial F_{\mu\nu}}{\partial X_\nu} = 0$, (equation 6). This can be shown with the aid of the variational principle which was mentioned before. It can be done by assuming that a definite Lagrangian L corresponds to the electromagnetic field is regarded as a generalized dynamical system, such as $A_\mu(x) \to q$ generalized coordinates and $\dfrac{\partial A_\mu}{\partial X_\nu} \to p$ generalized momenta. From the electrodynamics theory, it is known that the electromagnetic field must possess gauge invariance property in order the Lagrangian L can contain only $F_{\mu\nu}$ and not the components of the four-potential A_μ directly. So from the field components, it is possible to construct only two independent invariants. They are:

1. $E^2 - H^2$,
2. $(\vec{E} \times \vec{H})^2$.

If it is required that the superposition principle has to be held, that is to say the field equations should be linear, the Lagrangian density can only be a quadratic function in the field components, i.e., $L \propto (E^2 - H^2)$.

Now consider

$$L = \frac{1}{2}(E^2 - H^2) = -\frac{1}{4}F_{\mu\nu}F_{\mu\nu} = -\frac{1}{2}\partial_\mu A_\nu \partial_\mu A_\nu \text{ --------------(13).}$$

Hence, $L \neq L(\frac{\partial A_\mu}{\partial X_\nu})$, directly. So $\frac{\partial L}{\partial A_\mu} = 0$.

The Euler-Lagrange equation (as found) is

$$\frac{\partial L}{\partial X_\nu}\frac{\partial L}{\partial(\frac{\partial A_\mu}{\partial X_\nu})} - \frac{\partial L}{\partial A_\mu} = 0. \text{ It becomes } \frac{\partial L}{\partial X_\nu}\frac{\partial L}{\partial(\frac{\partial A_\mu}{\partial X_\nu})} = 0. \text{ ------- (14).}$$

Now consider $L = -\frac{1}{4}F_{\mu\nu}F_{\mu\nu} = -\frac{1}{4}F_{\mu\nu}^2$,

so $\delta L = -\frac{1}{2}F_{\mu\nu}\delta F_{\mu\nu} = -\frac{1}{2}F_{\mu\nu}(\delta\frac{\partial A_\nu}{\partial X_\mu} - \delta\frac{\partial A_\mu}{\partial X_\nu})$,

and inverting indices with the fact that $F_{\mu\nu} = -F_{\nu\mu}$,

then $\delta L = \frac{1}{2}F_{\mu\nu}(-2\,\delta\frac{\partial A_\mu}{\partial X_\nu}) = F_{\mu\nu}\delta(\frac{\partial A_\mu}{\partial X_\nu}) \Rightarrow \frac{\partial L}{\partial(\frac{\partial A_\mu}{\partial X_\nu})} = F_{\mu\nu}$ - (15)

so that $\left\{\frac{\partial}{\partial X_\nu}\frac{\partial L}{\partial\frac{\partial A_\mu}{\partial X_\nu}} = \frac{\partial F_{\mu\nu}}{\partial X_\nu}\right\} = 0$ ---------------------------- (16)

as it is required (see Maxwell equations sections).

Next to show that $\Box A_\mu = 0$, and $-\frac{1}{2}\partial_\mu A_\nu \partial_\mu A_\nu = -\frac{1}{4}F_{\mu\nu}F_{\mu\nu}$.

Take that $\Box A_\mu = 0$, from equation 16, $\frac{\partial F_{\mu\nu}}{\partial X_\nu} = 0$, using the fact that $F_{\mu\nu} = \frac{\partial A_\nu}{\partial X_\mu} - \frac{\partial A_\mu}{\partial X_\nu}$, then $\frac{\partial F_{\mu\nu}}{\partial X_\nu} = \frac{\partial}{\partial X_\nu}(\frac{\partial A_\nu}{\partial X_\mu} - \frac{\partial A_\mu}{\partial X_\nu}) = 0$

Now, switching indices and recalling the fact that

$F_{\mu v} = - F_{v\mu}$ and $F_{ij} = - F_{ji}$, then one finds:

$$\frac{\partial}{\partial X_v}(-2\frac{\partial A_\mu}{\partial X_v})=0=(\frac{\partial^2 A_\mu}{\partial X_v^2})= \Box A_\mu; \text{ hence, } \Box A_\mu =0, \text{------(17)}.$$

(As required.)

To show $L=-\frac{1}{2}\partial_\mu A_v \partial_\mu A_v$!

Consider $L= -\frac{1}{4} F_{\mu v} F_{\mu v} =-\frac{1}{4}(\frac{\partial A_v}{\partial X_\mu} - \frac{\partial A_\mu}{\partial X_v})(\frac{\partial A_v}{\partial X_\mu} - \frac{\partial A_\mu}{\partial X_v})$. Remember:

$F_{\mu v}=(\frac{\partial A_v}{\partial X_\mu} - \frac{\partial A_\mu}{\partial X_v})$. Carry on the multiplication, remembering that

$\frac{\partial A_\mu}{\partial X_\mu} =0$, and $\frac{\partial A_\mu}{\partial X_v} = - \frac{\partial A_v}{\partial X_\mu}$, the result will be:

$$L=-\frac{1}{2}\frac{\partial A_v}{\partial X_\mu}\frac{\partial A_v}{\partial X_\mu} = - (1/2)\partial_\mu A_v \partial_\mu A_v, \text{----------------} (18).$$

(As it should be.)

Next is to obtain the energy-momentum tensor designated as $T_{\mu v}$, in a symmetric form for the electric Lagrangian $L_{el.}$

From equation 11, it is known that $T_{\mu v} =L \, \delta_{\mu v}-\frac{\partial L}{\partial(\frac{\partial A_\rho}{\partial X_v})}\frac{\partial A_\rho}{\partial X_\mu}$.

And from equation 15, $F_{\rho v}=\frac{\partial L}{\partial(\frac{\partial A_\rho}{\partial X_v})}$; hence, $T_{\mu v}$ becomes

$T_{\mu v}=L \, \delta_{\mu v} - F_{\rho v}\frac{\partial A_\rho}{\partial X_\mu} =L \, \delta_{\mu v} +F_{v\rho}\frac{\partial A_\rho}{\partial X_\mu}$. Remember, $F_{\rho v} = - F_{v\rho}$.

Now to obtain a symmetric form, let the proposed term $\frac{\partial A_\mu}{\partial X_\rho} F_{v\rho}$

be subtracted. Such term has the form $\frac{\partial \, \Psi_{ijk}}{\partial X_\ell}$; Ψ is antisymmetric

tensor, since $\frac{\partial(A_i F_{k\ell})}{\partial X_\ell} =\frac{\partial A_i}{\partial X_\ell} F_{k\ell}+A_i\frac{\partial F_{k\ell}}{\partial X_\ell} 0$. (See equation 16.)

Therefore, $\frac{\partial(A_i F_{k\ell})}{\partial X_\ell} =\frac{\partial A_i}{\partial X_\ell} F_{k\ell}$. After the change, the new tensor will also satisfy that $\frac{\partial T_{\mu\nu}}{\partial x_\nu}=0$, since it is identically $\frac{\partial^2 \Psi_{ik\ell}}{\partial x_k \partial x_i}=0$.

Now using equations 11, 16, and 17, the following is obtained:

$$T_{\mu\nu}=L\delta_{\mu\nu} +F_{\nu\rho}\left(\frac{\partial A_\rho}{\partial x_\mu} - \frac{\partial A_\mu}{\partial x_\rho}\right)=L\,\delta_{\mu\nu} +F_{\nu\rho}F_{\mu\rho}, \text{ but } L=-\frac{1}{4}F_{\sigma\tau}^2.$$

Therefore, $[T_{\mu\nu} =F_{\nu\rho}F_{\mu\rho}-\frac{1}{4}\delta_{\mu\nu}F_{\sigma\tau}^2]$ ------------------------ (19).

This is the symmetric form of the energy-momentum tensor $T_{\mu\nu}$.

Now the following step is to find the energy flux density or the so-called the pointing vector (S_k), the momentum field density (G), the energy momentum tensor(T_{jk}), and the energy density $[\omega =\frac{1}{2}(E^2 + H^2)]$. To work this out, then by $T_{\mu\nu} =F_{\mu\rho}F_{\nu\rho}$ - $\frac{1}{4}\delta_{\mu\nu}F_{\sigma\tau}^2$. Now consider $F_{\sigma\tau}^2=F_{\sigma\tau}F_{\sigma\tau} =F_{11}^2 + F_{12}^2 +F_{13}^2 + F_{14}^2 + F_{21}^2 + F_{22}^2 + F_{23}^2+F_{24}^2+F_{31}^2+F_{32}^2+F_{33}^2 + F_{34}^2+F_{41}^2+F_{42}^2 + F_{43}^2 + F_{44}^2$.
With the aid of equations 2, 3, and 4, it can be found that:

$$F_{\sigma\tau}^2=2(H_z^2 + H_y^2 - E_x^2 +H_x^2 -E_y^2 -E_z^2)= 2[(H_x^2 + H_y^2 +H_z^2) -$$
$(E_X^2 +E_Y^2 + E_Z^2)] \to$
$F_{\sigma\tau}^2 = 2[H^2 - E^2]$------------------------------------ (20).

Now $T_{44}=F_{4\sigma}F_{4\sigma} -\frac{1}{4}F_{\sigma\tau}^2$, $\delta_{44} =1$,

$$=F_{41}^2 + F_{42}^2 + F_{43}^2 + F_{44}^2-\frac{1}{4}[2(H^2 -E_2)]=\frac{1}{2} (H^2 + E^2) =\omega \to$$

$\omega=\frac{1}{2}(H^2 + E^2)$----energy density--------------------(21).

Next is to find the pointing vector (energy flux density) $\vec{S_k}$,

Take $T_{k4}=i\ S_k=F_{k\rho}F_{4\rho} -\frac{1}{4}\delta_{\mu\nu}F_{\sigma\tau}^2=F_{k\rho}F_{4\rho}$ -0 =
$F_{k\rho}F_{4\rho}= F_{k1}F_{41} + F_{k2}F_{42}+F_{k3}F_{43} +F_{k4}F_{44}$. Now remember that
$T_{14}=-i(H_y E_z - H_z E_y)$, $T_{24} = -i(E_x H_z - H_x E_z)$, and
$T_{34}=i(H_x E_y - H_y E_x)$. These will lead to $\vec{S}=-[\ \vec{H} \times \vec{E}]\to$
$\vec{S}= [\vec{E} \times \vec{H}]$, (pointing vector) ---------------- (22).

355

Now to find G_k and to show it is the pointing vector $\vec{S}$.

Take $T_{4k} = F_{4\rho}F_{k\rho} = F_{41}F_{k1} + F_{42}F_{k2} + F_{43}F_{k3} + F_{44}F_{k4} \rightarrow$

$T_{41} = i(E_y H_z - E_z H_y); T_{42} = i(E_z H_x - E_x H_z)$ and,

$T_{43} = i(E_x H_y - E_y H_x)$. Hence, $T_{4k} = i(\vec{E} \times \vec{H}), \rightarrow \vec{G_k} = (\vec{E} \times \vec{H})$, momentum field density $= i\vec{S}$, --------- (23).

$T_{\mu\nu} = T_{\nu\mu}$, so $\vec{S_k} = \vec{G_k}$,------------------------- (24).

Now it is possible to find a general formula to T_{jk},

$T_{jk} = F_{j\rho}F_{k\rho} - \frac{1}{4}\partial_{ik}F_{\sigma\tau}^2$, for $j \neq k$, $T_{12} = -(H_y H_x + E_x E_y)$, and

$T_{23} = -(H_z H_y + E_y E_z)$; $T_{13} = -(H_x H_z + E_x E_z)$

But for j=k, $T_{11} = \frac{1}{2}(E^2 + H^2) - H_x^2 - E_x^2$

$T_{22} = \frac{1}{2}(E^2 + H^2) - H_y^2 - E_x^2$

$T_{33} = \frac{1}{2}(E^2 + H^2) - H_z^2 - E_z^2$; therefore, in general

$T_{jk} = \frac{1}{2}[E^2 + H^2]\delta_{jk} - [E_j E_k + H_j H_k]$ ---------------(25).

As another exercise of application, find:

(1.) The field equations.
(2.) The Hamiltonian equation. Given
$L = -\frac{1}{2}\partial_\mu\varphi_i\partial_\mu\varphi_i - \frac{1}{2}\mathcal{H}^2\varphi_i\varphi_i$
(3.) Finding $\vec{S}$ within the continuity equation,
$\frac{\partial H}{\partial t} + \nabla \cdot \vec{S} = 0$.

The solutions:

(1.) The field equations: to fined the field equations, the
following should be used, $(\frac{\partial L}{\partial\varphi_{,v}} - \frac{\partial}{\partial x_v}(\frac{\partial L}{\partial\varphi_{,v}}))=0$, where
$\partial\varphi_{,v} = \frac{\partial\varphi}{\partial x_v}.\frac{\partial L}{\partial\varphi_{,v}} = -\mathcal{H}^2\varphi \rightarrow \frac{\partial\varphi_{,v}}{\partial x_v} - \mathcal{H}^2\varphi_v = 0 \rightarrow$

$\Box\,\varphi_\nu - \mathcal{H}^2\varphi_\nu = 0$. Where, $\Box = \dfrac{\partial^2\varphi}{\partial X_\nu^2} = \dfrac{\partial\varphi_{,\nu}}{\partial\nu}$, therefore,

$[\Box\,\varphi_\nu - \mathcal{H}^2\varphi = 0]$ ---------------------------- (26).

This is the field equation.

(2.) Now the Hamiltonian equation to be derived, take the Lagrangian $L = -\dfrac{1}{2}\partial_\mu\varphi_i\partial_\mu\varphi_i - \dfrac{1}{2}\mathcal{H}^2\varphi_i\varphi_i$.

The Hamiltonian H is given by $H = \Pi_i\varphi_i - L$, and $\Pi_i = \dfrac{\partial L}{\partial\varphi_i} = \varphi^{\cdot}$.

So L can be written as $L = -\dfrac{1}{2}[(\dfrac{\partial\varphi_i}{\partial t})^2 + \sum_k(\dfrac{\partial\varphi_i}{\partial X_k})^2 + \mathcal{H}^2\varphi_i\varphi_i]$, so $H = \Pi_i^2 - \dfrac{1}{2}\Pi_i^2 + \dfrac{1}{2}\sum_k(\dfrac{\partial\varphi_i}{\partial X_k})^2 + \dfrac{1}{2}\mathcal{H}^2\varphi_I\varphi_I$.

So accordingly, $H = \dfrac{1}{2}[\Pi_i^2 + \sum_k(\dfrac{\partial\varphi_i}{\partial X_k})^2 + \mathcal{H}^2\varphi_i^2]$ -------------- (27).

This is the Hamiltonian equation.

(3.) To find $S^{\rightarrow}$, start with the Hamiltonian

$H = \Pi\varphi^{\cdot} - L$. ---- -(i).

Where $\Pi = \dfrac{\partial L}{\partial\varphi^{\cdot}}$ ------ (ii), and $\dfrac{\partial L}{\partial\varphi} = \dfrac{\partial L}{\partial X_\nu}(\dfrac{\partial L}{\partial\varphi_{,\nu}})$ -------- (iii).

And $L = L(\varphi_{,\nu},\varphi)$ --------------------------- (iv).

Now $\dfrac{\partial L}{\partial t} = \dfrac{\partial L}{\partial\varphi_{,\nu}}\varphi_{,\nu} + \dfrac{\partial L}{\partial\varphi}\varphi^{\cdot}$, by combining equations i and ii, and having $H = \dfrac{\partial L}{\partial\varphi^{\cdot}}\varphi^{\cdot} - L$, then, $\dfrac{\partial H}{\partial t} = \dfrac{\partial L}{\partial\varphi^{\cdot}}\varphi^{\cdot\cdot} + \dfrac{\partial}{\partial t}(\dfrac{\partial L}{\partial\varphi^{\cdot}})\varphi^{\cdot}$

Using equation iv, the following is obtained:

$\dfrac{\partial H}{\partial t} = \dfrac{\partial L}{\partial\varphi^{\cdot}}\varphi^{\cdot\cdot} + \dfrac{\partial}{\partial t}(\dfrac{\partial L}{\partial\varphi^{\cdot}})\varphi^{\cdot} - \dfrac{\partial L}{\partial\varphi_{,\nu}}\varphi_{,\nu} - \dfrac{\partial L}{\partial\varphi}\varphi^{\cdot}$, substitute for $\dfrac{\partial L}{\partial\varphi}$ from equation iii. It is obtained that

$$\frac{\partial H}{\partial t} = \frac{\partial L}{\partial \varphi^{\cdot}} \varphi^{\cdot\cdot} + \frac{\partial}{\partial t}\left(\frac{\partial L}{\partial \varphi^{\cdot}}\right)\varphi^{\cdot} - \frac{\partial L}{\partial \varphi_{,v}}\varphi_{,v}^{\cdot} - \frac{\partial}{\partial X_v}\left(\frac{\partial L}{\partial \varphi_{,v}}\right)\varphi^{\cdot}$$

By noticing the first line of the right hand side of this equation, it is recognized that it is the negative fourth component of the second line. Therefore, it is possible to rewrite the equation as

$$\frac{\partial H}{\partial t} = -\left[\frac{\partial L}{\partial \varphi_k}\varphi_k^{\cdot} + \frac{\partial}{\partial x_k}\left(\frac{\partial}{\partial \varphi_{,k}}\right)\varphi^{\cdot}\right] = -\frac{\partial}{\partial X_k}\left(\frac{\partial L}{\partial \varphi_{,k}}\varphi^{\cdot}\right) \rightarrow$$

$$\frac{\partial H}{\partial t} + \frac{\partial}{\partial X_k}\left(\frac{\partial L}{\partial \varphi_k}\varphi^{\cdot}\right) = 0, \quad \text{since } S^{\rightarrow} \text{ is given as } \left(\frac{\partial L}{\partial \varphi_{,k}}\varphi^{\cdot}\right), \text{ so}$$

$$\frac{\partial H}{\partial t} + \text{Div } S^{\rightarrow} = 0 \text{ ---------------------------- (28).}$$

This is the required continuity equation.

****

Appendix A

δ − Function

1. *Introduction and Definition:-*

In mathematics, it is known that sometimes a problem might meet the mathematicians or the mathematical physicists, where some of the integrals and their integrands have singularities at points, over such points that do not exist in a proper sense. So in this case, the evaluation of such integrals requires certain limiting processes. Therefore, it is important to find a way to get rid of such singularity by eliminating such point from the region of integration.

As an example, consider the function $(\frac{1}{x})$, which clearly has a singularity at x=0 (the origin). Suppose the integration of this function is taken over interval including the origin (x=0). Say the integration interval is –a ≤x≤a. This singularity may be eliminated as follows:

$$\int_{-a}^{+a}\frac{dx}{x}=limit_{\epsilon\to0}\left[\int_{-a}^{-\epsilon}\frac{dx}{x}+\int_{\epsilon}^{a}\frac{dx}{x}\right]=limit_{\epsilon\to0}\{\log\frac{\epsilon}{a}+\log\frac{a}{\epsilon}\}=1.$$

Note that if the integration is carried out by the usual rules without recourse to the limiting process, the result will be:

$$\int_{-a}^{a}\frac{dx}{x}=\log x]_{-a}^{a}=\log a\text{-}\log(-a)=\log(\frac{a}{-a})=\log(-1)=i\pi.$$

From this result, it is seen that the integration might be carried out by the usual rules without recourse of the eliminating process of the uses of the formula

$\frac{1}{x}=\frac{d}{dx}\log x\text{-}i\pi\delta(x)$, taking the advantages of δ(x) properties,

$\int_{c}\delta x)dx$ =1, and $\int_{c}-\delta(x)dx$ =0, ---------------------- (1).

Where c is any interval on the real axis that includes the point $x=0$, while c^- is any interval on the real axis that does not include the origin ($x\neq0$). Hence, the integration will be

$$\int_{-a}^{a}\frac{dx}{x}=\{\int_{-a}^{a}\frac{d}{dx}\log x-i\pi\delta(x)\}dx=i\pi- i\pi \int_{-a}^{a}\delta(x)dx$$

$=i\pi - i\pi=0$. Hence, the $\delta(x)$-function provides a shorthand method to evaluating integrals over singular points without recoursing to the limiting argument, which sometimes is quite tedious in some cases. Because the use of delta-function is very convenient, some of its properties shall be studied next. If the singular point is at $x=x_0$, delta function will take the form $\delta(x - x_0)$ which might be employed. This form has the following defining properties:

$$\left.\begin{array}{l} \int_c \delta(x - x_0)dx=1, \text{ if } c \text{ contains the point } x=x_0. \\ \int_{c^-}\delta(x - x_0)dx =0 \text{ if } c^- \text{not containing the point } x=x_0 \end{array}\right\} \text{--- (2)}.$$

The δ-function is defined by integral properties, and formulas involving delta function are to be understood as being true, when the two sides of the equations are integrands. To clarify the picture of the improper function $\delta(x)$, imagine a function of x that vanishes everywhere, except in small interval *of width* ϵ. When it is integrated over this interval, it gives unity. The limit of this function as $\epsilon \to 0$ (very, very narrow width) gives the delta function; it was developed mathematically by Dirac, and this is why it is called Dirac delta function.

2. *Representation of Dirac δ-Function.*

Mathematically, there are some simple functions which are in the limiting case that show the properties of delta function.

Some of these functions will be considered below.

(1.) The function $\sqrt{c/\pi}\, e^{-cx^2}$; for all values of c, this function has the property that $\int_{-\infty}^{\infty} \sqrt{c/\pi}\, e^{-cx^2} = 1$, as c is allowed to become very large. The function becomes very small at all points, except at the origin ($x=0$). Its value will be $\sqrt{c/\pi}$. So as $c \to \infty$, this function

$\to \delta$-function, i.e., $\sqrt{c/\pi}\,e^{-cx^2} \to \delta(x)$.

or $limit_{\,c\to\infty}\sqrt{c/\pi}\,e^{-c(x-x_0)^2} = \delta(x - x_0)$ ----------- (3).

(2.) Consider the function $f(x) = \dfrac{1}{c\pi}\dfrac{sin^2(cx)}{x^2} = \dfrac{c}{\pi}\dfrac{sin^2(cx)}{(cx)^2}$; for all c,

this function has the property $\int_{-\infty}^{\infty} \dfrac{1}{c\pi}\dfrac{sin^2(cx)}{x^2} = 1$ ---- (4).

(3.) Consider the function $f(x) = \dfrac{1}{\pi}\dfrac{sin(cx)}{x} = \dfrac{c}{\pi}\dfrac{sin(cx)}{(cx)}$,. This also

has the property $\int_{-\infty}^{\infty} f(x)dx = 1$, for all c

$f(x) \to 0$ as $c\to \infty$, $x \neq 0$

Thus, $limit_{\,c\to\infty}\dfrac{1}{\pi}\dfrac{sin(cx)}{x} = \delta(x)$. ------------------------------- (5).

The functions given by the equations 3, 4, and 5 serve as delta function $\delta(x)$.

(4.) As an example of three-dimensional delta function, it might consider, $limit_{\varepsilon\to 0}[\dfrac{1}{\pi^2}\dfrac{\epsilon}{(\epsilon^2+r^{\to 2})}] =$

$\delta(r^{\to}) = \delta(x^{\to})\delta(y^{\to})(z^{\to}) = \delta(x-x_0)\,\delta(y-y_0)\delta(z-z_0)$.

This function has the following properties:

(i) As $\epsilon \to 0$, the function approaches zero everywhere, except at $r^{\to} = 0$.

(ii) $\int_{all\ space} \dfrac{1}{\pi^2}\dfrac{\epsilon}{\epsilon^2+r^{(-)2}}\, dr^- = \dfrac{\epsilon}{\pi^2}\cdot 4\pi \int_0^{\infty}\dfrac{r^2 dr}{(\epsilon^2+r^2)} = 1$.

3. Use of Delta Function to Evaluate Certain Important Integrals

As stated before, Dirac delta function is quite important in evaluating certain integral. The following are some examples given:

(i) $\int_{-\infty}^{\infty} e^{iax} \, dx = limit_{L \to \infty} \int_{-L}^{+l} e^{iax} dx = L_{\to\infty} \frac{1}{ia} [e^{iaL} - e^{iaL}] =$

$2 limit_{L \to \infty} \frac{\sin (aL)}{a} = 2\pi \{ limit_{L \to \infty} \frac{\sin)aL)}{\pi a} \} = 2\pi\delta(a),$ ---- (6).

(ii) $\int_{all \, space} e^{ia^{\to} \cdot r^{\to}} dr^{\to}$, for three dimension case.

This can be written as $\int_{-\infty}^{\infty} e^{ia_x x} \, dx \int_{-\infty}^{\infty} e^{ia_y y} dy \int_{-\infty}^{\infty} e^{ia_z z} dz$

By using equation 6,

$= (2\pi)^3 \delta(a_x)\delta(a_y)\delta(a_z) = (2\pi)^3 \delta(a^{\to})$--------------(7).

(4.) *Some Special Properties of Dirac Delta Function* from the so far stated about the mathematical functioning of $\delta - function$, it can be deduced that some equations might represent its main properties and uses. These properties are demonstrated by certain equations. If they are carefully noticed, their two sides are equal in case they are considered as integrands over any defined interval. The following are those properties:

(1.) $f(x)\delta(x-x_0) = f(x_0) \, \delta \, (x-x_0)$
(2.) $\delta(x) = \delta(-x)$
(3.) $\frac{d\delta(x)}{dx} = -\delta(-x)$
(4.) $\delta(ax) = \frac{1}{a}\delta(x)$
(5.) $\delta(x^2 - a^2) = \frac{1}{2a}[\delta \, (x-a) + \delta(x+a)]$
(6.) $\int \delta(a-x)\delta(b - x) \, dx = \delta(a-b)$
(7.) $f(x)\delta^-(x-x_0) = f^-(x_0)\delta(x-x_0)$
(8.) It should also be noted that
$\int_{-\infty}^{+\infty} \delta[f(x)]dx = \frac{1}{[f^-(x)]_{f(x)=0}}$, where $f^- = \frac{df}{dx}$.

Appendix B

Solution of WhittekerˢEquation.

This equation is of important use in mathematical physics in general; its solution leads to very important mathematical functions which are used largely in physics.

Therefore, it finds the necessity for its detail solution separately. So to be started with this equation, it is

$$\frac{d^2 w}{dt^2} + \{-\frac{1}{4} + \frac{P}{X} + \frac{\frac{1}{4} - \lambda^2}{X^2}\} \, w = 0, \text{-----------------} (1).$$

Where w is a function given by $w = X^{\lambda + \frac{1}{2}} e^{-\frac{X}{2}} F$.

Now by substituting this function into equation 1, the following equation is obtained:

$$X \frac{d^2 F}{dX^2} + (2\lambda + 1 - X)\frac{dF}{dX} - ((1/2 + \lambda - P)F = 0, \text{-------------}(2).$$

Consider the function $f(\alpha, \beta, \gamma)$ that it satisfies the equation:

$$X \frac{d^2 f}{dX^2} + (\beta - X)\frac{df}{dX} - \alpha f = 0. \text{------------------------------------} (3).$$

This equation is bound at the origin, so its solution is:

$$f = \sum_{j=0}^{\infty} a_j X^j \text{--} (4),$$

By substituting equation 4 into equation 3, the result will be as follows:

$$\sum_{j=0}^{\infty} \{j(j-1) \, a_j X^{j-1} + \beta_j a_j X^{j-1} - (j+\alpha)a_j X^j\} \text{----- --------} (5).$$

By setting the coefficients of X^k equal to zero (k replaced j), it is obtained that $\frac{a_{k+1}}{a_k} = \frac{k+\alpha}{(k+1)(k+\beta)},$ \text{---------------------------}(6).

Equation 2 is known as confluent hypergeometric equation, whose solution is given by:

$$F(\alpha,\beta,X)=1+\frac{\alpha}{\beta}\cdot\frac{X}{1!}+\frac{\alpha(\alpha+1)}{\beta(\beta+1)}\cdot\frac{X^2}{2!}+\frac{\alpha(\alpha+1)(\alpha+2)}{\beta(\beta+1)(\beta+2)}\cdot\frac{X^3}{3!}+\cdots.$$

The values of F $(\alpha,\beta.X)$ are given by special tables according to the values of the parameters $\alpha,\beta,and\ X$.

Now $\frac{a_{k+1}}{a_k}\to\frac{1}{k}$, as $k\to\infty$, so $e^{PX}=\sum_{k=0}^{\infty}\frac{p^k x^k}{k!}\to\frac{a_{k+1}}{a_k}=\frac{P}{k}\}\to$ F$\to e^x$, as x$\to\infty$. The question is, what is the condition for the solution of a polynomial? For α is zero $(\alpha=0),or\ a\ negative\ integer$ (-n), then, $a_{k+1}=\frac{(k+\alpha)}{(k+1)(k+\beta)}a_k$, and F is a polynomial of degree (n).

The linearly independent solution for Whittaker equation is

$$W_{p,\lambda}(x)=X^{\lambda+\frac{1}{2}}e^{\frac{-x}{2}}F(1/2+\lambda-p,2\lambda+1,x)$$

Where $\alpha=1/2+\lambda$ -p, and $\beta=2\lambda+1$. The condition for the solution is $2\lambda+1\neq0,-1,-2,---,$ and

$$W_{p,-\lambda}(x)=X^{-\lambda+\frac{1}{2}}e^{\frac{-X}{2}}F(1/2-\lambda-p,-2\lambda+1,x),-2\lambda+1\neq-n\ ;\ n=0,\ 1,$$
2,----. Recursion relation for $F(\alpha,\beta,x)$ is

$$\frac{dF(\alpha,\beta,x)}{dx}=\frac{\alpha}{\beta}F(\alpha+1,\beta+1,x),\ F(\alpha,\beta,x)=e^xF(\beta-\alpha,\beta,-x).$$

There is certain relation between the associated Legendre function and F (α,β,x), it is represented as follows: $F(\alpha,\beta,x)=C_{\alpha\ell}L_{\beta-\alpha-1}^{\beta-1}(x)$, or $L_a^b(x)=C_{ab}$ F(α,β,x), when $a\ and\ b\ are\ integers\ as\ b\le a$, when b-a is zero or negative integer, then $L_b^a(x)$ is a polynomial (associated Legendre polynomial). For this mathematical polynomial, a concept of generating relation can be illustrated as follows:

Introducing $L_a^b(x) \equiv L_a(x)$ (Laquere polynomial), where

$L_a^b(x) = \dfrac{d^b}{dx^b} L_a(x)$, and $L_a^b(x) = L_a(x) \, e^x \dfrac{d^a}{dx^a} X^a e^{-x}$,

$L_2^2(x) = 2, L_0(x) = 1 = L_0^0(x), L_1^0 = 1-x, L_1^1(x) = -1$,

$L_2^0(x) = x^2-4x+2, L_2^1(x) = 2x-4$.

The asymptotic form of $F(\alpha, \beta, x)$, if $-3\pi/2 \leq$ arg. $\leq \pi/2$, is

$F(\alpha, \beta, x) \Rightarrow, [\text{as } |x| \rightarrow \infty], \dfrac{\Gamma(\beta)}{\Gamma(\beta-\alpha)}(-X)^{-\alpha} + \dfrac{\Gamma(\beta)}{\Gamma(\alpha)} e^x X^{\alpha-\beta}.]$

Where $\Gamma(x)$ is gamma function, it is an analytic memorphic function of x with a simple poles at x=-1, for ℓ=0, 1, 2,--, and the corresponding residues $\dfrac{(-)^\ell}{\ell!}$. The properties of gamma function are well known to the mathematicians and theoretical and mathematical physicists, which represented by

$\Gamma(1) = 1, \Gamma(x + 1) = x\Gamma(x)$.

It is real and positive when x is real and positive. There is a special table for $\Gamma(x)$ value for a given value of x. For more mathematical details, refer to *Formulas and Theorems for the Functions of Mathematical Physics* by W. Magnus and F. Oberhettinger.

Appendix C
Some Solved Problems

Problem 1: Given hydrogen molecules as shown in the diagram below:

H distance (d) H ⬚————$\lambda<<d$————⬚

Compare the microscopic concept with that of the macroscopic one, under the classical and the quantum theories.

The solution: The energy of the molecule as a system in motion about the common center of its atoms is given by

$$E = \frac{1}{2} KT + \frac{1}{2} m\omega^2 \text{---} (1).$$

Where K is Boltzmann constant, m is the molecule mass, and ω is the angular velocity. Now $\hbar\omega = 0.5$ eV, this gives $\omega = \frac{0.5eV}{\hbar} = \frac{5\ eV}{1.054\times6.666\times10^{20}\times10^{-34}eVsec\cdot} \cong 7\times 10^{13} sec^{-1}$?

KT=0.025 eV; it is the thermal energy.

T=300K^0, using these data, you find K$\cong 1.33 \times 10^{-23}$ J/k^0

The energy 1/2 KT will be about 0.013eV.

Now for the energy 0.5 eV, the angular velocity ω is given by
$\frac{0.5\ eV}{\hbar} = \frac{5\times1.6\times10^{-19}J}{1.054\times10^{-34}J\cdot sec} \approx 8\times 10^{13}/$ sec.

Calculate 1/2 $m_H\omega^2$. Remember, m_H=1 amu=939 MeV$\approx$1.5$\times$ 10^{-18} J. So Eω=1/2 $m_H\omega^2 \approx 5.1\times 10^{11}$ J.sec^{-2}

E=0.013   eV+5.1 $\times 10^{11}$J.sec^{-2} =0.013$\times$1.6$\times 10^{-19}$
J+5.1$\times 10^{11}$ J/sec$^2 \cong$ 5.1$\times 10^{11}$J/sec^2

$$\lambda = \frac{h}{\sqrt{2m_{HE}}} = \frac{6.62\times10^{-34}\ J.sec}{\sqrt{2\times1.6\times10^{-13}\ J\times5.1\times10^{11}/sec^2}} \approx 2.1 \times 10^{-33}$$

$= 10^{-25} A^0$. So the wavelength is too very small to be measured compared with the distance between the atomic molecules ($1A^0$).

Suppose the molecule is in a heavy atom, say 200 m$_H$, then λ will be about 1.9$\times 10^{-34}$, which is much smaller. So $\lambda \propto \frac{1}{m}$.

Even if the temperature is too low, say 10 k^0, then λ is still too small to be measured with the best available technology of the measurement. This is the classical approach to deal with the systems. Now if the system (molecule) is described from the quantum mechanical point of view, it will be considered in its motion as a linear harmonic oscillation. This system is described by a wave function given by

$$\psi_n = \frac{e^{i\gamma\mu}}{\sqrt{2^n\ n!\ a\sqrt{\pi}}}\ e^{\frac{-(\frac{x}{a})^2}{2}}\ H_n(s)\ e^{\frac{iE_n t}{\hbar}} \text{--------------(1).}$$

If the ground state is considered, then the wave function is

$$\psi_0 = \sqrt{a\sqrt{\pi}}\ e^{\frac{-1(\frac{x}{a})^2}{2}},\ \text{where}\ a = \sqrt{\frac{\hbar}{m\omega}}.\ \text{So}\ \psi_0 = Ae^{\frac{-1(\frac{x}{a})^2}{2}}$$

A is the amplitude $= \frac{1}{\pi^{\frac{1}{4}}\ (\frac{\hbar}{m\omega})^{\frac{1}{4}}} \approx \sqrt{\pi} \times 10^{-9}$ cm $= 0.17\ A^0$.

It is comparable with the distance between the molecules, which is measurable with the current available technical measurement for atomic distances. These two ways of measuring a microscopic system show clearly the difference between the classical and the quantum physics. (QED)

Problem 2. Suppose $\delta(x) = limit_{\epsilon \to 0} \dfrac{\epsilon}{x^2 + \epsilon^2} \cdot \dfrac{1}{\pi}$,

And $\delta(x) = limit_{\epsilon \to 0} \dfrac{e^{\frac{(x)^2}{\epsilon}}}{\sqrt{\epsilon}} \dfrac{1}{\sqrt{\pi}}$. Show that: (i.) $\delta(x)=0$ for $x \neq 0$, $=\infty$ for x=0; (ii.) $\int \delta(x)dx = 1$.

The solution: First step, it is necessary to picture of $\delta(x)$, as stated by Dirac himself. Take a function of a real variable x vanishes everywhere, except inside a small domain of length, say ϵ), that is surrounding the origin x=0. It is quite large inside this domain where its integral over this domain is unity. The exact shape of the function inside the domain itself does not matter, provided that there are no unnecessarily wild variations, (for example, provided that the function is always of order ϵ^{-1}). Then in the limit $\epsilon \to 0$, this function will go over into $\delta(x)$. For illustration, see the figure.

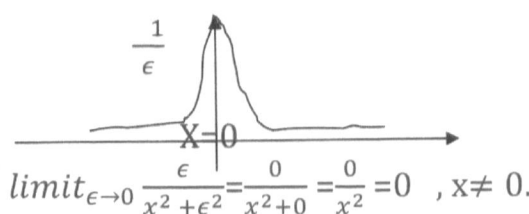

$$limit_{\epsilon \to 0} \frac{\epsilon}{x^2 + \epsilon^2} = \frac{0}{x^2 + 0} = \frac{0}{x^2} = 0 \ , x \neq 0.$$

$limit_{\epsilon, x \to 0} \dfrac{\epsilon}{x^2+\epsilon^2} = \dfrac{0}{0}$ (undetermined quantity), so, the derivative of each one has to be taken in this case, the result is, $limit_{\epsilon, x \to 0} \dfrac{1}{2x+2\epsilon} = \dfrac{1}{0+0} = 1/0 = \infty$, for x=0

Therefore,

$limit_{\epsilon \to 0} \dfrac{\epsilon}{x^2 + \epsilon^2}$

$\left. \begin{array}{c} \\ \\ \\ \end{array} \right\}$

0, if $x \neq 0$

∞ if x=0

So, $\delta(x)=\begin{cases} 0, \text{ if } x \neq 0 \\ \\ \infty, \text{ if } x=0 \end{cases}$

Now to show $\int \delta(x)dx=1$.

Take $\delta(x)=limit_{\epsilon \to 0}\frac{\epsilon}{x^2+\epsilon^2}\cdot\frac{1}{\pi}$, so $\int \delta(x)dx=\frac{1}{\pi}\int_{-\infty}^{\infty}\frac{\epsilon dx}{x^2+\epsilon^2}$

$=\frac{2}{\pi}limit_{\epsilon \to 0}\int_0^{\infty}\frac{\epsilon}{x^2+\epsilon^2}\,dx=\frac{2}{\pi}limit_{\epsilon \to 0}[\,tan^{-1}\frac{x}{c}]_0^{\infty}=\frac{2}{\pi}(\frac{\pi}{2})=1.$

$\therefore \int \delta(x)dx=1.$

Now if $\delta(x)=limit_{\epsilon \to 0}e^{\frac{-x^2}{\epsilon}}\frac{1}{\sqrt{\epsilon}}\cdot\frac{1}{\sqrt{\pi}}$, then $\int \delta(x)dx=1$.

The proof: $\int \delta(x)dx=\frac{1}{\sqrt{\pi}}limit_{\epsilon \to 0}[\int_{-\infty}^{\infty}\frac{e^{\frac{-x^2}{\epsilon}}}{\sqrt{\pi}}dx=$

$\frac{1}{\sqrt{\pi}}limit_{\to 0}[\frac{\sqrt{\epsilon\pi}}{\sqrt{\epsilon}}]=\frac{\sqrt{\pi}}{\sqrt{\pi}}=1, \therefore \int \delta(x)dx=1.$

Note that $\frac{\epsilon}{x^2+\epsilon^2}$ and $\frac{e^{\frac{-x^2}{\epsilon}}}{\epsilon}$ are continuous functions; therefore, the limit can be taken inside or outside the integral. These functions can be represented graphically as shown below.

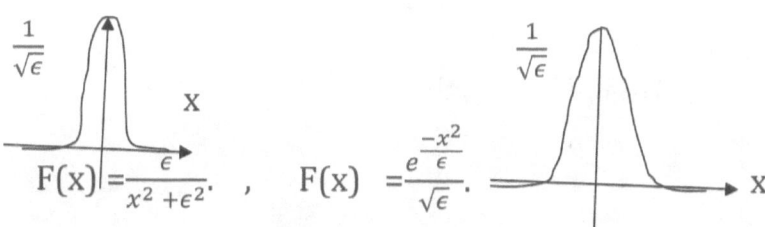

$\frac{1}{\sqrt{\epsilon}}$ X $F(x)=\frac{\epsilon}{x^2+\epsilon^2}$, $F(x)=\frac{e^{\frac{-x^2}{\epsilon}}}{\sqrt{\epsilon}}$. $\frac{1}{\sqrt{\epsilon}}$ X

Problem 3: Prove that $\delta(ax)=\frac{1}{a}\delta(x)$.

The proof: By definition $\delta(x) = \frac{1}{2\pi}\int_{-\infty}^{\infty} e^{ixy}\,dy$; therefore,

$\delta(a) = \frac{1}{2\pi}\int_{-\infty}^{\infty} e^{iaxy}\,dy$, let $ay = Z \to dy = \frac{dZ}{a}$, so if $a < 0$,

$y \to \infty, Z \to -\infty$, or $y \to = -\infty, Z \to \infty$. This leads to

$\delta(ax) = \frac{1}{2\pi}\int_{\infty}^{-\infty} \frac{e^{ixZ}}{(-a)}\,dz = \frac{-1}{2\pi a}\int_{\infty}^{-\infty} e^{ixZ}\,dz =$

$(\frac{1}{a})\cdot[\frac{1}{2\pi}\int_{-\infty}^{\infty} e^{ixZ}\,dz] = \frac{1}{a}\delta(x)$. For $a > 0$, it is directly that

$\delta(ax) = \frac{1}{a}\delta(x)$. $\therefore \delta(ax) = \frac{1}{a}\delta(a)$, as required.

Now let again assumed that $a < 0$, and let it equal to (-b) where b>0, so, $\delta(-bx) = \frac{1}{2\pi}\int e^{-bxy}\,dy$. Again, let

$(-by) = Z \to dy = \frac{dZ}{-b}$, $\to y = +\infty, Z \to -\infty, y = -\infty, Z \to +\infty$

Hence, $\delta(-bx) = \frac{1}{2\pi}\int_{+\infty}^{-\infty} \frac{e^{ixZ}}{-b}\,dZ = \frac{1}{b}[\frac{1}{2\pi}\int_{-\infty}^{\infty} e^{ixZ}\,dZ] = \frac{1}{b}\delta(x)$.

$\therefore \delta(-bx) = \frac{1}{b}\delta(x)$. Since $a = -b$, so $b = -a$; therefore, for that

$\delta(ax) = \delta(-b) = \frac{1}{(-a)}\delta(x) = (1/|a|)\delta(x)$ for $a < 0$.

But for $a > 0$, it is $\delta(ax) = \frac{1}{a}\delta(x)$.

Problem 4: Show that

(i.) $\delta(-x) = \delta(x)$
(ii.) If $\int f(x)\,\delta(x)\,dx = f(0)$, then $\int f(x)\delta(x - a)\,dx = f(a)$.

The proof:

(i.) Take

$$\delta(-x) = \frac{1}{2\pi}\int_{-\infty}^{\infty} e^{-ixy}\,dy = \frac{1}{2\pi}\,limit_{R \to \infty}\int_{-R}^{R} e^{ixy}\,dy =$$
$$\frac{1}{2\pi}\,limit_{R \to \infty}\frac{e^{-ixy}}{-ix}]_{-R}^{R}\frac{1}{2\pi}\,limit_{R \to \infty}\frac{-2i\,sinRx}{-ix}$$

$=\frac{1}{\pi}[limit_{R\to\infty}\frac{sinRx}{x}]=\delta(x)$, by definition. So

$\delta(-x)=\delta(x)$, as required.

(ii.) *If $\int f(x)\delta(x)dx$ =f(0),then, $\int f(x)\delta(x-0)dx=$ f(0).*

The proof: $\int f(x)\delta(x)dx$ can be written as

$\int f(x)\delta(x-0)dx$ =f(0).
Let x-0=y, or x=y+0, dx =dy,
then $\int f(y+0)\,\delta(y)$ dy=f(0),
so $\int f(x)\delta$(x-a)dx can be written as
$\int f(y+a)\,\delta(y)$ dy=f(a).

Assuming x-a =y or x =y+a, dx =dy., $\therefore$ $\int f(x)\delta(x-a)dx$=f(a),$\to$
$\int f(x)\,\delta(x-x_0)$dx=f(x$_0$).

And so on. It is property of Dirac delta function.

Problem 5): Show that <p>=$\int \psi^*$(r, t)$\mathbb{P}_{op}\psi$(r, t) dr, where the momentum operator $\mathbb{P}_{op}$ =$-i\hbar\frac{\partial}{\partial r}$.

The proof: It is known that $\mathbb{P}= \hbar\mathbb{k}$, where $\mathbb{P}$ *and* $\mathbb{k}$ are vectors. The average value of $\mathbb{P}$ is by <$\mathbb{P}$>=$\hbar < k$>=
$\hbar \int \phi^*$(k, t) $\mathbb{k}\phi(k,t)dk$. Using Fourier transformation, $\phi(k,t)$
$=\frac{1}{(2\pi)^{3/2}} \int \psi(r,t)e^{-ik\cdot r}$dk, so ϕ^*(k, t) $=\frac{1}{(2\pi)^{3/2}} \int \psi^*(r^-,t)\,e^{ik\cdot r}$dr.

Therefore,
$<P>=\frac{\hbar}{(2\pi)^3} \iiint \psi^*(r^-,t)\,\psi(r,t)\mathbb{k}e^{ik\cdot r^-}e^{-ik\cdot r}$dr- dr dk

Using $\mathbb{k}$ =$-\frac{i\partial}{\partial r}$, operator, then
$<P> = =\frac{-i\hbar}{(2\pi)^3} \iint_{-\infty}^{\infty} d\mathbb{k}\,dr^-\,\psi^*(r^-,t)e^{ik\cdot r^-} \int_{-\infty}^{\infty} \psi(r,t)\frac{\partial}{\partial r}e^{-ik\cdot r}$dr.

Now take $\int_{-\infty}^{\infty} \psi(r,t) \frac{\partial}{\partial r} e^{-i\,\mathbb{k}\cdot r} dr$, integrate it by parts and you get $\psi(r,t) e^{i\,\mathbb{k}\cdot r}]_{-\infty}^{\infty} - \int_{-\infty}^{\infty} e^{-i\mathbb{k}\cdot r} \frac{\partial}{\partial r} \psi(r,t) dr$

$= \int_{-\infty}^{\infty} e^{-i\,\mathbb{k}\cdot r} \frac{\partial}{\partial r} \psi(r,t) dr$, the first term $=0$ due to periodic condition. Hence, $<P> = \int \psi^*(r,t)\, \mathbb{P}_{0p} \psi(r,t) dr$. QED.

Or $<P>=$

$\frac{-i\hbar}{(2\pi)^3} \iint_{-\infty}^{\infty} dr\, dr^- \psi^*(r^-,t) \frac{\partial}{\partial r} \psi(r,t) \int_{-\infty}^{\infty} e^{ik\cdot r^- -r)} dk=$

$\frac{-i\hbar}{(2\pi)^3} \int_{-\infty}^{\infty} \psi^*(r^-,t) \frac{\partial}{\partial r} \psi(r,t) dr \int_{-\infty}^{\infty} (2\pi)^3\, \delta(r^- -r) dr^-$

Using the property of Dirac delta function, it is obtained that

$<P>=-i\hbar \int_{-\infty}^{\infty} \psi^*(r,t)(\frac{\partial}{\partial r})\psi(r,t) dr \quad = \int_{-\infty}^{\infty} \psi^*(r,t)(-i\hbar \frac{\partial}{\partial r})\psi(r,t)$
$= \int_{-\infty}^{\infty} \psi^*(r,t) \mathbb{P}_{0p}\psi(r,t) dr.$
So $<P> = \int_{-\infty}^{\infty} \psi^*(r,t) \mathbb{P}_{0p}\psi(r,t) dr$, as required. QED.

Problem 6: Given $\psi(r,t)=\frac{1}{(2\pi\hbar)^{3/2}} \int_{-\infty}^{\infty} \phi(p,t)\, e^{\frac{ip\cdot r}{\hbar}} dp$, find $\phi(p,t)$ in terms of $\psi(r,t)$. Where $\psi(r,t)$ *is the spacial* wave function and $\phi(p,t)$ is the momentum space wave function.

By Fourier transformation, it is known that $\phi(p,t)$ is given by
$\phi(p,t) = \frac{1}{(2\pi\hbar)^{3/2}} \int_{-\infty}^{\infty} \psi(r,t)\, e^{\frac{-ip\cdot r}{\hbar}} dr$ and

$\psi(r,t)=\int_{-\infty}^{\infty} \phi(p,t)\, e^{\frac{-ip\cdot r}{\hbar}} dp$. Now substitute for $\psi(r,t)$ to find $\psi(p,t)$. Hence,

$\phi(p,t) = \frac{1}{(2\pi)^3} \int_{-\infty}^{\infty} \phi(p,t) dp \int_{-\infty}^{\infty} e^{\frac{i(p-p^-)\cdot r}{\hbar}} dr$

$= \int_{-\infty}^{\infty} \phi(p,t)\, \delta(p - p^-) dp = \phi(p^-,t)$

$\therefore \phi(p,t) = \phi(p^-,t) \rightarrow \phi(p,t)=\frac{1}{(2\pi\hbar)^{3/2}} \int_{-\infty}^{\infty} \psi(r,t) e^{\frac{-ip\cdot r}{\hbar}} dr.$

QED.

Problem 7: Prove that the energy eigenvalue is discrete.
The proof: Take the simplest case: a particle is in a potential
well, where V(r)=0, inside the well (finite range for x, but it is
infinity outside. As it is shown in the figure below.

Schrodinger Schrödinger equation for this is

$\psi'' + (E - V)\psi = 0$. But for V=0, it is

$\psi'' + E\psi = 0$. The solution is give by

$\psi(x) = A \sin\sqrt{E}\, x$, $0 \leq x \leq L$,

So at x=0, $\psi(0) = 0$, at x=L, $\psi(L) = A \sin L$.

V=∞ |

V=0

x=0 |_____| x=L

Using boundary conditions, then $\psi(0)$=A sin0=0 and $\psi(L)$=
$A\sin\sqrt{E}L$=0. This implies that $\sqrt{E}\, L \neq 0$, but $=n\pi$, so $\sin n\pi=0$,
$\sin 2\pi=0$, ---; hence, $\sqrt{E}\, L = n\pi \to E_n L^2 = n^2\pi^2 \Rightarrow$

$E_n = \dfrac{n^2\pi^2}{L^2}$ -- (1).n=0, 1, 2, 3,---

$\psi_n = A \sin\dfrac{n\pi}{L}$ ------------------------------------ (2).

Equation 1 is the eigenvalue of the system (particle), and
equation 2 is its eigenstate.

Therefore, $E_1 = \dfrac{\pi^2}{L^2}$, $\psi_1 = A \sin\dfrac{\pi}{L}$,   this clearly indicates that

$E_2 = \dfrac{4\pi^2}{L^2}$, $\psi_2 = A \sin\dfrac{2\pi}{L}$,   the energy eigenvalue is

------------------------------   discrete. QED.

$E_n = \dfrac{n^2\pi^2}{L^2}$, $\psi_n = A \sin\dfrac{n\pi}{L}x$,

Problem 8: Show that for a central potential, the lowest energy
level is always the S-level ($\ell=0$).

The solution and the proof: The radial equation for a central potential is given:

$$\frac{d^2 R_{n\ell}}{dr^2} + \frac{2}{r}\frac{dR_{n\ell}}{dr} + \frac{2m}{\hbar^2}[E - \{V(r) + \frac{\hbar^2\ell(\ell+1)}{2mr^2}\}]R_{n\ell}, \text{------------- (1)}.$$

Where $V(r) + \frac{\hbar^2\ell(\ell+1)}{2mr^2}$ is the effective potential $V_{eff.}$, $V(r)$ is the central potential, m is the mass of the particle moving under the effect of $V(r)$, $\frac{\hbar^2\ell(\ell+1)}{2mr^2} = \frac{L^2}{2mr^2}$ is the centrifugal potential ($V_{cf.}$), and L is the angular momentum. Now take the radial equation for $\ell = 0$, i.e., for the lowest state (S-state), then equation 1 takes the form

$$\{\frac{d^2}{dr^2} + \frac{2}{r}\frac{d}{dr} + \frac{2m}{\hbar^2}[E - V(r)]\}R_{n1} = 0, \text{------------------ (2)}.$$

This can be written for $\ell=1, 2, 3, --- \ell$, so let $\ell=N$, N=1, 2, 3,-- Hence, equation 1 takes the form

$$\frac{d^2}{dr^2} + \frac{2}{r}\frac{d}{dr} + \frac{2m}{\hbar^2}[E - (V(r) + \frac{\hbar^2 N(N+1)}{2mr^2})]R_{nN} = 0, \text{--------(3)}.$$

To simplify things, assume that u=kr,

where k=$\frac{2m}{\hbar^2}$ [E-V(r)]. ---------------------------- (4).

So equation 1 becomes $[\frac{d^2}{du^2} + \frac{2}{u}\frac{d}{du} + (1 - \frac{\ell(\ell+1)}{u^2})]R_{n\ell}$ --- (5).

The known solution to this equation has two linearly independent solutions; they are

$\{j_\ell(u), n_\ell(u)\}$
or $\{h_\ell^{(1)}(u), h_\ell^{(2)}(u)\}$----------------------------(6),
where $h_\ell^{(1)}(u) = j_{\ell(u)} + n_\ell(u)$ -----------------(7)
And $h_\ell^{(2)}(u) = j_\ell(u) - n_\ell(u)$-------------- (8).

$j_\ell(u)$ is Bessel function, $n_\ell(u)$ is Newman function, and $h_\ell^{(n)}$ are Hangle functions.

Therefore, the solution of equation 5 is

$R_{n\ell}(r) = A_1 j_\ell(kr) + A_2 n_\ell(kr)$, for r<a.

$R_{n\ell}(r) = B_1 h_\ell^{(1)}(kr) + B_2 h_\ell^{(2)}(kr)$, for r>a.

Where a is the radius of the central potential well.

Since R(r) must be finite for r=0, then, $A_2 = B_2 = 0$, $k_0 = \sqrt{\dfrac{2m\,E}{\hbar^2}}$,

E=-ω , E<0. Now let $k_0 = ik^-$, where $k^- = \sqrt{\dfrac{2m(-\omega)}{\hbar^2}}$.

$\therefore R_{\ell n}(r) = A j_\ell(kr)$, for r $\leq$ 0.

$R_{n\ell}(r) = B\, h_\ell^{(1)}(kr)$, for r $\geq$ 0.

By applying the join conditions at r=a, it is found for the S-state ($\ell = 0$), that cot. ka $= -\dfrac{k}{k^-}$, ------------------(9).

It gives the energy eigenvalues. This can be represented graphically below

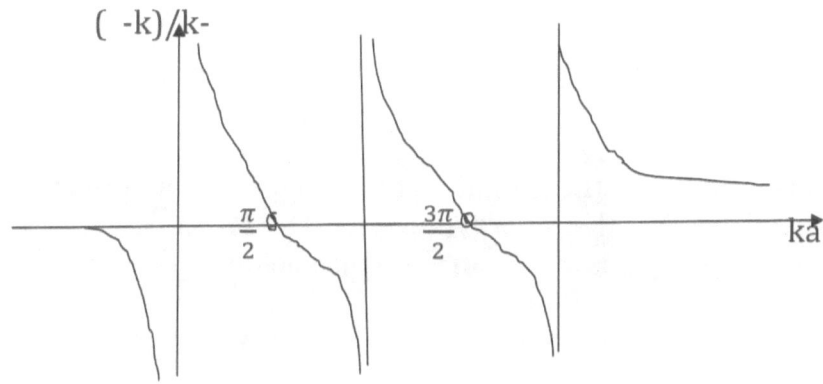

A shown in the figure above, the zero points are $\frac{\pi}{2}, \frac{3\pi}{2}$,---that give the eigenvalue, i.e.,

Cot.ka=0 leads to ka $=\frac{\pi}{2}$, so $(\frac{2\pi}{\hbar^2}(E-V)^{1/2}a=\pi/2$, this gives

$\frac{2m}{\hbar^2}(-\omega -V) a^2 =\frac{\pi^2}{4} \rightarrow [E_n=V(r)+\frac{\pi^2\hbar^2}{8ma^2}]$, n=0, 1, 2, 3,---, (a=a).

Or it can be said that a new level appears whenever $\frac{k}{k^-}$=0,

Or cot ka =∞(notice the figure), it occurs at ka=π, 2π,-----.

Thus, there is no energy level with ℓ=1,

when V a²=V₀ a²$\leq \frac{\pi^2\hbar^2}{2m}$; V(r) =V₀, for r<a.

So if $\frac{\pi^2\hbar^2}{2m} \leq V_0 a^2 \leq \frac{\pi^2\hbar^2}{m}$, there is one bound state with $\ell = 1$, etc. It might be noticed that the smallest value of (V₀ a²) for which there exists a bound state with ℓ=1 is in fact greater than the value of (V₀a²) for $\ell = 0, i.e., the S - state.$

Therefore, S- state is the lowest energy level in the central potential well. It means S-state is necessarily the ground state. Hence, from the energy eigenvalue it can be noticed that E₀<E₁< E₂<E₃, --------< E_L, L=$\sqrt{\ell(\ell + 1)}\hbar$, $\ell < L$.

So it canbe put as a theorem that states: for a central potential, the lowest energy level is always S-level, ($\ell = 0$). QED.

Problem 9: Show the bound state is related to the number of zeroes, discuss the energy levels, etc., for a square well potential (s.w.p), as shown in the figure below:

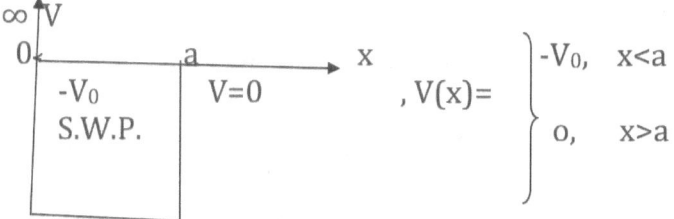

The solution:
Schrödinger equations for this problem are

$$\frac{-\hbar^2}{2m}\frac{d^2\psi}{dx^2}-V_0\psi=E\psi,\text{---------------x<a -----------(1)}$$

$$\frac{-\hbar^2}{2m}\frac{d^2\psi}{dx^2}=E\psi,\text{--------------------x>a ------------(2)}$$

One can notice the following in this problem:

(1.) The energy scale is lowered everywhere by the amount V_0.
(2.) The domain of x is $0\rightarrow\infty$.
(3.) The boundary condition that the wave function does not become infinite at x =0.

As it is known from the primarily quantum mechanics, the solutions to equations 1 and 2 are given by:

$\psi(x)$ =A sin αx +B cosαx, for x<a, ------------ (3).
$\psi(x)$ =C $e^{-\beta x}$, for x>a,------------------------(4).

Where $\alpha=\sqrt{\frac{2m\,(V_0-|E|)}{\hbar^2}}$, $\beta=\sqrt{\frac{2m\,E}{\hbar^2}}$.

Now the interesting thing in this problem is the bound state energy levels for which E<0; therefore, it is required the wave function $\psi(x)$ be finite at x=0, so B in the second term of equation 3 has to be set to zero, Hence, the solutions become,

$\psi(x) = A\sin\alpha x$, for x<a ------------------- (5).

$\psi(x) = Ce^{\beta x}$, for x>a -------------------- (6).

By applying the join condition at x=a, i.e.,

$$\frac{1}{\psi_{x<a}}\frac{d\psi_{x<a}}{dx}\Big]_{x=a} = \frac{1}{\psi(x)_{x>a}}\frac{d\psi_{x>a}}{dx}\Big]_{x=a} =>$$

$\alpha\mathrm{Cot}\alpha a = -\beta$ --------------------- (7).

The solutions at x-0, a are:

$A\sin\alpha a + B\cos\alpha a = 0$

$Ce^{-\alpha\beta} = 0$

$\mathrm{Cos}0 = 1 \neq 0$, $e^{-\alpha\beta} = 0$, therefore B=0 and C=0.

Now it is not plausible that both A and B are zero because this would give physically uninteresting solution, $\psi(0)=0$ everywhere. Also, it is not possible to make both $\sin\alpha a$ and $\cos\alpha a$ equal to zero, for a given value of α or E. Therefore, there are two possible classes of solutions:

    (1.)   The first class solution is A=0 and $\cos\alpha a=0$.

    (2.)   The second class solution is B=0 and $\sin\alpha a=0$.

These imply that $\alpha a = \frac{n\pi}{2}$ =>

$\psi(x) = B\cos\frac{n\pi}{2a}x$, -------------odd n, ------------------- (8).

$\psi(x) = A\sin\frac{n\pi}{2a}x$, --------------even n, ------------------ (9).

Next is to discuss equation 7, $\alpha\cot\alpha a = -\beta$,

For simplicity, let $\alpha a = \zeta$ and $\beta a = \eta$, then $\zeta\cot\zeta = -\eta$.

$\therefore \eta = -\zeta\cot\zeta$, --------------------------------------- (10).

Substituting the values for α and β, the following formula is obtained: $\zeta^2 + \eta^2 = (\frac{2m}{\hbar^2} V_0 a^2)$. $= R^2$. Geometrically, this represents a circle of radius R, where $R = \sqrt{\frac{2m\,(V_0 a^2\,)}{\hbar^2}}$.

η and ζ are restricted to positive value. $\eta = -\zeta \cot \zeta$.

The energy levels can be found from the intersections in the first quadrant of the curve of ($\zeta \cot \zeta$) plotted against ζ, with the circle of known radius (R), shown above.

The graphical solution of $\alpha \cot \alpha a = -\beta$ is illustrated below. For $V_0 a^2$ between zero and $\frac{\pi^2 \hbar^2}{2m}$, there is just one energy level of the first class (equation 8). For $V_0 a^2$ between $\frac{\pi^2 \hbar^2}{2m}$ and four times the value, there is one energy level of each class (equations 8 and 9) or the two altogether. As $V_0 a^2$ increases energy levels appear successively. First is of the first class and then followed by the other. From equation 3, $\psi(x) = A \sin \alpha a + B \cos \alpha a$; it can be seen that when ordered according to increasing of the eigenvalues, the nth eigenfunction has n nodes. Then it follows from this discussion that there is no energy level unless $V_0 a^2 > \frac{\pi^2 \hbar^2}{8m}$.

There is one bound state if $(\frac{\pi^2 \hbar^2}{8m}) < V_0 a^2 \leq \frac{9\pi^2 \hbar^2}{8m}$, etc.

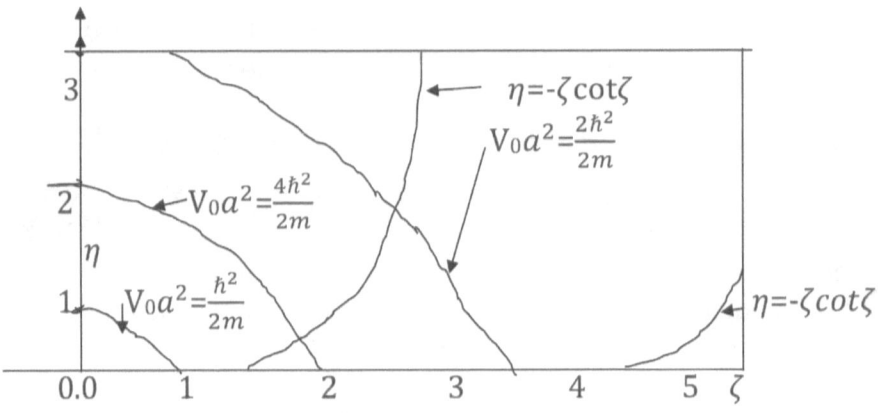

Schematic diagram η *versus* ξ

QED

Problem 10: Show that the Transmission (T) and the Reflection (R) are the same for diagrams (a and b) as shown below.

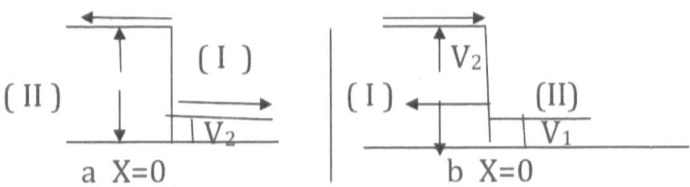

a X=0   b X=0

The solution: Take the diagram a, the wave function is

$\psi_I = R_1 e^{-ik_1 x} + R_2 e^{ik_1 x}$, for $x \geq 0$------------ (1).

$\psi_{II} = S_1 e^{-ik_2 x} + S_2 e^{ik_2 x}$, for $x \leq 0$, ---------------- (2).

Since there is no coming particle from left to write in a, $S_2 = 0$. Hence, equation 2 becomes $\psi_{II} = S_1 e^{-ik_2 x}$

Where $k_1 = \sqrt{\dfrac{2m\,(E-V_{10})}{\hbar^2}}$, $k_2 = \sqrt{\dfrac{2m\,(E-V_2)}{\hbar^2}}$ --------(3).

Now $\psi_i = R_1 e^{-ik_1 x}$, -------incident wave.

$\psi_R = R_2 e^{ik_1 x}$, --------reflected wave.

$\psi_T = S_1 e^{-ik_2}$,---------transmitted wave. $\left.\right\}$ ----- (4).

Now the current of incident wave J_i is given by

$J_i = \frac{\hbar}{2mi}(\psi_i^* \nabla \psi_i - \psi_i \nabla \psi_i^*) = |R_1|^2 v_1$, v_1 is the current velocity.

The reflected current is

$J_r = \frac{\hbar}{2mi}(\psi_R^* \nabla \psi_R - \psi_R \nabla \psi_R^*) = |R_2|^2 v_1$, where $v_1 = \frac{\hbar k_1}{2m}$.

Also, the transmitted current is

$J_t = \frac{\hbar}{2mi}(\psi_T^* \nabla \psi_T - \psi_T \nabla \psi_T^*) = |S_1|^2 v_2$, where $v_2 = \frac{\hbar k}{2m}$.

Now, the reflection coefficient is R$=\frac{J_r}{J_i}$, ------------ (5).

The transmission coefficient is T$=\frac{T_t}{J_i}$, ------------- (6).

Therefore, T+R$=\frac{J_r + J_t}{J_i}$ =1; it is the probability current.

By applying the join boundary conditions on ψ_I and ψ_{II}, the result will be $S_1 = \frac{2k_1}{k_1 + k_2}$, $R_2 = \frac{k_1 - k_2}{k_1 + k_2}$.

Now R$= \frac{J_r}{J_i} \frac{|R_2^2|}{|R_1^2|} \neq R_2|^2$, $R_1 = 1$, $\therefore$ R $= 1\frac{k_1 - k_2}{k_1 + k_2}|^2$

T=1-R, substitute for k_1 and k_2 from equation 3, T will be,

T$= \frac{4\sqrt{E - V_2}}{\left|1 + \sqrt{\frac{E - V_2}{E - V_1}}\right|}$ 2------------------------- (7). This for diagram a.

For diagram b, follow the same procedure, where,

$\psi_I = S_1 e^{ik_1 x} + S_2 e^{-ik_1 x}$, x$\leq$ 0. $\left.\right\}$ -----------(8).

$\psi_{II} = R_1 e^{ik_2 x} + R_2 e^{-ik_2 x}$, x$\geq$ 0. Notice from diagram b there is no incident particle from right to left, so R_2 =0. Also, k_1

$= \sqrt{\dfrac{2m(E-V_{10})}{\hbar^2}}$, and $k_2 = \sqrt{\dfrac{2m(E-V_2)}{\hbar^2}}$. Follow the other same steps in a

the T is given by [$T = \dfrac{4\sqrt{E-V_2}}{\left|1+\sqrt{\dfrac{E-V_2}{E-V_1}}\right|^2}$].----------------(9).

Hence, equations 7 and 9 are the same as required.
QED.

Problem 11: Given below a potential well, where

$$U(x) \quad = \quad \left\}\begin{array}{l} 0, \qquad x>L \\ V_0\ (>0),\ 0<x<L, \end{array}\right.$$

discuss the problem for,

(1.) $\mathcal{E}<U_0$
(2.) $\mathcal{E}>U_0$ Find the transmission coefficient T, and the transit time t. This problem is graphically represented below.

The solution: The Schrödinger equation in general is

$\psi^{"} -(\mathcal{E}\text{-}U)\psi =0,$ \quad for \quad $0 \leq x \leq L.$
$\psi^{"}\text{-}\mathcal{E}\psi=0,$ for $x>L.$

The solution for the case $x>L$ is quite well known as

$\psi = e^{-i\sqrt{\mathcal{E}}x} + Re^{i\sqrt{\mathcal{E}}x},$ ------------------------- (1).

And for the case x<0, the solution is

$\psi = S e^{-i\sqrt{\mathcal{E}}\,x}$, -------------------- (2).

$e^{+i\sqrt{\mathcal{E}}x}$. Here is not physically accepted. See the diagram, according to the physical condition of ψ. Hence,

$$\psi(x) = \begin{cases} e^{-i\sqrt{\mathcal{E}}x} + R\, e^{i\sqrt{\mathcal{E}}x}, & \text{for } x>L \\ \\ S\,e^{-\sqrt{\mathcal{E}}x}, & \text{for } x<0 \end{cases} \quad \text{--------- (3).}$$

For the case $0 \geq x \leq L$, there are two cases:

(1.) If $\mathcal{E} < U_0$, the solution is,

$\psi = A e^{kx} + B e^{-kx}$, $k = \sqrt{U_0 - \mathcal{E}}$,----------- (4).

(2.) If $\mathcal{E} > U_0$, the solution is $\psi = C e^{ikx} + D e^{-ikx}$,----- -(5).

$k = \sqrt{\mathcal{E} - U_0}$. Hence, the solutions altogether are,

$$\psi(x) = \begin{cases} e^{-i\sqrt{\mathcal{E}}x} + R\, e^{i\sqrt{\mathcal{E}}x}, & \text{for the case } x>L \\ C e^{ikx} + D\, e^{-ikx}, & \text{for the case } 0<x<L \\ S e^{-i\sqrt{\mathcal{E}}x}, & \text{----for the case } x<L \end{cases} \quad \text{----(6).}$$

Now the physical meaning of S and R can be inferred by substituting the solutions of (x>L and x<0) into one dimensional form of the probability of current density given by $j(x)=v(\ |R|^2 -1)$, x>L, $j(x) = v|C|^2$, $v = p/m = \frac{\sqrt{\mathcal{E}}}{m}$. Speed of the particle, for $\mathcal{E} > U_0$, the solution is inside as the second in equation 6. By applying the joining conditions at x=0 and x=L, and eliminating C and D it is found that

$$R=\frac{(\mathcal{E}-k^2)(1-e^{2ikL})}{(\sqrt{\mathcal{E}}+k)^2-(\sqrt{\mathcal{E}}-k)^2\,e^{2ikL}}$$

$$S=\frac{4k\sqrt{\mathcal{E}}}{(k+\sqrt{\mathcal{E}})^2-(\sqrt{\mathcal{E}}-k)^2e^{2ikL}}\;e^{i(k-\sqrt{\mathcal{E}})L}$$

$$\left.\vphantom{\begin{matrix}1\\1\\1\\1\end{matrix}}\right\} \quad\text{---------- (7).}$$

Now to find the transmission coefficient, the transmitted current, and the total current must be founded. Therefore, taking the transmitted wave, $\psi = Se^{-i\sqrt{\mathcal{E}}x}$ to find the transmitted current as $j_t = \frac{\hbar}{2m}[\psi^*\nabla\psi - \psi\nabla\psi^*] = \frac{\hbar\sqrt{\mathcal{E}}}{m}|S|^2$.--- (8).

Also, the incident wave $\psi = e^{-i\sqrt{\mathcal{E}}x}$ to be used by the same way to find the total current as, $j_{tot.} = \frac{\hbar\sqrt{\mathcal{E}}}{m}$. So T= $\frac{j_t}{j_{tot.}} = |S|^2$.

So from equation 7, (S), T= $\frac{4k^2\,\mathcal{E}}{4k^2\,\mathcal{E} + (\mathcal{E}-k^2)^2 \sin^2 kL}$, but

k= $(\mathcal{E}-U_0)^{1/2}$, so by substituting for k, T will take the form

$$T=\left[\frac{4\mathcal{E}(\mathcal{E}-U_0)}{4\mathcal{E}(\mathcal{E}-U_0)+U_0^2\,\sin^2 kL}\right]\;\text{------------for } \mathcal{E}>U_0.\;\text{-------- (9).}$$

For $\mathcal{E}<U_0$, the solution is, equation 4, i.e.,

$\psi(x) = Ae^{\hbar x} + Be^{-\hbar x}$, $0<x<L$, $\hbar = \sqrt{U_0 - \mathcal{E}}$

This case can be obtained by replacing k by $(i\hbar)$ in the solution of $\mathcal{E} > U_0$.　　Because　　k　　= $\sqrt{\mathcal{E} - U_0}$　　= $\sqrt{(-(U_0 - \mathcal{E})}$ = $\sqrt{-1}\sqrt{(U_0 - \mathcal{E})}$ = $i\sqrt{U_0 - \mathcal{E}}$ = $i\hbar$.

Also, it is known from complex variable in mathematics that $\sin i\hbar = i\sinh\hbar$; therefore,

$$T=\frac{4k^2\mathcal{E}}{4k^2\mathcal{E}+(\mathcal{E}-k^2)^2\sinh^2 kL}=$$

$$[T=\frac{4\mathcal{E}(U_0-\mathcal{E})}{4\mathcal{E}(U_0-\mathcal{E})+U_0^2\sinh^2 kL}]\text{-------- for } \mathcal{E}<U_0 \left.\phantom{\begin{array}{c}a\\a\\a\end{array}}\right\} \text{--------(10)}.$$

$$\text{And }\{T=\frac{4\mathcal{E}(\mathcal{E}-U_0)}{4\mathcal{E}(\mathcal{E}-U_0)+U_0^2\sin^2 kL}\}\text{ ------for } \mathcal{E}>U_0$$

Now to find out, what is the transit time? It is already found

that $S=\{\dfrac{4k\sqrt{\mathcal{E}}\,e^{i(k-\sqrt{\mathcal{E}})L}}{(k+\sqrt{\mathcal{E}})^2-(\mathcal{E}-k)^2 e^{2ikL}}\}=$

$$=\frac{1}{\frac{(k+\sqrt{\mathcal{E}})^2}{4k\sqrt{\mathcal{E}}}e^{i(\sqrt{\mathcal{E}}-k)}-\frac{(\mathcal{E}-k)^2}{4k\sqrt{\mathcal{E}}}e^{i(k-\sqrt{\mathcal{E}})}\;e^{i(k-\sqrt{\mathcal{E}})L}}=1/\mathcal{T},; \text{ therefore,}$$

$\psi=1/\mathcal{T}e^{-i\sqrt{\mathcal{E}}x}$, the transmitted wave.

Now there is a wave remains concentrated inside region II during a time that is much greater than a classical time splitting into a transmitted and reflected waves. This time is called a transit time; it is given as t= $\frac{2m}{\hbar}(\Delta$ x)by spreading of distribution at a time measured by

$$\Delta x_t=\sqrt{(\Delta x)^2+(\frac{\hbar}{2m}\frac{t}{\Delta x})^2}$$

Now comparing this with the amplitude of $\psi= S\,e^{-i\sqrt{\mathcal{E}}\,x}$,

it is found that $(\Delta x)^2=\frac{(k+\sqrt{\mathcal{E}})^2}{4k\sqrt{\mathcal{E}}}$. Hence, the transit time is

$$t=[\frac{2m}{\hbar}\frac{(k+\sqrt{\mathcal{E}})^2}{4k\sqrt{\mathcal{E}}}]=[\frac{2m}{\hbar}\frac{(\sqrt{\mathcal{E}-U_0}-\sqrt{\mathcal{E}})^2}{4\sqrt{\mathcal{E}}(\sqrt{\mathcal{E}-U_0})}]$$

By investigating the transmission coefficient T, it is clear that its function of the energy($\mathcal{E}$) remains very small.

Problem 12:

Given $\psi(x,0)=\frac{1}{(\pi\xi^2)^{1/4}}e^{\frac{ip_0 x}{\hbar}-\frac{x^2\xi_0^2}{2}}$, at t=0. --------- (A)

Show its time development and its specific change after t>0.
The solution: With the aid of Fourier transformation (FT), $\psi(x,t)$ can be written as

$$\psi(x,t)=\frac{1}{\sqrt{2\pi\hbar}}\int f(p)\, e^{\frac{ipx}{\hbar}}\, e^{\frac{ip^2 t}{2m}}\, dp\text{----------------} (1)$$

So $\psi(x,0)=\frac{1}{\sqrt{2\pi\hbar}}\int f(p)\, e^{\frac{1px}{\hbar}} dp$. Also using FT, $f(p)$ is

$$f(p)=\frac{1}{\sqrt{2\pi\hbar}}\frac{1}{(\pi\,\xi_0^2)^{1/4}}\int e^{\frac{-i(x(p-p_0))}{\hbar}}\exp.(\frac{-x^2}{2\xi_0^2})\,dx.\text{-------}(2).$$

Now add the exponential quantities and complete the square, letting $2\hbar^2\,\xi_0^2$ =constant to be 1 for simplicity, then

$$f(p)=\frac{1}{\sqrt{2\pi\hbar}}\frac{1}{(\pi\,\xi_0^2)^{1/4}}\,e^{\frac{-\xi^4(p-p_0)^2}{\hbar^2}}\int e^{-(x+c)^2}dx\text{-------}(3).$$

Where c= $\frac{i\xi_0^2}{\hbar}$(p-p_0). Again for simplicity, let (x+c)=y,=>dx=dy, $\therefore$
$$f(p)=\frac{1}{\sqrt{2\pi\hbar}}\frac{1}{(\pi\xi_0^2)^{1/4}}$$

Now substitute for $f(p)$ in equation 1 to become $\psi(x,t)$,

$$\psi(x,t)=\frac{1}{\sqrt{2\pi\hbar}}\frac{1}{(\pi\xi_0^2)^{1/4}}\int e^{\frac{1px}{\hbar}}\,e^{\frac{\xi_0^4(p-p_0)}{\hbar^2}}\,e^{\frac{-ip^2 t}{2m}}.$$

Now take the integral, then add the exponential, and complete the square. Also take the integral $\psi(x,t)$; it will be

$$\psi(x,t)=\frac{1}{2\hbar\sqrt{\pi}}\frac{1}{(\xi_0^2+\frac{\hbar^2 t^2}{m^2\xi^2})^{1/4}}\,e^{\frac{-\xi_0^4 p_0^2}{\hbar^2}}\,e^{\frac{1\mathcal{M}}{4}}\text{----------}(B),\text{ where}$$

$$\mathcal{M}=[\frac{\frac{ix}{\hbar}+\frac{2\xi_0^4 p_0}{\hbar^2}}{\frac{i}{2m}t+\xi_0^4/\hbar^2}]^2,\text{ all squared.}$$

Now compare equation A and equation B; the new width of the wave after time t will be $\xi = \xi_0^2[1+\frac{\hbar^2 t^2}{m^2 \xi_0^2}]$. Therefore, ψ remains in the Gaussian shape during the time evaluation; only the width increases, as shown above. If t=0, it is clear $\xi^2=\xi_0^2$.
QEQ.

Problem 13: Given $F(A)=(A- a_n)\, g_n\,(A)$,------------(1).

Show that: (i.) $F(A)=0$, P_n (projector operator)$=g_n(A)/g_n(a_n)$; (ii.) $F(A)$ is a function of observable N eigenvalues.

The solution:

(i.) Using the eigenvalue equation that

$F(A)|a_n>=F(a\,)|a_n>$, $|a_n>$ is an eigenstate. And $f(a_n)$ is a function of eigenvalues $a_1, a_2, a_3,$------a_n.

Using equation 1, then $F(a_n) =(a_n-a_n)g_n(a_n)=0$. Therefore, $F(A)|\,a_n>=0$. But $|\,a_n> \neq 0$, it is a given state,. Hence, $\{F(A)=0\}$ ----------------------- (2).

Now $F(A)\,|a_n>=(A-a_n)g_n(A)|\,a_n>$, from (1). But $F(A)=0$. So,

$A\,g_n(A)|\,a_n> =a_n\,g_n(A)|\,a_n>$, using the fact $A|a_n>=a_n|a_n>$ and
$Ag_n(A)|a_n>=a_n g_n(A)|a_n>$
$g_n(A)\,A|a_n> =a_n\,g_n(a_n)\,|\,a_n> =>$
$g_n(A)a_n\,|\,a_n>=a_n\,g_n(a_n)\,|\,a_n>$.

If a projector operator as $P_n=|\,a_n><a_n|$ is introduced, it is possible to write that $g_n(A)a_n\,|\,a_n>=g_n(a_n)a_n\,|a_n><a_n|a_n>1$ $= g_n(a_n)a_n|\,a_n>P_n \Rightarrow P_n=g_n(A)/g_n(a_n)$, as required.

(ii). Now if $F(A)=(A-a_n)g_n(A)=0$, (see equation 2), then $(A-a_n)=0, \Rightarrow A=a_1, a_2, a_3,------a_n$, i.e., $A=a_n$. So A is a function of observable of N eigenvalues, as required.

QED.

Problem 14: Given lu><vl show that, $T_r$lu><vl=<vlu>.

The solution: Define first the trace matrix of an operator, say A, it is $T_r A=\sum_i A_{ii}$, ----------------------------------- (1).

Now the matrix of the given operator lu><ul with respect to two states, li> and lj>, is given by (l u><ul)$_{ij}$ =
(<ulj><ilu>)-------------Scalar product----------- (2).

The diagonal elements are (lu><u>)$_{ii}$ =(<uli><ilu>), so
T_r =lu><ul=$\sum_i (u><u)_{ii}=\sum_i <u|i><i|u>=$
$\sum_i |i><i|<u|u> = $=<ulu>$\sum_i |i><i|$. Since $\sum_i |i><i|$=1

Therefore, [T_r lu><ul=<ulu>] ----------------- (3).
Now to find $T_r$lu><vl=<ulv>?

Following the same way, write (lu><vl)$_{ij}$ =(<ilu><vlj>)=
(<vlj><ilu>), so the diagonal elements are
(lu><vl>)$_{ij}$ =(<v li><ilu>). Hence, $T_r$lu><vl=
$\sum_i$ <vli><ilu>=<vlu>$\sum_i |i><i|$=<vlu> because the closure relation is 1. (C-R.)

So [$T_r$lu><vl=<vlu>]. QED.

Problem 15: If H is a positive definite, then <ulH u>$\geq$ 0, for any lu>, and <v lH lv>$\geq$ 0, for any lv>. If T_r H=0$\rightarrow$H=0.

The proof: Since H is Hermitian operator, i.e., $H=H^*$, it can be written that

<ulHlu><vlHlv>=H²<ulu><vlv>.--- --------------(1).

Using Schwartz inequality, it gives
<ulu><vlv>≥ l<ulv>l², --------------------------------(2).

Multiplying equation 2 by H², it becomes
H²<ul u><vlv>≥ H²l<ulv>l²⇒
<ulHlu><vlHlv>≥l<ulHl v>l²,
∴|<ulHlv>l²≤<ulHlu><vlHlv>,--------------------(3).

It is given that <u lHl u>≥0, positive definite.
∴ H<ulu>≥ 0, since H=H*, Hermitian.

Now if <ulHlu>=0, and <ulu> is norm≠0, →Hlu>=0.
The matrix element of H is <ilHlj>=H_{ij}, so the diagonal elements
are H_{ii}=<ilHli>, T_r H=$\sum_i H_{ii}$=$\sum_i$ <$|i H|i$ > or
T_rH=$\sum_i$ < $u_i|H|u_i$ >, since <$u_i|$H$|u_i$>≥0, →T_rH≥ 0,→
$\sum_i H_{ii}$ ≥ 0, so if T_r H=0, it implies that H=0.
Q.D.

Problem 16: Given A as a linear operator, show that:

(1.) $T_r(A^+ A)$=$\sum$ lAl²≥0
(2.) If $T_r A^+A$=0, then A=0.

The solution: Let A an operator operating on a certain
eigenstate, say ψ, then

$A\psi = \psi$-, a new state, which also can be written as
$\psi^- = \psi A^+$, then (ψ^-, ψ^-)= $(\psi A^+A \psi)$,
But (ψ^-, ψ^-)=1≥ 0, normalized, so $(\psi A^+A \psi)$=1⇒
$A^+ A$=1; hence, A is a unitary operator >0.→<ψ^-, ψ^-)≥ 0.

It is a positive definitive. And $A^+A=1$ is also a positive definitive and Hermitian operator.

T_rA^+ $A=\sum_i(A^+A)_{ii}$, where ($A^+A)_{ii}$ is the diagonal element of A^+A matrix. Hence, $(A^+A)_i =<\psi_i|A^+A|\psi_i>=$
$<\psi_i A |A\psi_i>=<A\psi_i|A\psi_i>=<\psi_i|\psi_i> |A|^2=|A|^2$,
$<\psi_i|\psi_i>=1$, normalized.
$\therefore [T_r (A^+A) =\sum_i(A^+ A)_{ii} =\sum |A|^2 \geq 0]$.

Now if $T_r(A^+A)=0 \rightarrow (\psi_i|A^+ A|\psi_i)=0$, but $(\psi_i,\psi_i)=1$, so A should be zero (A=0). Therefore, $T_r0=0$.
QED.

Problem 17: Prove that $(\frac{\ell+L\cdot\sigma}{2\ell+1})$ is a projector operator onto state of total angular momentum $J=\ell+1/2.(\ell \uparrow \ s \uparrow)$. And that $\frac{\ell-L\cdot\sigma}{(2\ell+1)}$ is a projector operator onto state of $J=\ell-1/2. (\ell \uparrow \ s \downarrow)$

The proof: It is known that any projector operator, P satisfies the following properties:

(1.) $P^2=P$, so P=0 or 1.
(2.) $P\psi^\wedge =\psi^\wedge$.

This indicates that P is projector operator on $\psi^\wedge$.
Now an investigation should be made of the projector operator P under the properties 1 and 2.

(a.) $(\frac{\ell+1+L\cdot\sigma}{2\ell+1})^2=\frac{\ell^2+\ell L\cdot\sigma+\ell+1+L\cdot\sigma+\ell L\cdot\sigma+L\cdot\sigma+L^2-L\cdot\sigma}{(2\ell+1)^2}$

By using the identity, $(L\cdot \sigma)(L\cdot \sigma) =L^2+i\sigma(L^\rightarrow \times L^\rightarrow)$
And $L^\rightarrow \times L^\rightarrow = iL^\rightarrow$, $L^2=\ell(\ell + 1)$ in $\hbar$ unit; the result will be
$=\frac{\ell+1+L\cdot\sigma}{(2\ell+1)}$, i.e., $P^2=P$, property 1 is satisfied.

Now using the fact that $\sigma=2S$, Pauli matrices, operate on the state lj$=\ell + 1/2$, $1/2 >$; therefore, $(\frac{\ell+1+L\cdot\sigma}{(2\ell+1)})$l $\ell + \frac{1}{2}$, $1/2>=\frac{\ell+1+2L\cdot S}{(2\ell+1)}$lj,$1/2>$, but $J^2=L^2 + S^2+2LS$,

so $2L\cdot S=J^2 - L^2\text{-}S^2$. And remember that $J^2=j(j+1)$ in $\hbar$ unit. $S^2=s(s+1)$ in $\hbar$ unit, $L^2 =\ell(\ell + 1)$ in $\hbar$ unit, also $s=(1/2)\hbar$.

Using all these information, $[\frac{(\ell+1+J^2-L^2-S^2}{(2\ell=1)}]$lj,$1/2>=$lj,$1/2>$ is obtained This satisfies property 2, i.e., $P\psi^\wedge=\psi^\wedge$.

(b.) For the case of the state lj$= \ell -\frac{1}{2} >$, operated by $\frac{\ell-L\cdot\sigma}{2\ell+1}$, and following the same way and steps, the result will be $[\frac{\ell-L\cdot\sigma}{2\ell+1)}]^2=\frac{\ell-L\cdot\sigma}{2\ell+1}$. It satisfies $P^2=P$, i.e., $P=1$ or 0.

Now operate on l j$=\ell - 1/2$, $1/2>$. Show it equals to lj, $1/2>$. Following the same way, using the mentioned information, the result will be $(\frac{\ell-J^2+L^2+S^2}{2\ell+1})^2$ lj,$1/2>=(\frac{2\ell+1}{2\ell+1})$lj,$1/2>=$ lj, $1/2>$, for j$=\ell$-1/2. Again, it satisfies $P\psi^\wedge=\psi^\wedge$.

Also, it can be shown that $(\frac{\ell-L\cdot\sigma}{2\ell+1})lj=1+1/2$, $1/2>=0$, (Show it.) Hence, it is shown that $(\frac{\ell-L\cdot\sigma}{2\ell+1})$ is a projector operator onto state of total angular momentum j$=\ell$-1/2, $(\ell \uparrow s \downarrow)$. QED.

Problem 18: Show that $J^2\{T_q^k\}=k(k+1)T_q^k$.
Where $T^{(k)}$ is tensor of rank (k).
The proof: By definition, we have:
$J^2\{A\}=[J_x,[J_x\cdot A]]+[J_y[J_y,A]]+[J_z[J_z,A]]$. It can be written as
$J^2\{T_q^k\}=[J_x[J_x,T_q^k]]+[J_y[J_y, A]]+[J_z[J_z,A]]$ --------- (1).

The following useful information are already known:

(a.) $[J_{\pm}, T_q^k] = \sqrt{k(k+1) - q(q \pm 1)} T_q^k$.

(b.) $[J_z, T_q^k] = q\, T_q^k$.

(c.) $J_+ = J_x + iJ_y$,

(d.) $J_- = J_x - iJ_y$

(e.) $J_+ J_- = J_x^2 + J_y^2 J_z^2$

(f.) $J_x^2 + J_y^2 = J_+ J_- - J_z^2$

Now using the identity [A, B+C]=[A, B]+[A, C], equation 1 can be written as:

$$J^2\{T_q^k\} = [J_+ J_-, T_q^k] - [J_z, T_q^k] + [J_z, [J_z, T_q^k]] \text{----------------- (2).}$$

Now $[J_+ J_-, T_q^k] = [J_+, [J_-, T_q^k]] = [k(k+1) - q(q-1)] T_q^k$ -----(3).

Substituting equation 3 into equation 2 with the use of the relation b, one gets $J^2\{T_q^k\} = [k(k+1) - q(q-1)] T_q^k$ $-q(q-1) T_q^k$ $= k(k+1) T_q^k$.

So $J^2\{T_q^k\} = k(k+1) T_q^k$, as required.

Here, T_q^K could be spherical harmonic function $Y_L^M(\theta, \varphi)$.

QED.

Problem 19: From the following relations:

$[a_r, a_s] = [a_r^+, a_s^+] = 0$, $[a_r a_s^+] = \delta_{rs}$

$[a_1, a_1^+] = 1$, $S = 1/2[a_1^+ a_1 + a_2^+ a_2]$, $J_1 = 1/2[a_2^+ a_1 + a_1^+ a_2]$.

$J_2 = i/2 [a_2^+ a_1 - a_1^+ a_2]$, $J_3 = 1/2[a_1^+ a_1 - a\frac{+}{2} a_2]$. ----------- (1).

Show that: $[J_1, J_2] = iJ_3$, $[J_1, J_3] = i\, J_2$, $[J_2, J_3] = iJ_1$, in $\hbar$ unit.

The proof: From this information, it is easy to find that $a_1 a_1^+ = 1 + a_1^+ a_1$; $a_2 a_2^+ = 1 + a_2^+ a_2$; $a_1 a_2 = a_2 a_1$, and $a_1^+ a_2^+ = a_2^+ a_1^+$.

Now all these in mind the proof can be done, take $[J_1, J_2]$, substitute for J_1 and J_2 from the above information. With some mathematical manipulation, you find that:

$[J_1, J_2]=i/2[a_1^+ a_1 - a_2^+ a_2]=i\,J_3$, as required.---------(2).

Following the same steps, the others will be proven as:

$[J_2,J_3]=iJ_1$---------------------------------- (3).
$[J_1, J_3]=IJ_2$ ----------------------------- (4).

Also, it can be shown that $J \times J=iJ$. Remember that $J=e_x^{\rightarrow}J_x^{\rightarrow} + e_y^{\rightarrow}J_y^{\rightarrow} + e_z^{\rightarrow}J_z^{\rightarrow}$, J is a vector. And $J^2=J_x^2+J_y^2+J_z^2$.

If J-components are subtracted and working out the mathematical relations, $J^2=s(s+1)$ in $\hbar$ unit. Therefore,

$[s, J]=[sJ\text{-}Js]=(s\sqrt{s(s+1)} -\sqrt{s(s+1)}\, s\,)=0$. Hence $[s, J]=0$.

So s commutes with J. Also, one can show that $[J^2, J_3]=0$, i.e., J_3 commute with J^2, which means they have the same eigenfunction. This can be easily proved as follows:

Already, it is shown that $J=s(s+1)$ in $\hbar$ unit. And

$J_3 =1/2(a_1^+ a_1 - a_2^+ a_2)$, $S=1/2[a_1^+ a_1 + a_2^+ a_2]$ are given in the information above. Hence, $J_3 +S=a_1^+ a_1 \rightarrow J_3 =a_1^+ a_1\text{-}S$.

Now $[J^2, J_3]=0$, substitute and work the mathematical steps.
$\therefore$ $[J^2, J_3]=0$, as required.
QED.

Problem 20: Given the figure as shown below:
$r^{\rightarrow} \cdot r^{--\rightarrow}=rr^-\cos\alpha=r\,r^-.\,r^\wedge.r^{-\wedge}$

Where r^, $r^{-\wedge}$ are unit vector,

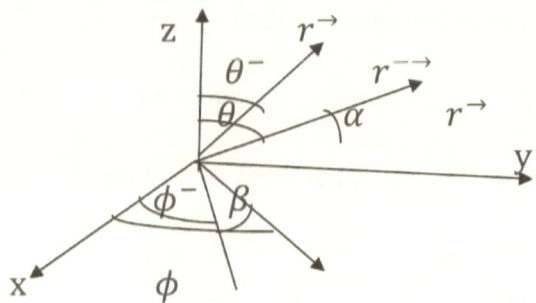

Show that $P_\ell(\cos\alpha) = \frac{4\pi}{(2\ell+1)} \sum_{m=-\ell}^{\ell} (-1)^m\, Y_\ell^m(\Omega)\, Y_\ell^{-m}(\Omega^-)$.

The solution: Geometrically, it can be shown that

$r^\rightarrow . r^{-\rightarrow} = \cos\theta\, \cos\theta^- + \sin\theta\, \sin\theta^- \cos(\phi - \phi^-)$ ---------- (1).

$\therefore r^\rightarrow . r^{-\rightarrow} = rr^-\cos\alpha = rr^-[\cos\theta\,\cos\theta^- + \sin\theta\,\sin\theta^- \times$
$\cos(\phi - \phi^-)]$ --(2)

So $\cos\alpha = \cos\theta\,\cos\theta^- + \sin\theta\sin\theta^-\cos(\phi - \phi^-)$ ------- (3).

Now consider $r^{-\rightarrow}$ as a fixed vector in space. The Legendre polynomial $P_\ell(\cos\alpha)$ is a function of the angles θ and ϕ, and the angles θ^- and ϕ^- are as parameters. Since the angular momentum is conserved, invariant under rotation, then $L^2 = L^{(-)^2}$, and $\ell = \ell^-$. Since r^- can be taken as a fixed vector in space, so it might be taken along Z-axis; hence, the angle α becomes the usual polar angle, which implies that $P_\ell(\cos\alpha)$ satisfies the equation:

$\nabla^{(-)^2} P_\ell\text{-}(\cos\alpha) + \frac{L^{(-)^2}}{r^2} P_\ell\text{-}(\cos\alpha) = 0$, in $\hbar$ unit. ---------- (4).

Mathematically, it is known that $\nabla^2 V = \nabla \cdot \nabla V$. It is an operator and scalar product. But it is known that the scalar products are invariant under rotation; hence, $\nabla^{(-)^2} = \nabla^2$, r unchanged.

Therefore, equation 4 can take the unprimed form

$$\nabla^2 P_\ell(\cos\alpha)+\frac{L^2}{r^2}P_\ell(\cos\alpha)=0,\text{---------------------------(5)}.$$

From this equation, $P_\ell(\cos\alpha)$ can be calculated. It is the solution of the differential equation 5. So $P_\ell(\cos\alpha)$ is the eigenfunction of the total angular momentum, L^2 and $L^{(-)^2}$.

The system is invariant under rotation, but also it is known

$$L^{(-)^2}=\frac{1}{\sin\theta^-}[\frac{\partial}{\partial\theta^-}(\sin\theta^-\frac{\partial}{\partial\theta^-})]+\frac{1}{\sin^2\theta^-}\frac{\partial^2}{\partial\phi^{(-)2}}=\nabla^2_{\theta^-,\phi^-}.$$

$$L^2=\frac{-1}{\sin\theta}[\frac{\partial}{\partial\theta}(\sin\theta\frac{\partial}{\partial\theta}]+\frac{1}{\sin^2\theta}\frac{\partial^2}{\partial\phi^2}=\nabla^2_{\theta,\phi}, \text{ in unit of } \hbar$$

But again, the primed system is equal to the unprimed system. The components of the angular momentum are given by

$$\ell_x=-i[\cos\phi\frac{\partial}{\partial\theta}-\cos\phi\cot\alpha n\,\theta\frac{\partial}{\partial\phi}],\text{--------------(6a)}$$

$$\ell_y=-i[\cos\phi\frac{\partial}{\partial\theta}-\sin\phi\cot\alpha n\,\theta\frac{\partial}{\partial\phi}],\text{------------(6b)}.$$

$$\ell_z=-i\frac{\partial}{\partial\phi},\text{--- (6c)}.$$

The same valid for the primed system. So $\ell_x^-, \ell_y^-\ell_z^-$ are

$$\left.\begin{array}{l}\ell_x^-=-i[\sin\phi^-\frac{\partial}{\partial\phi^-}-\cos\phi^-\cot\alpha n\theta^-\frac{\partial}{\partial\phi^-}.\\[2mm]\ell_y^-=-i[\cos\phi^-\frac{\partial}{\partial\theta^-}-\sin\phi^-\cot\alpha n\,\theta^-\frac{\partial}{\partial\phi^-}\end{array}\right\}\quad\text{----------(7)}.$$

$$\ell_z^-=-i\frac{\partial}{\partial\phi^-}.$$

Using equations 6abc and 7, it can be shown that:

$$(\ell_x+\ell_x^-)P_\ell(\cos\alpha)=0$$
$$(\ell_y+\ell_y^-)P_\ell(\cos\alpha)=0.$$

Or in general, $(\ell_i + \ell_i^-)P_\ell(\cos\alpha)$----------------------- 8)
$(\ell_z + \ell_z^-)P_\ell(\cos\alpha)=0$

So far, it has been found that:

$$\left.\begin{array}{l} L^2 P_\ell(\cos\alpha) = L^{-2} P_{\ell^-}(\cos\alpha) = \ell(\ell+1)P_\ell(\cos\alpha) \\ (\ell_i + \ell_i^-)P_\ell(\cos\alpha)=0 \end{array}\right\} - - - (9).$$

The next step is to show $P_\ell(\cos\alpha)=\frac{4\pi}{2\ell+1}\sum_{m=-\ell}^{\ell}(-1)^m Y_\ell^m Y_\ell^{-m}$.

First, it is useful to write down the relation between P_ℓ and the known spherical harmonic function $Y_\ell^m(\alpha)$. It is well known as

$Y_\ell^m(\alpha)=\sqrt{\frac{(2\ell+1)(\ell-m)!}{4\pi\,(\ell+m)!}}(-1)^m e^{im\phi} P_\ell (\cos\alpha)$.

Where α depends on θ, ϕ, θ^- and ϕ^- as shown in the figure.

Now if m=0, then $Y_\ell^0(\alpha)=\sqrt{\frac{2\ell+1}{4\pi}} P_\ell(\cos\alpha)$,------------(10).

And if $\alpha=0$, then $Y_\ell^0(0)=\sqrt{\frac{2\ell+1}{4\pi}} P_\ell(0)=\sqrt{\frac{2\ell+1}{4\pi}}$,-------- (11).

$P_\ell(0)=1$. Therefore, $P_\ell(\cos\alpha)=\sqrt{\frac{4\pi}{2\ell+1}} Y_\ell^0(\alpha)$. ---------- (12).

Now multiply equation 12 by equation 11, the result is
$P_\ell(\cos\alpha)=(\frac{4\pi}{2\ell+1}) Y_\ell^0(\cos\alpha)Y_\ell^0(0)$,--------------------(13).

Now use the D-matrix that connects between $Y_\ell^m(\alpha)$ and $Y_\ell^m(\theta,\phi)$ under rotation case. This relation is
$Y_\ell^{m^-}(\alpha,\beta) =\sum_{m^-=-\ell}^{\ell} D_{mm^-}^\ell(\alpha,^-\beta,^-\gamma^-)Y_\ell^m(\theta,\phi)$
α,β,γ are the well-known Euler angles.

Set $m^-=0$. Then $Y_\ell^0(\alpha,0)=\sum_{m=-\ell}^{\ell} Y_\ell^m(\theta,\phi)D_{mo}^\ell(\alpha,^-\beta,^-\gamma^-)$.

But $D^\ell_{m0}(\alpha^-,\beta^-,\gamma^-)=\sqrt{\frac{4\pi}{2\ell+1}}\,Y^{m*}_\ell(\theta^-,\phi^-)$, [See Edmond, A.M.]

But $Y^{m*}_\ell(\Omega^-)=(-1)^m Y^{-m}_\ell(\Omega^-)$.

$\therefore [P_\ell(\cos\alpha)=\frac{4\pi}{(2\ell+1)}\sum_{m=-\ell}^{\ell}(-1)^m\, Y^m_\ell(\Omega)\, Y^{-m}_\ell(\Omega^-)]$.---(14)

QED.

Problem 21: Given $Q=\frac{(S.r)^2}{r^2}$. Show that Q is a projector.

The solution: If Q is a projector. It has to satisfy that

$1\text{-}Q^2=Q$, $(2)\text{-}Q\psi=\psi$, $Q=1$ or 0.

Now S=s_1+s_2, so $Q=\frac{[(s_1+s_2).r]^2}{r^2}=\frac{(s_1.r+s_2.r)(s_1.r+s_2.r)}{r^2}=$

$$\frac{(s_1.r)^2+(s_2.r)^2+2(s_2.r)(s_2.r)}{r^2}$$

$s_1=(1/2)\sigma_1$, $s_2=\frac{1}{2}\sigma_2$, σ_i(i=1, 2, 3) is Pauli matrices.

Remember that s=(1/2) in $\hbar$ unit. Therefore, $\searrow$ r^2 sin0=0.

$(s_1\cdot r)^2=1/4(\sigma_1.r)(\sigma_1.r)=1/4[r2+\sigma_1(r\times r)]=(1/4).r^2$

By the same way, $(s_2.r)^2=(1/4).r^2$,

Therefore, $[\,Q=1/2+\frac{2(s_1.r)(s_2.r)}{r^2}\,]$------------------(1).

Now find $Q^2=[1/2+[\frac{2(s_1.r)(s_2.r)}{r^2}]^2$. Complete the product steps. Remember r$\times$ r =r^2 sin r^.r^ =r^2 sin0=0.

The result will be $Q^2=Q$, so first property is proved.
Now show that $Q\psi=\psi$?
It is known S_{12} =2(3Q - S^2), a tensor operator.

Now it is needed to show that this satisfies the identity

$[S^2_{12}=4S^2-2S_{12}]$. Take $S^2_{12}=4(9Q^2+S^2S^2-3Q\,S^2-3S^2Q)=$

$4(9Q^2-3s(s+1)Q-3Qs(s+1)+s(s+1)S^2)$

S=1, triplet, or 0, singlet. Therefore, $S_{12}^2 = 4S^2 - 4(3Q - S^2) =$
$4S^2 - 2S_{12}$. So S_{12}^2 $4S^2$ -2 S_{12}; therefore, this is satisfied by $S_{12} = 2(3Q - S^2)$, as required. For triplet case.

Now take the tensor operator $S_{12} = 2[3 \frac{(S.r)^2}{r^2} - S^2]$, but

$S^2 = (s_1 + s_2)^2 = s_1^2 + s_2^2 + 2 s_1.s_2 = s_1(s_1+1) + s_2(s_2+1) + 2s_1.s_2 = 3/4$
$+ 2s_1.s_2$ -------(2).

No $S_{12} = 2(3Q - S^2) = 2[6 \frac{(s_1.r)(s_2.r)}{r^2} - 2 s_1.s_2]$ ---------(2).

For singlet state S=0, $\rightarrow$ $s_1 = -s_2$; hence,

$S_{12} = 2[6. \frac{9(s_1.r)^2}{r^2} + 2s_1^2] = 2[6 \frac{-(\sigma_1. r)(\sigma_1.r)}{4r^2} + 2S_1^2]$

Remember that $(\sigma_1 . r)(\sigma_1 . r) = r^2 + i\sigma$ ~~r×r~~ = r²
$S_1^2 = 3/4$. Substitute for S_{12}, the result is $S_{12} = 0$, for singlet case.
For triplet case, S=1.
$S_{12} = 2(3Q - S^2)$, if Q=0 $\rightarrow S_{12} = -2s(s+1) = -4$

If Q=1 $\rightarrow$ $S_{12} = 2(3-2) = 2$; therefore, the only possible values to S_{12} are 0, 2, -4. Singlet, triplet Q=1, Q=0.

Now to see the action, as example, of both Q and S_{12} on $\mathcal{Y}_{LSJ}^M$.
Since there are two cases: singlet, s=0; and triplet, s=1, so $\mathcal{Y}_{LSJ}^M$ takes the form $\mathcal{Y}_{L0J}^M$ for the singlet case and $\mathcal{Y}_{L1J}^M$ for the triplet case. Therefore, let $\mathcal{Y}(0)$ *for the singlet case and* $\mathcal{Y}(1)$ for the triplet case, then $\mathcal{Y}(0) = 0 \, \mathcal{Y}(0) = 0, S_{12} \mathcal{Y}(0) = 0$

Also, Q $\mathcal{Y}(1) = \mathcal{Y}(1)$, $S_{12}\mathcal{Y}(1) = 0 \, \mathcal{Y}(1) = 0$
So $Q\mathcal{Y}(1) = \mathcal{Y}(1)$, and $S_{12}\mathcal{Y}(1) = 0$, as required.
QED.

Problem 22: Given the diagram below, show that

$$(1/|r_1-r_2|)= 4\pi \sum_{\ell=0}^{\infty} \sum_{m=-\ell}^{\ell} \frac{1}{2\ell+1} \frac{Y_L^\ell}{r_>^{\ell+1}} Y_{\ell m}^*(\theta^-\phi^-)Y_{\ell m}(\theta,\phi).$$

The proof: It is mathematically known that

$$(1/|r_1 - r_2|)=\frac{1}{[r_1^2+r_2^2-2r_1r_2\,\cos\alpha]^{1/2}}, \text{----------------------(1).}$$

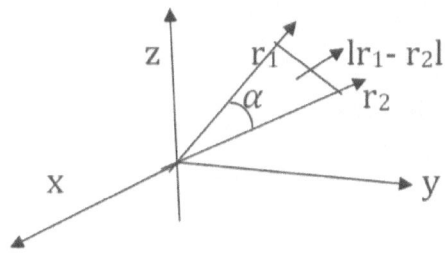

Equation 1 can be written in the form:

$$\frac{1}{|r_1-r_2|} =\frac{1}{r_2}\left[(\tfrac{r_1}{r_2})^2 +1-2\tfrac{r_1}{r_2}\cos\alpha\right]^{-1/2} \text{--------------(2).}$$

By expanding equation 2 in powers of $\frac{r_1}{r_2}$ and $\frac{r_2}{r_1}$, according to $\frac{r_1}{r_2}<<1$ or $\frac{r_2}{r_1}<<1$. So equation 2 becomes according to the region of convergence as follows:

$$\frac{1}{|r_1-r_2|} =\frac{1}{r_1}\sum_{\ell=0}^{\infty}(\tfrac{r_2}{r_1})^\ell P_\ell(\cos\alpha)=\frac{1}{r_2}\sum_{\ell=0}^{\infty}(\tfrac{r_1}{r_2})^\ell P_\ell(\cos\alpha).$$

$$=\sum_{\ell=0}^{\infty} \frac{r_2^\ell}{r_1^{\ell+1}} P_\ell(\cos\alpha), \text{-------- } r_2<r_1.$$

$$\text{Or}=\sum_{\ell=0}^{\infty} \frac{r_1^\ell}{r_2^{\ell+1}} P_\ell(\cos\alpha), \text{-------}r_1<r_2.$$

Now substitute for $P_\ell(\cos\alpha)$ from equation 14 of problem 20, the result is:

$$\frac{1}{|r_1-r_2|}=4\pi \sum_{\ell=0}^{\infty} \sum_{m=-\ell}^{\ell} \left(\frac{1}{2\ell+1}\right) \frac{Y_L^\ell}{r_>^{\ell+1}} Y_{\ell m}^*(\theta^-,\phi^-) Y_{\ell m}(\theta,\phi)$$

QED.

Problem23: Show that

(a.) $\int_0^{2\pi} d\varphi \int_0^{\pi} \sin\theta \, d\theta \, Y_{lm}^*(\theta, \varphi) \, Y_{\ell-m-}(\theta, \varphi) = \delta_{\ell\ell-}\delta_{mm-}$

(b.) $\sum_{\ell=0}^{\infty} \sum_{m=-\ell}^{\ell} Y_{\ell m}^*(\theta, \varphi) Y_{\ell m}(\theta^-, \varphi^-) = \dfrac{\delta(\theta-\theta^-)\delta(\varphi-\varphi^-)}{\sin\theta} \equiv$

$\delta(\Omega - \Omega^-)$.

The solution:
Take equation 1. Remember that $Y_{\ell m}(\theta, \varphi)$ is given by,

$$Y_{\ell m}(\theta, \varphi) = \sqrt{\frac{(2\ell+1)(\ell-|m|!)}{4\pi(\ell+|m|!)}} \, P_\ell^m(\cos\theta) e^{im\varphi} \text{---------------(1)}.$$

$$Y_{\ell-m-}(\theta, \varphi) = \sqrt{\frac{(2\ell^-+1)(\ell^--|m^-|)!}{4\pi(\ell^-+|m^-|)!}} \, P_{\ell^-}^{m^-}(\cos\theta) e^{im^-\varphi} \text{-----(2)}.$$

Now substitute these into a, do the mathematical manipulations. Remember that $\int_{-\ell}^{\ell} P_\ell^m(x) \, P_{\ell^-}^{m^-}(x) dx = $

$$\sqrt{\frac{4(\ell+|m|)!(\ell^-+|m^-|)!}{(2\ell+1)(2\ell^-+1)(\ell-|m|)!(\ell^--|m^-|)!}} \, \delta_{\ell\ell-}\delta_{mm-} \quad \text{set x=}\cos\theta.$$

$$dx = d(\cos\theta) = -\sin\theta d\theta.$$

Go back and substitute this into the previous result; the answer is the required a. So

$$[\int_0^{2\pi} d\varphi \int_0^{\pi} \sin\theta \, d\theta \, Y_{\ell m}^*(\theta, \varphi) Y_{\ell-m-}(\theta, \varphi) = \delta_{\ell\ell-}\delta_{mm-}] \text{.Q.E.D.}$$

(b.) To show that, use the closure relation:

$\sum_{\ell=0}^{\infty} u_\ell^*(Q) u_\ell(Q^-) = \delta(Q - Q^-)$; therefore,

$\sum_{\ell=0}^{\infty} \sum_{m=-\ell}^{\ell} Y_{\ell m}^*(\theta, \varphi) Y_{\ell m}(\theta^-, \varphi^-) =$

$\delta(\varphi - \varphi^-) \, \delta(\cos\theta - \cos\theta^-) = \dfrac{\delta(\theta-\theta^-)\delta(\varphi-\varphi^-)}{\sin\theta} = \delta(\Omega - \Omega^-)$.

Where $d\Omega = \sin\theta \, d\theta \, d\varphi$

$[\because \sum_{\ell=0}^{\infty} \sum_{m=-\ell}^{m} Y_{\ell m}^{*}(\theta, \varphi) Y_{\ell m}(\theta^{-}, \varphi^{-}) = \delta(\Omega - \Omega^{-})], \text{QED}$

Problem 24: Given $\psi(x, 0) = \dfrac{1}{(2\pi\xi)^{1/4}} e^{\frac{ip_0 x}{\hbar} - (\frac{x}{2\xi})^2}$, --------(1).

It describes a free electron (V=0). Show its time evaluation, and it is a normalized wave function. Also find $\varphi(p)$ showing it is normalized too.

The solution: It is known by Fourier transformation $\psi(x, t)$ is given by $\psi(x, t = \int \varphi(p) \dfrac{e^{\frac{ipx}{\hbar}}}{\sqrt{2\pi\hbar}} e^{-\frac{ip^2 t}{2m}} \, dp,$ --------------(2).

So $\psi(x, 0) = \int \varphi(p) \dfrac{e^{\frac{ipx}{\hbar}}}{\sqrt{2\pi\hbar}} \, dp,$ -------------------(3).

Also, by Fourier transformation $\varphi(p)$ is given by

$\varphi(p) = \int \psi(x, 0) \dfrac{e^{\frac{-ipx}{\hbar}}}{\sqrt{2\pi\hbar}} \, dx,$ ------------------- (4).

Substitute for $\psi(x, 0)$ as given in equation 1, the result is:

$\varphi(p) = \dfrac{1}{\sqrt{2\pi\hbar}} \dfrac{1}{\sqrt[4]{2\pi\xi}} \int_{-\infty}^{\infty} e^{\frac{-ix(p_0 - p) - (\frac{x}{2\xi})^2 \hbar}{\hbar}} \, dx,$ --------- (5).

Now to find the integral, first complete the square of the exponential $(\dfrac{-ix(p_0 - p) - (\frac{x}{2\xi})^2 \hbar}{\hbar})$, which gives the following:

$\dfrac{1}{4\xi^2}[\underbrace{[(x^2 + \frac{4\xi^2 i}{\hbar}(p_0 - p)x]^2 - \{\frac{2i\xi^2}{\hbar}(p_0 - p)\}^2]}_{Y^2} \Big\downarrow =$

$\dfrac{-1}{4\xi^2} Y^2 \cdot \frac{\xi^2}{\hbar^2}(p_0 - p)^2, \; dx = dy.$

Hence, $\varphi(p)$ takes the form

$$\varphi(p)= \frac{1}{\sqrt{2\pi\hbar}} \frac{e^{\frac{-\xi^2(p_0-p)^2}{\hbar^2}}}{\sqrt[4]{2\pi\xi^2}} \int_{-\infty}^{\infty} e^{\frac{-1Y^2}{4\xi^2}} dy \text{----------------- (6)}.$$

Now by using the standard integral, $\int_{-\infty}^{\infty} e^{bx^2} dx = \sqrt{\pi/b}$ ---(7).

So by comparison, $b \equiv (-1/4\xi^2)$, the integral will be $= \sqrt{4\xi^2 \pi}$

$= 2\xi\sqrt{\pi}$; therefore, $\varphi(p) = \frac{\sqrt{2\xi}}{\sqrt[4]{\hbar^2 2\pi}} e^{\frac{-\xi^2[p_0-p]^2}{\hbar^2}}$

Hence, $[\varphi(p) = (\frac{4\xi^2}{\pi\hbar^2})^{1/4} e^{\frac{-i[p_0-p]^2}{\hbar^2}}]$ -------------(8).

Now $\psi(x) = \frac{1}{\sqrt[4]{2\pi\xi^2}} e^{\frac{ip_0x-\hbar(\frac{x}{2\xi})^2}{\hbar}}$. From equation 1.

So $\psi^*(x) = \frac{1}{\sqrt[4]{2\pi\xi^2}} e^{\frac{-ip_0x-\hbar(\frac{x}{2\xi})^2}{\hbar}}$. Complex conjugate.

So $\int \psi^*(x) \psi(x) dx = \frac{1}{\sqrt{2\pi\xi^2}} \int_{-\infty}^{\infty} e^{\frac{-x^2}{2\xi^2}} dx$, using the property of the standard integral (see equation 7), the result will be $\int |\psi(x)|^2 dx = 1$. $\therefore \psi(x)$ is normalized.

Now to show that the momentum space wave function $\varphi(p)$ is normalized too, i.e., $\int_{-\infty}^{\infty} \varphi^*\varphi dp = 1$

Following the same way, the result is $[\int |\varphi(p)|^2 dp = 1]$.
QED.

Problem 25: From the given $\psi(x)$ and $\varphi(p)$ in problem 24. Find <x>, <x²>, <p> and <p²>.

The solution:
Finding <x>, the average value of x. By definition, it is,

$$<x>=\int \psi^*(x)\,\psi(x)\,dx=\frac{1}{(2\pi\hbar)}\iint \phi^*(p^-)\,dp^-\,dx\int x\phi(p)\,e^{\frac{i(p-p^-)x}{\hbar}}\,dx\,dp$$

$$=\frac{1}{2\pi\hbar}\iint \phi^*(p^-)\,dp^-\,dx\,e^{-ip^-x}\int \frac{\hbar}{i}\phi(p)\frac{\partial}{\partial p}e^{\frac{ipx}{\hbar}}\,dp=$$

$$\frac{-i\hbar}{2\pi\hbar}\iint \phi^*(p)\,dp^-\,dx\,e^{\frac{-ip^-x}{\hbar}}\int \phi(p)\frac{\partial}{\partial p}e^{\frac{ipx}{\hbar}}\,dp.$$

Now integrate the third integral by parts to find that:
It equals to $\{-\int e^{\frac{ipx}{\hbar}}\frac{\partial}{\partial p}\phi(p)\,dp\}$, show that!

Therefore,

$$<x>=\frac{i\hbar}{2\pi\hbar}\iint \phi^*(p-)\frac{\partial}{\partial p}\phi(p)\,dp\,dp^-\int e^{\frac{i(p-p^-)x}{\hbar}}\,dx$$

$$=\frac{i\hbar}{2\pi}\iint \phi^*(p^-)\frac{\partial}{\partial p}\phi(p)\,(2\pi)\,\delta(p^- - p)\,dp\,dp^-\ so,$$

$$<x>=\iint \phi^*(p)\left(i\hbar\frac{\partial}{\partial p}\right)\phi(p)\,dp. \text{------------------ (1).}$$

Substitute for $\phi(p)$ and $\phi^*(p)$, into equation 1, and the result will be

$$<x>=i\left(\frac{2\xi^2}{\pi}\right)^{1/2}\left(\frac{2\xi^2}{\hbar^2}\right)[\int (p_0\quad -p)\ e^{\frac{-2\xi^2(p_0-p)^2}{\hbar^2}}\,dp=0,\quad \text{odd-even}$$

integral]$\rightarrow$<x>=0, as required and should be since there are +x and−x.

Or it can be shown by another way; let $p_0-p=y$, so $dy=-dp$, let $\frac{2\xi^2}{\hbar^2}=a$, so $-\int ye^{-ay^2}\,dy$, with the assumption that a $y^2=x$, and if x= ∞, y=-∞, x=-∞, y=∞ and $ydy=dx/2a$, takes the form, $\left(-\frac{1}{2a}\right)\int_{-\infty}^{\infty}e^{-x}\,dx=\frac{-1}{2a}[e^{-x}]_{-\infty}^{\infty}=0$, as it should be.

$\therefore$<x>=0.

Now to find< x²>, $<x^2>=\int \psi^*(x)\psi(x)\,dx$

Substitute for ψ^* and ψ as

$$\psi(x) = \frac{1}{\sqrt{2\pi\hbar}} \int e^{\frac{ipx}{\hbar}} \varphi(p)\, dp, \quad \psi^*(x) = \frac{1}{\sqrt{2\pi\hbar}} \int e^{\frac{-ip^-x}{\hbar}} \varphi(p^-)\, dp^-$$

Following the same way, the result will be:

$$<x^2> = (\frac{4\xi^2}{\pi^2}) \int (p_0 - p)^2 e^{\frac{-\xi^2(p_0-p)^2}{\hbar^2}}\, dp.$$ Following the same way, $[<x^2> = \xi^2]$. Hence, $<x> = 0$ and $<x^2> = \xi^2$.

Next: to find $<p>$ and $< p^2>$,
$<p> = \int \varphi^*(p)\, p\, \varphi(p) dp$, remember $p_{op.} = p$ in p-space.

Use spatial space $\psi(x)$, where $p_{op} = -i\,\hbar\frac{\partial}{\partial x}$; therefore,

$<p> = \frac{1}{2\pi\hbar} \iiint \psi^*(x^-)\, p_{op}\psi(x)e^{\frac{ip(x^- - x)}{\hbar}} dx^- dx\, dp = p_0$, carry on the details as in finding $<x>$.

So $< P> = p_0$.

For finding $<p^2>$, the operator is $(-i\,\hbar)^2 (\frac{\partial}{\partial x})^2$ in spatial space.

Hence, $< p^2> = \int \psi^*(x)((-i\,\hbar)^2 (\frac{\partial}{\partial x})^2 \psi(x)\, dx$.

Now substitute for $\psi^*(x)$ and $\psi(x)$, the result is

$$<p^2> = \frac{(-i\hbar)^2}{\sqrt{2\pi\,\xi^2}} \int e^{\frac{-ip_0\,x - \hbar x^2/4\xi^2}{\hbar}} (\frac{\partial}{\partial x})^2 e^{\frac{-ip_0\,x - \hbar x^2/4\xi^2}{\hbar}} dx$$

Carry on the calculations as done before; the result is

$[<p^2> = p_0^2 + \frac{\hbar^2}{4\xi^2}]$.

QED.

Problem 26: Use the minimum uncertainty principle to build up a minimum wave packet.

The solution: Consider x and p_x as two dynamical variables described by such minimum wave packet, where $\nabla x \nabla p \geq \frac{\hbar}{2}$ is valid. The minimum uncertainty means that $\nabla x \nabla p = \hbar$.

Mathematically, the following two important inequalities are to be used:

(1.) Schwartz inequality, $\| \alpha^- \psi \| \, \| \beta^- \psi \| \geq | \alpha^- \psi, \beta^- \psi |$.

(2.) Triangle inequality,
$$| (\psi, \alpha^- \beta^- \psi) | + | (\psi, \beta^- \alpha^- \psi) | \geq | (\psi, (\alpha^- \beta^- - \beta^- \alpha^-) \, \psi) |.$$

It should be noted that $(\psi, \alpha^- \beta^- \psi)$ is a scalar product; it can be represented by the integral form as $\int \psi^* \alpha^- \beta^- \psi \, d\gamma$.

And $\| \alpha^- \psi \| =$ length of a vector, but $\| \alpha^- \psi \|^2 = (\alpha^- \psi, \alpha^- \psi)$.
One might introduce, $\alpha^- \psi = c \, \beta^- \psi$, ----------------(1).

As a condition for a minimum wave packet, where c is scalar. (It is a complex number.)

Now.
(A.) Take $\alpha^- \psi = c \beta^- \psi$
A) (B.) Inquire $| (\psi, \alpha^- \beta^- \psi) | + | (\psi, \beta^- \alpha^- \psi) | =$
$| (\psi, \alpha^- \beta^- \psi) - (\beta^- \alpha^- \psi) |$. Also, it should be noted $(\psi, \beta^- \alpha^- \psi)$
$= c(\psi, \beta^{(-)^2} \psi) = c(\beta^- \psi, \beta^- \psi)$.

And $(\psi, \alpha^- \beta^- \psi) = (\alpha^- \beta^- \psi, \psi)^+ = c^+ (\beta^- \psi, \beta^- \psi)$, (see equation 1.), $\alpha^- = c \, \beta^-$. --------------- (2).

Substitute equation2 into B. The following is obtained:
$2|c| = |c - c^+|$ $\Rightarrow$ $c = ib$, --------------- (3).
So $\alpha^- \psi = (ib) \beta^- \psi$, as in(1) -------------------------(4).

Now let $\alpha = x$, $\alpha^- = x - x^- = x - x_0$,
$\beta = p_x$ and $\beta^- = p_{op} = \frac{-i\hbar\partial}{\partial x}$, then the condition for minimum wave packet becomes

$(x - x_0)\psi = ib[(\frac{-i\hbar\partial}{\partial x}) - \hbar k_0]\psi$, where $k_0 = \frac{p_0}{\hbar}$, -----------(5).

Equation 5 is just an ordinary differential equation; its solution is given as $\psi(x, 0) = Ae^{ik_0 x}e^{\frac{1}{ib}(x-x_0)^2}$ ---------(6).

Using the following auxiliary condition:

$\int \psi^*(x)\,\psi(x)\,dx = 1$
$(\Delta x)^2 = \int \psi^*(x)(x-x_0)^2\psi(x)\,dx$ $\left.\right\}$ ---------------(7).

Using equation 7, A and B in equation 6 can be determined; hence,

$\int_{-\infty}^{\infty}|A|^2 e^{\frac{1}{b\hbar}(x-x_0)^2}\,dx = 1$, normalization condition. $\rightarrow$

$|A|^2(\pi b\hbar) = 1$, $=> A = \frac{1}{\sqrt{2b\hbar}}$, ------------------(8).

$\therefore (\Delta x)^2 = \int_{-\infty}^{\infty}|A|^2 e^{\frac{1(x-x_0)^2}{b\hbar}}(x - x_0)^2\,dx$, let $x - x_0 = y$,

$dx = dy$, the result is $|A|^2 = \frac{\sqrt{b\hbar}}{2\sqrt{\pi}} \Rightarrow \sqrt{b\hbar} = 2\sqrt{\pi}\,(\Delta x)^2$

$\rightarrow b\hbar = 4\pi(\Delta x)^4$; therefore, $A = \frac{1}{2\pi(\Delta x)^2}$. Hence,

$\psi(x) = \frac{1}{2\pi(\Delta x)^2}[e^{ik_0 x - \frac{(x-x_0)^2}{8\pi(\Delta x)^4}}] = \frac{1}{2\pi\xi^2}[e^{\frac{1p_0 x}{\hbar} - \frac{(x-x_0)^2}{8\pi\xi^4}}]$-------(9).

Where $\xi^2 = (\Delta x)^2$, $k_0 = p/\hbar$,

equation 9 is the minimum wave packet for which $\Delta x\,\Delta p = \frac{\hbar}{2}$.
QED.

Problem 27: Find:

(1.) $<r> = \int \phi^*(p,t)(\frac{i\hbar\partial}{\partial p})\,\phi(p,t)\,dp$

(2.) $<p^2> = \int \phi(p,t)l^2 p^2 dp = \int \psi^*(r,t)(\frac{-i\hbar\partial}{\partial r})^2 \psi(r,t)dr.$

(3.) $< F(r,p)> = \int \psi^*(r,t)(F(r,p=\frac{i\hbar\partial}{\partial r})\psi(r,t)\,dr.$

The solution: The solution of these three problems is quite clear and direct, using Fourier transformation which connects between the spatial wave function and its corresponding momentum space wave function, i.e., between $\psi(r,t)\ and\ \phi(p,t)$ as mentioned above. Also, you might need sometimes to use the well-known integration by parts. Therefore, the first will be solved as an example; the other two are to be left as exercises. So take the first to show that $<r> = \int \psi^*(r,t)r\psi(r,t)\,dr$ ---------- (1).

By Fourier transformation (F-T),

$\psi(r,t) = \dfrac{1}{\sqrt[2]{2\pi\hbar}^3}\int e^{\frac{ip\cdot r}{\hbar}}\,\phi(p)\,dp$, p- is a momentum vector

And $\psi^*(r,t) = \dfrac{1}{\sqrt{2\pi\,\hbar}^{3/2}}\int e^{\frac{-ip^-\cdot r}{\hbar}}dp^-.$

Now substitute these into equation 1, and with well-known mathematical manipulations, $<r>$ will be as

$<r> = \dfrac{-i\hbar}{(2\pi\hbar)^3}\iint \phi^*(p^-,t)dp^-\,dr\,e^{\frac{i(p^-.r)}{\hbar}}\int \underline{\phi(p,t)\dfrac{\partial}{\partial p}e^{\frac{ip.r}{\hbar}}dp}.$

Now take the underlined integral; integrate it by parts, applying the periodic condition $(\phi(\infty) = \phi(-\infty))$, the result is
$\int \phi(p,t)\dfrac{\partial}{\partial p}e^{\frac{ip.r}{\hbar}}dp = \overbrace{\phi(p,t)e^{\frac{ip.r}{\hbar}}}^{}\big]_{-\infty}^{\infty} = 0$

$\int e^{\frac{ip.r}{\hbar}}\dfrac{\partial}{\partial p}\phi(p,t)dp. = -\int e^{\frac{ip.r}{\hbar}}\dfrac{\partial}{\partial p}\phi(p,t)dp;$ therefore,

407

$$<r>=\frac{-i\hbar}{(2\pi\hbar)^3}\iint \phi^*(p^-,t)dp^-dr \; e^{\frac{i(p^-.r)}{\hbar}}\int e^{\frac{ip.r}{\hbar}}\frac{\partial}{\partial p}\phi(p,t)dp$$

$$=\frac{-i\hbar}{(2\pi\hbar)^3}\iint \phi^*(p^-,t)\frac{\partial}{\partial p}\phi(p,t)dp^-dp\underline{\int e^{\frac{i(p-p^-).r}{\hbar}}\,dr}$$

The underlined integral is $(2\pi)^3\delta(p-p^-)=(2\pi)^3\,\delta(p^--p)$.

So $<r>=\int i\hbar\frac{\partial}{\partial p}\phi(p,t)dp\;\underline{\int \phi^*(p^-,t)\delta(p^--p)dp^-}$

The underlined integral gives $\phi^*(p,t)$, (see Appendix A).

Hence, $<r>=\int \phi^*(p,t)\,(i\hbar\frac{\partial}{\partial p})\phi(p,t)dp.$ -------- (2).

So, equation 2 is the proof of equation 1.
Work equations 2 and 3 as an exercise, follow the same way.
QED.

Problem 28: If two observables commute, they possess a complete orthonormal set of common eigenfunction.

The solution: Assume A and B as two commuting observables, so [A B]=0. Let $|\psi_a>$ be an eigenfunction of A and a is its eigenvalue. Therefore, $|\psi_a>$ can be expanded into a complete orthonormal set of eigenfunctions of B. Hence, it can be written such as $|\psi_a>=\sum_m \varphi(a, b_m)>$,------------(1).

Where $|\varphi (a, b_m)>$ is an eigenfunction of B corresponding to the eigenvalue b_m. It can always be arranged matters in such a way that every function that is occurring in the sum (1) corresponds to a different eigenvalue, i.e., $|\psi>=\sum_p \psi_p>$. Now the expansion $|\psi>=\sum_{p.r} c_p(r)|\varphi_p(r)>$ can be used instead of $\sum_p |\psi_p>$.

Let us show that $|\varphi_m^\wedge>\equiv (A-a)|\varphi(a, b_m)>=0$, -------- (2).

Since A and B commutes, then

$B|\varphi_m> \equiv (A-a)B|\varphi(a, b_m)>$, but $B|\varphi\, (a, b_m)>=$

$b_m\, |\varphi(a, b_m)>$, where b_m is the eigenvalue associated with B. Therefore, $B|\hat{\varphi}> \equiv (A-a)B|\varphi(a, b_m)>=b_m\, |\hat{\varphi}_m>$.-----(3).

Hence, the functions $|\hat{\varphi}_m>$ are eigenfunctions of B. Also, since the corresponding eigenfunctions are all different, they are linearly independent.

However, $\sum_m |\hat{\varphi}_m> = (A-a)|\psi_a> = 0$, ------------------ (4).

Equation 4 is only possible if each of the functions $|\hat{\varphi}_m>$ vanishes, i.e., $|\varphi(a, b_m)>$ are simultaneously eigenfunctions of A and B.

Now, consider $|\psi_n^{(r)}>$ as a complete orthonormal set of eigenfunction of A, then $A|\psi_n^{(r)}> = a_n\, |\psi_n^{(r)}>$.

As it has been done before, it can be written that

$|\psi_n^{(r)}> = \sum_n \varphi|^{(r)}>$, where $\varphi^{(r)} = \varphi^{(r)}(a_n; b_m)$.

For example, $|\psi_n^{(r)}> = \sum_m \psi^{(r)}(a_n; b_m)>$.

Therefore, the functions $|\varphi^{(r)}(a_n, b_m)>$ are common functions to A and B. The ensemble of the functions $|\varphi^{(r)}(a_n, b_m)>$ are corresponding to the pair eigenvalues a and b_m and may not be linearly independent. Therefore, by means of orthogonalization processes, it is possible to choose a set of orthonormal functions $|X^{(r)}(a_m; b_m)>$ corresponding to the same pair of eigenvalues, such that the functions $|\varphi^{(r)}(a_n; b_m)>$ are linear combinations of these chosen functions, i.e.,

$|\varphi^{(r)}(a_n; b_m)> = \sum_s c_{rs}|X^{(r)}(a_m; b_m)>$.---------- (5).

The ensemble {l $\mathcal{X}$>} of all these functions constitutes an orthonormal set of eigenfunctions common to A and B. Moreover, it is a complete set since any wave function lψ> can be expanded in a series of l $\mathcal{X}$>; to form this expansion, it suffices to expand l ψ> in a series of functions of the complete set of {l $\psi^{(r)}$>} and then to substitute for the function l $\psi_n^{(r)}$> in this expansion. Hence,

l $\psi_n^{(r)}$>=$\sum_m$ l$\varphi^{(r)}(a_n, b_m)$ >, but l$\varphi^{(r)}(a_n, b_m)$>=

   $\sum_s c_{rs}$ l $\mathcal{X}^{(s)}(a_m, b_m)$>.

Therefore, [l$\psi_n^{(r)}$>=$\sum_m \sum_s c_{rs}$ l$\mathcal{X}^{(s)}(a_m, b_m)$>]. QED.

Problem 29: Show that

<$\tau$$J$$a$ lk$_u$l τ Jb>=<$\tau$$JalJ_ul\tau$$J$b>$\frac{<J \cdot K>}{j(j+1)}$,
where <JK>=<$\tau$$JalJ \cdot K$l $\tau$$J$a>.

The proof: It is known that the commutation relations between J and K are

[J_i, K_i]=0, [J_l, K_j]=iK_k ----------------- (1) or, more explicit,
[J_x, K_x]=0, [J_x, K_y]=iK_z, [J_x, K_z]=i K_y, [J_y, K_z]=iK_x ------------- (2).

Now take the commutation relations

[J^2, K]=$J \cdot$ [J, K]=-i[$K \times J$-$J \times K$] or, more explicit,
=$J^2 K_x$−$K_x J^2$=[J_y^2 +J_z^2] K_x−K_x[J_y^2 +J_z^2].

Carry on the mathematical manipulation using the correlations properties in equations 1 and 2, the result is

$=-i[J_yK_z+K_zJ_y]$, i.e., $[J_y^2, K_x]=-i[J_y K_z+ K_z J_y]$, and,

$[J_z^2, K_y]=i[J_zK_y+K_yJ_z]$, or in general,

$[J^2, K]=i[K\times J-J\times K]$,------------------(3).

Now take the commutation

$[J^2, [J^2, K]]=-2i[J^2, J\times K-iK]=-2i[-2iJ\times(J\times K-iK)-i(J^2K-K J^2)]$
$=2(J^2K+KJ^2)-4J(J K)$

$\therefore [J^2, [J^2, K]]=2(J^2K+KJ^2)-4J(J\cdot K)$, -----------(4).

Equation 4 can be written as

$[J^2, [J^2, K]]=J^4K-2J^2KJ^2+KJ^4\Rightarrow$ from equation 4,

$[-4J(J\cdot K)=J^4K-2J^2 KJ^2+KJ^4-2(J^2 K+KJ^2)]$,-----------(5).

Now if the matrix element of equation 5 is taken in $|\tau Ja>$ representation the following will be obtained:

$-4<\tau^- J^- a^- |J(J\cdot K)|\tau Jb>=$
$<\tau^- J^- a^- |J^4K-2J^2KJ^2+KJ^4-2(J^2K+KJ^2)]| \tau Jb>$------ (6).

With the use of the eigenvalue equation,

$J^2|\tau^- J^- a^->=j^-(j^-+1)|\tau^- J^- a^->$
$J^2|\tau Jb>=j(j+1)|\tau Jb>$, and the fact J and K are Hermitian operators, (considering $j=j^-$), then the following mathematical calculations take place to reach the result:

$-4<\tau^- J^- a^- |J(J K)|\tau J b>=-4j(j+1)<\tau^- J^- a^- |k\tau Jb>$.

But since J_u and K_u are in the same direction and $|\tau^- J^- a^->$ and $|\tau J b>$ belong to the same substate $\mathcal{E}(\tau, J)$ {see Messiah, Volume II}, then $|\tau^- J^- a^- > \equiv |\tau Jb>$; therefore, the result is $-4<\tau^- J^- a^- |J(J K)|\tau Jb>=-4j(j+1)<\tau Jalk_u|\tau J b>$

So this leads to the final required result:

$$j(j+1) <\tau \; Jalk_u l\tau \; J \; b> = <\tau^- J^- a^- \; lJ_u(J_u \cdot K)l\tau \; Jb> =$$
$$<\tau \; JalJ_u l\tau \; Jb>(J \; K) \rightarrow$$

$$\left[\begin{array}{l} [<\tau \; Jalk_u l\tau \; J \; b> = <\tau \; JalJ_u l\tau \; J \; b> \dfrac{(J \cdot K)}{j(j+1)}] \\[2mm] \text{Where } (J \cdot K) = <\tau \; JalJ \cdot Kl\tau \; J \; b>. \end{array} \right]$$

QED.

Problem 30: Prove the following:

A.

(1.) $[A, B] = -[B, A]$
(2.) $[A, B+C] = [A,B] + [A, C]$.
(3.) $[A, BC] = [A, B]C + B[A, C]$
(4.) $[A, B^n] = \sum_{s=0}^{n-1} B^s [A, B] B^{n-s-1}$

B.

Using the above properties, find:

(1.) $[L_x, L_y] = iL_z$, in $\hbar$ unit.
(2.) $[L_z, L_x] = iL_y, = = =$.
(3.) $[L_y, L_z] = iL_x, = = =$. .
(4.) $[L^2, L] = 0$
(5.) $[L^2, L_z] = 0$.

The proof:

(1.) $[A, B] = AB - BA$, by definition.

Take [A, B]=-AB+BA=BA-AB=[B, A], multiply by (-), so
[A, B]=-[B, A].

(2.) [A, B+C]=[A(B+C)-(B+C)A]=AB+AC-BA-CA=

AB-BA+AC-CA=[A, B]+[A, C].

(3.) [A, BC]=(ABC-BCA), LHS $\longrightarrow$ 0

RHS is ABC-BAC+B$\underline{AC\text{-}BCA}$=ABC-BCA=[A, BC].
∴ [A, BC]=[A, B]C+B[A, C].

(4.) $[A, B^n]=\sum_{s=0}^{n-1} B^s[A, B]B^{n-s-1}$

By induction, set n=1, so it is [A, B]
Let n= $\hbar$+1
Consider $[A, B^{\hbar+1}]=[A, B^{\hbar}B]=B^{\hbar}[A, B]+[A, B^{\hbar}]B$
$=B^{\hbar}[A, B]+(\sum_{s=0}^{\hbar-1} B^s[A, B]B^{\hbar-s})B=\sum_{s=0}^{(\hbar+1)-1} B^s[A, B]^{\hbar+1-s-1}$
∴$[A, B^n]=\sum_{s=0}^{n-1} B^s[A, B]^{n-s-1}$. This might be solved by another way. Try it.

B. Using the fact that $L^{\rightarrow} = (r^{\rightarrow} \times p^{\rightarrow})$, angular momentum, it can be shown that:

L_X=[yp$_z$-zp$_y$] ---------(i).
L_Y =[zp$_x$-xp$_z$] --------(ii).
L_Z=[xp$_y$-y p$_x$] ------- (iii).

Also, the identity $[x,P_x]=[y,p_y]=[z,p_z]=i\hbar$, is useful to use.
Now $[L_X, L_Y]=[L_XL_Y-L_YL_X]$. Use equations i and ii, and the results will be

(1.) $[L_x, L_y]= i\hbar\ L_z$. Following the same way, you also get:
(2.) $[L_z, L_x]= i\hbar\ L_y$

(3.) $[L_y, L_z] = i\hbar\, L_x$

(4.) $[L^2, L] = [L^2L - LL^2] = [L^3 - L^3] = 0$

(5.) $[L^2, L_z] = [L_X^2 L_z] + [L_Y^2, L_z] + [L_Z^2, L_z] = 2L_X[L_X.L_z] + L_Y[L_Y, L_z] +$
$2L_z[\overline{L_z, L_z}] = -2i\,\hbar L_X L_Y + 2\,i\hbar\, L_Y L_X = 0.$

So $[L^2, L_z] = 0$.

QED.

Problem 31: Show the following:

(1.) $[I, A] = 0$

(2.) $[\mathbb{O}, A] = 0$, where, $\mathbb{O} = \lambda 1$

(3.) $[AB]^{-1} = B^{-1}A^{-1}$

(4.) $<\psi_a \,|BC|\psi_b> \equiv (<\psi_a|B)(C|\psi_b> = <\psi_a|(B(C|$
$\psi_b >) = (<\psi_a|B)C|\psi_b>$

The proof:

(1.) $[I, A] = [I\,A - AI]$, where $I = = A^{-1}A$;
therefore, $[AAA^{-1} - AA^{-1}A] = [A-A] = 0$, so $[I, A] = 0$, as required.

(2.) $[\mathbb{O}, A] = [\lambda 1 A - A\,\lambda 1] = [\lambda AA^{-1}A - A\,\lambda AA^{-1}] = [\lambda A - A\lambda] = 0$, as required.

(3.) $[AB]^{-1}[=B^{-1}A^{-1}AB(AB)^{-1}$, $A^{-1}A = I$, $IB = B$, where I is identity operator. Hence,
$[AB]^{-1} = B^{-1}I\,B[AB]^{-1} = B^{-1}A^{-1}BB^{-1} = B^{-1}A^{-1}$
So $[[AB]^{-1} = B^{-1}A^{-1}$, as required.

(4.) $<\psi_a|BC|\psi_b> \equiv (<\psi_a|B)(C|\psi_b> = (<\psi_a|(B(C|\psi_b>$
$= (<\psi_a|B)C)|\psi_b>$. QED

Problem 32: Show Schmidt's process of orthogonalization using Dirac notation for the eigenfunctions.

The solution: Define $|\varphi_1>$ as $|a_1\varphi_1> = |\psi_1>$,-----ket ----(1).

Its complex conjugate is $<a_1\varphi_1| = <\psi_1|$, ----------bra----- (2).

Now multiply equation 1 by equation 2 from the left; the result is $<a_1\varphi_1|a_1\varphi_1> = a_1^2<\varphi_1|\varphi_1> = <\psi_1|\psi_1>$; a_1 is constant.

Now $|a_1|^2$ can be adjusted such that $<\varphi_1|\varphi_1>=1$,
namely, $|a_1|^2 = <\psi_1|\psi_1>$, $\therefore <\varphi_1|\varphi_1>=1$,------------------(3).

By the same logic, define $|\varphi_2>$ by $|a_2\varphi_2>$,
such that $|a_2\varphi_2> = |\psi_2> - |\varphi_1><\varphi_1|\psi_2>$ ------------------ (4).

So $<a_2\varphi_2| = <\psi_2| - <\varphi_1|<\psi_2|\varphi_1>$ ----------------(5).

Multiply equation 4 by equation 5, the result is:

$|a_2|^2<\varphi_2|\ \varphi_2> = <\psi_2|\psi_2> - <\psi_2|\varphi_1><\varphi_1\ |\psi_2>$ ⟶ 0 -
$\leq\psi_2|\varphi_1><\psi_1|\varphi_1\ \geq 0$ $=<\psi_2|\psi_2>\rightarrow$
$|a_2|^2<\varphi_2|\ \varphi_2> = <\psi_2|\psi_2>$. Now a_2 to be adjusted in such a way
so that $<\varphi_2|\ \varphi_2>=1$; therefore, $|a_2|^2 = <\psi_2|\psi_2>$, so

$<\varphi_2|\ \varphi_2>=1$, --------------------------------------- (6).

By the same procedure, define $|\varphi_3>$ as $|a_3\varphi_3> =$
$|\psi_3> - |\varphi_1><\varphi_1|\psi_3>.0 \rightarrow |\psi_3>\rightarrow$
$<a_3\varphi_3\ |a_3\varphi_3> = <\psi_3|\psi_3>. \rightarrow |a_3|^2<\varphi_3|\varphi_3> = <\psi_3|\psi_3>$.

Again, a_3 to be adjusted such as $<\varphi_3|\ \varphi_3>=1$; therefore,
$|a_3|^2 = <\psi_3|\psi_3> \rightarrow [<\varphi_3|\ \varphi_3>=1]$, as required.

Now take $|c_1\varphi_1> = |\psi_1>$------------------------------- (7).
$<c_2\varphi_2| = <\psi_2|$ --------------------------- (8).

Multiply equation 7 by equation 8. The result is
$<c_2\varphi_2|c_1\varphi_1> = <\psi_2|\psi_1> \to 0$ (orthogonal). $=0 \to$
$<c_2\varphi_2| c_1\varphi_1> = 0$; hence, $c_2^* c_1 <\varphi_2| \varphi_1> = 0$, but $c_2^* c_1 \neq 0$, so
$<\varphi_2| \varphi_1> = 0$. $\therefore$ $<\varphi_2| \varphi_1>$ are orthogonal.

Now consider the general case: $|c_\ell \varphi_\ell> = |\psi_\ell>$, ------- (9).
And $c_m \varphi_m| <= <\psi_m|$ ---------------------------------- (10).

Multiply equation 9 by equation 10, the result is:

$$<c_m \varphi_m| c_\ell \varphi_\ell> = <\psi_m|\psi_\ell> \to c_m^* c_\ell <\varphi_m|\varphi_\ell> = <\psi_m|\psi_\ell>,$$

$c_m^* c_\ell$ can be adjusted to be Kroneker delta $\delta_{m\ell}$, such that

$$<\varphi_m| \varphi_\ell> = \delta_{m\ell} = \begin{cases} 1, & m = \ell \\ \\ 0, & m \neq \ell. \text{ QED.} \end{cases}$$

Problem 33: Show that $P_r = \frac{\hbar}{i}(\frac{1}{r}\frac{\partial}{\partial r}r) = \frac{\hbar}{i}(\frac{\partial}{\partial r} + 1)$ satisfies,

$P_r = \frac{1}{2}[\frac{\mathbb{r}}{r} \mathbb{P}. + \mathbb{P}.\frac{\mathbb{r}}{r}]$, where $\mathbb{P} = m\mathbb{r} = mrr^\wedge = P_r r^\wedge$,
$\mathbb{r} = rr^\wedge$, $r^\wedge$ is a unit vector. P_r is Hermitian.

The solution: Take $\frac{1}{2} [\frac{\mathbb{r}}{r} \mathbb{P}. + \mathbb{P}.\frac{\mathbb{r}}{r}]$, substitute for $\mathbb{P}$, and $\mathbb{r}$. Do the mathematical consolation. The result is

Take $\frac{1}{2} [\frac{\mathbb{r}}{r} \mathbb{P}. + \mathbb{P}.\frac{\mathbb{r}}{r}] = \frac{2P_r}{2} = P_r$. Remember, $r^\wedge \cdot r^\wedge = 1$
So $[P_r = \frac{1}{2} [\frac{\mathbb{r}}{r} \mathbb{P}. + \mathbb{P}.\frac{\mathbb{r}}{r}]$, as required.

Now substitute for $\mathbb{P} = P_r r^\wedge = \frac{\hbar}{i}(r\frac{\partial}{\partial r}r)r^\wedge$.

Hence, after some mathematical manipulation, the result is

$$\frac{1}{2}\left[\frac{\mathbb{r}}{r}\cdot\mathbb{P}.+\mathbb{P}.\frac{\mathbb{r}}{r}\right]=\frac{\hbar}{i}(\frac{1}{r}\frac{\partial}{\partial r}\ r)=P_r, \therefore P_r=\frac{1}{2}\left[\frac{\mathbb{r}}{r}\ \mathbb{P}.+\mathbb{P}.\frac{\mathbb{r}}{r}\right]=\frac{\hbar}{i}(\frac{1}{r}\frac{\partial}{\partial r}\ r).$$ But $(\frac{1}{r}\frac{\partial}{\partial r}\ r)=(\frac{\partial}{\partial r}+1/r)$; therefore, $P_r=\frac{\hbar}{i}(\frac{\partial}{\partial r}+1/r)$ too.

To show P_r is Hermitian, it must show that $<\psi|P_r|\psi>=<\psi|P_r|\psi)^*$, where $\psi(r)$ is any square integrable function. This can be argued and shown that P_r is Hermitian, but the problem now is to show it can never be argued with $(\frac{\hbar}{i}\frac{\partial}{\partial r})$ and $(\frac{\hbar}{i}\frac{\partial}{\partial \theta})$, since $(\frac{\hbar}{i}\frac{\partial}{\partial r})$ is not Hermitian because $(\frac{\hbar}{i}\frac{\partial}{\partial r})^*$ $=(-\frac{\hbar}{i}\frac{\partial}{\partial r})\neq(\frac{\hbar}{i}\frac{\partial}{\partial r})$. It means

$<\psi,(\frac{\hbar}{i}\frac{\partial}{\partial r})\ \psi>\neq<\psi\ (\frac{\hbar}{i}\frac{\partial}{\partial r}),\psi>$, i.e., $\int \psi^*(\frac{\hbar}{i}\frac{\partial}{\partial r})\ \psi(r)d\mathbb{r}$ - $\int[(\psi(r)(\frac{\hbar}{i}\frac{\partial}{\partial r})]^*\ \psi(r)\neq 0$. Where $d\mathbb{r}=dr\ d\Omega$, $d\Omega=\sin\theta d\phi$. So $\frac{\hbar}{i}\int_0^\pi \sin\theta d\theta\int_0^{2\pi}d\phi\ [\int_0^\infty \psi^*(r)\frac{\partial}{\partial r}\psi(r)\ dr-\frac{i}{\hbar}\int_0^\infty \psi^*(r)(\frac{i}{\hbar}\frac{\partial}{\partial r})^*dr$ $4\pi\frac{\hbar}{i}[(\int_0^\infty \frac{\partial}{\partial r}|\ \psi(r)|^2dr\neq 0)-(\int_0^\infty(\frac{\partial}{\partial r})^{*}|\ \psi(r)|^2dr\neq 0)]=0$.

So, it is clear that, no arguing with $((\frac{\hbar}{i}\frac{\partial}{\partial r})$, since it is not Hermitian operator, and also the same with $(\frac{\hbar}{i}\frac{\partial}{\partial \theta})$.

Problem 34: Find the orthogonality relation and the closure relation for Bessel function $j_\ell(kr)$. Also, show that $k(\frac{2}{\pi})^{1/2}Y_{\ell m}(\theta,\varphi)\ j_\ell(kr)$ form an orthonormal and complete set.

The solution: First, the relation must be found:
$\int_0^\infty j_\ell(kr)\ j_\ell(k^-r)r^2dr=\frac{\pi}{2k^2}\delta(k-k^-),$-----------------(1).

By using the known relation, $j_\ell(kr)=\sqrt{\frac{\pi}{2kr}}J_{\ell+\frac{1}{2}}(kr)$, one gets
$\int_0^\infty j_\ell(kr)\ j_\ell(k^-r)\ r^2\ dr=\frac{\pi}{2\sqrt{kk^-}}\int_0^\infty r\ j_{\ell+\frac{1}{2}}(kr)j_{\ell+\frac{1}{2}}(k^-r)dr.$

But $\int_0^\infty x\, j_m(kx)\, j_m(k^- x)\ dx = \frac{1}{k}\delta(k - k^-)$. (See Jackson, ED. Page 77).

Therefore, $\frac{\pi}{2\sqrt{kk^-}}\int_0^\infty r\, j_{\ell+\frac{1}{2}}(kr)\, j_{\ell+\frac{1}{2}}(k^- r)\ dr = \frac{\pi}{2k^2}\delta(k - k^-)$; hence, $[\int_0^\infty j_\ell(kr)\, j_\ell(k^- r)\ r^2 dr = \frac{\pi}{2k^2}\delta(k - k^-)]$, C. relation - (2)

Now take $\psi_{n\ell m}(r,\theta,\varphi) = k\sqrt{\frac{2}{\pi}}\, Y_{\ell m}(\theta,\varphi)\, j_\ell(kr)$, normalize it. The result is $\int \psi^*_{k\ell m}(r,\theta,\varphi)\, \psi_{k^- \ell^- m^-}(r,\theta,\varphi) r^2 \sin\theta\, d\theta\, d\varphi\, dr$

$= \int_0^\pi \int_0^{2\pi} \int_0^\infty \frac{2}{\pi} kk^-\, Y^*_{\ell m}(\theta,\varphi) Y_{\ell^- m^-}(\theta,\varphi) j_\ell(kr) \times j_\ell(k^- r) \sin\theta\ d\theta\ d\varphi\, r^2 dr =$

$\frac{2}{\pi} kk^- [\int_0^\infty j_\ell(kr)\, j_\ell(k^- r)\ r^2\ dr \delta_{\ell - \ell} \delta_{mm^-}$. Using the fact that,

$[\ \int_0^\pi \int_0^{2\pi} Y^*_{\ell m}(\theta,\varphi) Y_{\ell^- m^-}(\theta,\varphi)\sin\theta\ d\theta\ d\varphi = \delta_{\ell - \ell} \delta_{mm^-}\ ]$. Then using equation 1, the following will be obtained:

$[\int \psi_{k\ell m}(r,\theta,\varphi)\qquad \psi_{k^- \ell^- m^-}(r,\theta,\varphi) r^2 dr \sin\theta\ d\theta\ d\varphi = \delta(k - k^-)\, \delta_{mm^-}]$ -------------------------------- (3).

It is the normalization condition, as required.
Now to show that $\int e^{ik\cdot r}\, Y_{\ell m}(\theta_k,\varphi_k) j_\ell(k^-\cdot r) dr$
$= \frac{2\pi^2}{k^2}(-i^\ell) Y_{\ell m}(\theta_k,\varphi_k)\, \delta(k - k^-)$.

It is known that $e^{ik\cdot r} = 4\pi \sum_{\ell,m} i^\ell\, j_\ell(kr) Y^*_{\ell m}(\theta,\varphi)\, Y_{\ell,m}(\theta,\varphi)$
So $e^{-ik\cdot r} = 4\pi \sum_{\ell,m}(-i)^\ell\, j_\ell(kr) Y_{\ell m}(\theta,\varphi)\, Y_{\ell,m}(\theta_k,\varphi_k)$

Hence, $\int e^{-ik\cdot r}\, Y_{\ell m}(\theta_k,\varphi_k) j_\ell(k^-\cdot r) dr =$
$4\pi \sum_{\ell,m}(-i)^\ell \int Y_{\ell m}(\theta,\varphi) Y_{\ell m}(\theta,\varphi) Y^*_{\ell m}(\theta_k,\varphi_k) \times j_\ell(kr) j_\ell(k^-\cdot r) dr.$ ----------------(4)

With the aid of the fact that

$$\sum_{\ell m} Y_{\ell m}^*(\theta, \varphi)\, Y_{\ell,m}(\theta^-, \varphi^-) = \frac{\delta(\theta-\theta^-)\delta(\delta\varphi-\varphi^-)}{\sin\theta}$$

And $\int_0^\infty j_\ell(kr)\, j_\ell(k^-r) r^2 dr = \frac{\pi}{2k^2}\delta(k-k^-)$, the result of equation

4 is $\frac{2\pi^2}{k^2}(-i)^\ell Y_{\ell m}(\theta_k,\varphi_k)\delta(k-k^-)$; therefore,

$[\int e^{-ik\cdot r}\, Y_{\ell m}(\theta_k,\varphi_k)\, j_\ell(k^-r) dr =$

$\quad \frac{2\pi^2}{k^2}(-i)^\ell Y_{\ell m}(\theta_k,\varphi_k)\delta(k-k^-)]$------------(4),

as required. QED.

Problem 35: For the radial equation given below, show:
The only solution for bound state is the Hankle function $h_\ell^+(i\hbar r)$.

$[-\frac{\hbar^2}{2m}\frac{d^2}{dr^2} + \ell(\ell+1)\frac{\hbar}{2mr^2} + V(r)-E]R(r)=0,$------------(1).

The solution: The solution is for the case that E<0, using the Hankle function.

Let $\rho=kr$, where $k=\sqrt{\frac{-2m(E-V_0)}{\hbar^2}}=i\hbar$ -------------------- (2).

$\hbar=\sqrt{\frac{2m(E-V_0)}{\hbar^2}}$, --------------------------------------- (3)

Hence, equation 1 becomes in the form

$[\frac{d^2}{d\rho^2}+\frac{2}{\rho}\frac{d}{d\rho}+(1-\frac{\ell(\ell+1)}{\rho^2})]R(r)=0,$ ----------------------- (4).

The solution of equation 4 has two linearly independent solutions, either

$h_\ell^{(-)}(\rho)=n_\ell(\rho)- ij_\ell(\rho)$ or

$h_\ell^{(+)}(\rho)=n_\ell(\rho)+ij_\ell(\rho),$ $\left.\begin{array}{c}\\\\\end{array}\right\}$ -------------------------(5).

The asymptotic form of $j_\ell(\rho)$, and $n_\ell(\rho)$ are used, given by:

$$\left. \begin{array}{l} j_\ell(\rho)(\text{as } \rho \to \infty) \sim \dfrac{\sin\left(\rho - \frac{\ell\pi}{2}\right)}{\rho}, \\[4mm] n_\ell(\rho)(\text{as } \rho \to \infty) \sim \dfrac{\cos\left(\rho - \frac{\ell\pi}{2}\right)}{\rho}. \end{array} \right\} \quad \text{----------------------(6)}.$$

Now substitute equation 6 into equation 5. The asymptotic Hangle functions are:

$$\left. \begin{array}{l} h_\ell^{(-)}(\rho) \ (\text{as } \rho \to \infty) \sim \dfrac{1}{\rho} e^{-i\left(\rho - \frac{\rho\pi}{2}\right)} \\[4mm] h_\ell^{(+)}(\rho) \ (\text{as } \rho \to \infty) \sim \dfrac{1}{\rho} e^{i\left(\rho - \frac{\rho\pi}{2}\right)} \end{array} \right\} \quad \text{----------------(7)}.$$

Since in this case $i k = k$, so, $kr = i k r \to \rho \to i\rho$.
Therefore, Hangle functions becomes:

$$\left. \begin{array}{l} h_\ell^{(-)}(ikr) \cong (r \to \infty) \dfrac{1}{ikr} e^{kr + \frac{i\ell\pi}{2}} \\[6mm] h_\ell^{(+)}(ikr) \cong (r \to \infty) \dfrac{1}{ikr} e^{-kr - \frac{i\ell\pi}{2}} \end{array} \right\} \quad \text{-------------- (8)}.$$

Therefore, the solution of equation 4 might be considered as $R(r) = A j_\ell(kr)$ $r < a$, interior solution.

Since $R(r)($ as $r \to 0) < \infty$, $R(r) = B h_\ell^{(+)}(ikr)$, $r > a$, is exterior solution.

Hence, the desired solution for $r > a$ is
$R(r) = B \ h_\ell^{(+)}(ikr)$, as $kr \gg \ell$, $h_\ell^{(+)}(ikr)$ can be written as
$h_\ell^{(+)}(ikr) \sim_{r \to \infty} e^{-kr} e^{\frac{i\pi\ell}{2}} = \dfrac{e^{-kr}}{ikr}\left(\cos\dfrac{\pi\ell}{2} - i\sin\dfrac{\pi\ell}{2}\right) =$
$e^{-kr} e^{\frac{-i\pi\ell}{2}} e^{\frac{i\pi}{2}}, \ (-i) = e^{\frac{-i\pi}{2}}.$

Therefore, $h_\ell^{(+)}(i k r) \sim_{r \to \infty} \frac{-1}{k r} e^{-k r} e^{\frac{i \pi(1-\ell)}{2}} = \frac{-1}{k r} e^{-k r} e^{\frac{i \pi(1-\ell)}{2}}$

$\therefore i^\ell h_\ell^{(+)}(i k r)$ (as $r \to \infty$) $\sim \frac{i^\ell}{k r} \frac{1}{i^\ell} e^{-k r}$ (as $r \to \infty$) $\sim$

$\frac{-1}{k r} e^{-k r}$. ------------------------(9).

Since R(r) is bound at infinity $[e^{-k r}$ (as r→ ∞)=0].

So the solution that is only bound (as r→ ∞)is $h_\ell^{(+)}(i k r)$.

It is $h_\ell^{(+)}(i k r) \sim \frac{1}{i^\ell k r} e^{-i k r}$, as r→ ∞, which approaches zero exponentially. More precisely $i^\ell h_\ell^{(+)}(i k r)$ is a real function. It is the product of a polynomial of degree ℓ in $\frac{1}{k r}$ and $e^{-i k r}$, but $h_\ell^{(-)}(i k r) \cong (r \to \infty) \frac{1}{i k r} e^{k r + \frac{i \ell \pi}{2}}$ is not bounded at ∞). So the only bounded state solution is $h_\ell^{(+)}(i k r)$, as required. QED.

Problem 36: Find the solution of the following differential equation by series:

$$\frac{d^2 \Theta(z)}{d z^2} - \frac{2 z}{1-z^2} \frac{d \Theta(z)}{d z} + \frac{\lambda}{1-z^2} - \frac{m^2}{(1-z^2)^2} \Theta(z) = 0, \text{--------------- (1).}$$

The solution: Try the solution:

$\Theta(z) = (1-z^2)^s (a_0+a_1 z+a_2 z^2+----+a_n z^n+--)$.

$\Theta^-(z) = \frac{d \Theta(z)}{d z} = -2s(1-z^2)^{s-1}((a_0+a_1 z+a_2 z^2 +----+a_n z^n+-)+(1-z^2)^s(a_1+2a_2 z+------+n a_n z^{n-1} + - -)$.

$\Theta^{''}(z) = \frac{d^2 \Theta(z)}{d z^2} = 4s(s-1)((1- z^2)^{s-2} (a_0+a_1 z +a_2 z^2 +----+a_n z^n)+(a_1+2a_2 z+-----+na_n z^{n-1})(4s(s-1)(1- z^2)^{s-2})-2s((1-z^2)^{s-1}(a_1+2a_2 z+------+na_n z^{n-1} + - -)+(1-z^2)^s \times (2a_2+--------+n(n-1)a_n z^{n-2}+--)$

Now substitute, these in equation 1 with the required mathematical manipulations. Equation 1 becomes

$(1-z^2)^{s-2}$ $[4s^2(a_0\ z^2+a_1z^3+a_n\ z^{n+2}+\text{---})-m^2(a_0\ +a_1z+a_2\ z^2\ +--$ $+a_nz^n+\text{--})+(1-z^2)^s(2a_2+\text{---}+n(n-1)a_nz^{n-2}+\text{---})+$
$(1-z^2)^{s-1}[a_n\ (\lambda+1)+(\ \lambda+5)a_1z_1+\text{---}+a_nz^n(4n+\lambda+1)]=0.$ (2).

Now let $\lambda = v(v + 1)$, so from the constant terms and the linear terms, it is obtained that $4s^2-m^2=0$, ------------------ (3)
$2\cdot 1a_2+[\ v(v + 1)\text{-lml(l ml+1)}]a_0 =0,\text{---------}(4).$

Or in general,

$(s+1)(s+2)\ a_{s+2}=\text{lml(lml+1)}-(v -s)(\ v +s+1)a_s,$ or
$$\frac{a_{s+2}}{a_s} = \frac{l\,m\,l(l\,m\,l+1)-(v-s)(v+s+1)}{(s+1)(s+2)}, \text{------------ (6).}$$

From equation 3, $s=\pm\frac{m}{2}$, neglect (-), [s=lml/2].-----(7).

Let s=0, then $\frac{a_2}{a_0}=\frac{l\,m\,l(l\,m\,l+1)-\lambda)}{(2\cdot 1)}$, ------------------(8).
$\therefore (2\ \cdot\ 1)a_2+[\lambda - l\,m\,l(l\,m\,l + 1))]a_0=0,\text{-----}(9).$

Let s=1, then
$(2\cdot 3\ )a_3+\{\ \lambda\text{-2 1-l}\,m\,l(l\,m\,l + 1)\}a_1=0,$ --------(10).
Let s=2, then $3\ 4\ a_4+[\lambda \text{-2 2-l}\,m\,l(l\,m\,l + 1) - 1\ 2]a_2=0$ (11).
For s=3, then $4\ 5\ a_5+\{\lambda\text{-2 3-l}\,m\,l(l\,m\,l + 1) - 2\cdot 3]a_3$
$=0,$ ----------------------------(12).
$(k +1)(k +2)\ a_{k+2} +[\lambda - 2k\text{-l}\,m\,l(l\,m\,l + 1) \text{-(k-1)k}\}a_k$
-----(13), k=0, 1, 2, 3,-------Q.E.D.

Problem 37: Show that:

(1.) $i\hbar \frac{d}{dt} |\psi''(t)> = H_1''(t)|\psi''(t)>$.

(2.) $i\hbar \frac{d}{dt} q_r''(t) = [q_r''(t), H_0(q'', p'')$.

(3.) $i\hbar \frac{d}{dt} p_r''(t) = [p_r''(t), H_0(q'', p'')]$.

(4.) $[q_r'', p_s''] = i\hbar \delta_{rs}$.

The solution: It is known that

$$\left.\begin{array}{l} |\psi''(t)> = U_0^+(t, t_0)|\psi(t)> \\ q_r''(t) = U_0^+(t, t_0)\, q_r\, U_0(t, t_0) \\ p_r''(t) = U_0^+(t, t_0) p_r U_0(t, t_0) \\ H = H_0 + H_1 \\ H_1'' = U_0^+(t, t_0) H_1\, U_0(t, t_0) \end{array}\right\} \quad \text{---------------(1).}$$

Take $i\hbar \frac{d}{dt} U(t, t_0) = HU(t, t_0)$,------------------(2).

It is convenient to write $U = U_0 U^-$, so

$i\hbar \frac{d}{dt}(U_0 U^-) = H(U_0 U^-)$,------------------------(3).

Carry on the differentiation and notice equations in equation (1.) The result is $i\hbar \frac{d}{dt} U^- = H_1'' U^-$ ---------------- (4).

Take $|\psi''(t)> = U_0^+(t, t_0)|\psi(t)>$, but $|\psi(t)> = U|\psi(t_0)>$. Therefore, $|\psi''(t)> = U_0^+(t, t_0)U_0|\psi(t_0)>$

But $U^- = U_0^+(t, t_0)\, U_0$, so $|\psi''(t)> = U^- |\psi(t_0)>$
Where $|\psi(t_0)>$ is time independent.

Therefore,

$$\frac{d\,|\,\psi''(t)>}{dt}=\frac{d}{dt}U^-|\psi(t_0)>,$$ substitute for $\frac{d}{dt}U^-$ from equation 4.

The result is $\{\frac{d\,|\,\psi''(t)>}{dt}=H_1''(t)|\,\psi''(t)>\}$, as required.

(2.) Take $q_r''(t)=U_0^+(t,\,t_0)\;q_r\;U_0(t,\,t_0)$. Differentiate both sides, $i\hbar\frac{d}{dt}q_r''(t)=\frac{d}{dt}(U_0^+(t,t_0)\;q_r\;U_0(t,\,t_0)$. Carry on the differentiation.

Remember $U_0^+(t,t_0)=e^{\frac{i}{\hbar}(t-t_0)H_0}$ and

$U_0(t,\,t_0)=e^{\frac{-i}{\hbar}(t-t_0)H_0}$. The result will be

$i\hbar\frac{d}{dt}q_r''(t)\;=[q_r''(t),\;H_0(q'',\;p'')+\frac{\partial q_r''(t)}{\partial t}$, but $q_r''(t)$ is not time dependent explicitly, $\frac{\partial q_r''(t)}{\partial t}=0$.

Hence, $[\;i\hbar\frac{d}{dt}q_r''(t)\;=[q_r''(t),\;H_0(q'',\;p'')]$, as required.

(3.) Follow the same procedures and the steps,

where $p_r''(t)=U_0^+(t,\,t_0)\;p_r U_0(t,\,t_0)$ and for the same reasoning $\frac{\partial p_r''(t)}{\partial t}=0$.

So $\{\;i\hbar\frac{d}{dt}p_r''(t)=[p_r''(t),\;H_0\;(q'',\;p'')]\}$, as required.

The References:

(1.) *Quantum Mechanics,* I. Schiff, 1955, McGRAW-Hill Book Company, Inc.

(2.) *Quantum Mechanics,* 3rd ed., Eugen Merezbacker, 1998, John Willy and Sons Inc.

(3.) *The Principle of Quantum Mechanics.* 4th ed., 1958, Oxford Clarendon Press.

(4.) *Elementary Quantum Mechanics,* David S. Saxon, 1968, Holden Day Series in Physics.

(5.) *Quantum Mechanics, Principles and Applications,* M. Alonso and Henry Valk, 1973, Addison Wessly Publication Company.

(6.) *Principles of Modern Physics,* R. B. Leighton, 1950, McGraw-Hill Book Company Inc.

(7.) *Quantum Mechanics.* Vol. 1 and 2, Messiah, 1962, Amsterdam North-Holland.

(8.) *Theoretical Nuclear Physics,* J. M. Blott and V. F. Weiskopf, 1952, John-Wiley.

(9.) Quantum Mechanics: Foundation and Applications, 2nd ed., Bohm Arno, 1994, Springer Verlog.

(10.) *Classical Electrodynamics,* 2nd ed., J. D. Jackson, 1975, John –Wilely.

(11.) Elementary Theory of Angular Momentum, M. E. Rose, 1957, John-Wiley.

(12.) *Classical Mechanics,* 2nd ed., H. Goldstein, 1980, Addison Wesley.

(13.) *Introduction to Quantum Mechanics.* Griffiths, 1995, Englewood, New Jersey, Prentice-Hall.

(14.) Relativistic Quantum Mechanics and Field Theory, Gross, Franz, 1993, John-Wiley.

(15.) *Quantum Mechanics,* L. D. Landau and Lifshitz, 1958, Addison Wesley.

(16.) *Quantum Field Theory,* Mandle and G. Shaw, 1984, John-Wiley.

(17.) *Understanding Quantum Mechanics,* Morrison and A. Micheal, 1990, Englewood, New Jersey, Prentice Hall.

(18.) *Method of Theoretical Physics.* Vols. I and II, P. M. Morse and H. Meshach, 1953, McGraw-Hill.

(19.) *The Theory of Atomic,* 3rd ed., Collisions. N. F. Mott and H. S. W. Massey, 1965, Oxford Clarendon Press.

(20.) *Mathematical Foundation of Quantum Mechanics,* J. von Newmann, 1955 (translated to English by Beyer), Princeton University Press.

(21.) *The Interpretation of Quantum Mechanics,* Omnes, Roland, 1994, Princeton, New Jersey, Princeton University Press.

(22.) The Quantum Vaccum:An introduction to Quantum Electrodynamics, Milonni, Peter, 1995, New York, Academic Press.

(23.) *Quantum Mechanics: Concepts and Methods,* Peres, Asher, 1995, Dordrecht-Kluwer, Academic Publisher.

(24.) *Introduction to High Energy Physics,* 2nd ed., Perkins, H. Donald, 1982, Reading, Massachusetts, Addison-Wesley.

(25.) The Theory of Groups and Quantum Mechanics, Hermann Weyle, (translated by H. P. Robertson), 1931, Dover Publications, Inc.

(26.) *The Classical Theory of Fields,* Landau and Lifshitz, (translated by Hmermish) 2nd ed., 1962, Addison – Wesley.

(27.) *Introduction to the Theory of Relativity,* 8th ed., Peter G. Bergmann, 1958, Prentice Hall, Inc., Englewood Cliffs, New Jersey.

(28.) *Problems in Quantum Mechanics.* V. I. Koglan and V. M. Galistskiy,1963, translated from the Russian by Scripta Technica, Inc.

(29.) *Complex Variables and Applications,* R. V. Churchill, 1960, McGraw-Hill Book Company, Inc.

(30.) *Studies in Particles and Fields,* edited by H. Aly (Southern Illinois University, Edwardsville, USA), 1980, Baghdad University Press.

(31.) *A Modern Approach to Quantum Mechanics,* Townsend, John S., 1992, New York, McGraw-Hill.

(32.) *Quantum Theory of Scattering*, To-You Wu and Takashi Ohmura, 1962, Prentice-Hall, Inc., Englewood Cliffs, New Jersey.

(33.) *The Conceptual Basis of Quantum Field Theory*, Gerard't Hooft, 2004, Handbook of the Philosophy of Science, Elsevier

(34.) *Quantum Field Theory*, C. Itzykson and J. B. Zuber, 1987, Addison-Wesley.

(35.) *An Introduction to Quantum Field Theory*, M. E. Peskin and D. V. Schroeder, 1995, Addison-Wesley.

(36.) *Quantum Theory of Fields*, Vol. 1, S. Weinberg, 1995, CUP.

(37.) *Relativistic Quantum Fields 1*, Mark Hindmarsh, 2002, http//www.pact.cpes..susk.ac.uk/users,markh/RQf$_1$/rqf$_1$. Pdf.

(38.) *Introductory Quantum Mechanics*, 2nd ed., Liboff, Richard L., 1992, Reading, Massachusetts, Addison-Wesely.

(39.) *Quantum Mechanics of Many Degrees of Freedom*, Olton, S. Daniel, and Judah M. Eisenberg, 1988, John-Wiley, New York.

(40.) *Topics in Advanced Quantum Mechanics*, Holstein, Barry, 1992, Reading, Massachusetts, Addison-Wesley.

(41.) *Second Quantization and Atomic Spectroscopy*, Judd, Brian, 1967, Baltimore, Maryland, Johns Hopkins press.

(42.) *Quantum Mechanics II*, L. D. Landau and H. Robin, 1990, New York, John Wiley.

(43.) *Scattering Theory*, Taylor and R. John, 1972, New York, John Wiley.

(44.) *Elements of Advanced Quantum Theory*, J. M. Ziman, 1969, London, Cambridge University Press.

(45.) *Modern Quantum Mechanics*, Revised Edition, edited by J. J. Sakurai and San Fu Tuan, 1994, New York, Benjamin Cummings.

(46.) *Atlas for Computing Mathematical Functions*, Thompson and J. William, 1997, New York, John Wiley.

(47.) *Lectures in Quantum Mechanics*, Michigan State University, 2001, www.nsc1.msu.edu.

(48.) *Angular Momentum in Quantum Mechanics,* 2nd ed., A. R. Edmonds, 1960, second printing 1963, Princeton University Press.

(49.) *Fourier Series and Boundary Value Problems,* 2nd ed., R. V. Churchill, 1963, McGraw-Hill Books Company, Inc.

(50.) *Formulas and Theorems for the Functions of Mathematical Physics,* Wilhelm Magnus and Fritz Oberwettinger, 1949, Chelsea Publishing Company, translated by John Wermer.

(51.) *Quantum Theory of Angular Momentum,* D. A. Varshalovich, et al., World Scientific, 1988.

(52.) *Angular Momentum,* W. J. Thompson, 1994, John Wiley and Sons, Inc.

(53.) *Angular Momentum,* D. M. Brink et al., 1968, Clarendon Press, Oxford.

(54.) *The 3-J and 6-J Symbols*, M. Rotenberg et al., 1959, (Technology Press MIT, Cambridge).

Index

Q

R

radiations, 11, 15–16, 19–20, 23–24, 278, 436

radius, 167, 187, 190, 213, 245, 247, 375

region, 141, 177, 182, 188, 197, 284, 359, 399

relativity, special, 266–67, 269, 274

resonances, 192–93

result, 46, 71, 151, 158, 204, 226, 268, 279, 281, 299, 322–23, 359, 391, 410–11, 415–16

rotation, 53, 99, 101–13, 120–23, 127, 129, 234, 319–23, 342, 394

S

scalar product, 59, 61, 124, 165, 388, 394, 405

scalar quantity, 108, 320

scattering, 167–68, 170, 174–79, 186–87, 191, 200–201, 211, 246, 276, 322, 328, 331, 338, 346, 427

scattering cross section, 173, 183, 209, 258

Schrödinger equation, 27–28, 35, 37, 39, 41, 64, 71–72, 74–76, 115, 150, 175, 178–79, 188–89, 201, 269–70

science, 12, 15, 66, 281, 427, 429, 432–33, 435, 437

set, complete, 35, 58, 87, 94, 99–100, 143, 152, 281, 410

singlet, 128, 133, 215, 234, 398

solution, 35, 37, 64, 81–82, 87–88, 140, 176–80, 188–89, 201–5, 328–29, 363–64, 377–78, 383–84, 386–89, 405–8

complete, 329

regular, 189

time-dependent, 35, 64

SP. *See* symmetrization postulate

space, 21–22, 41, 57, 60, 68–69, 101, 141–43, 146–47, 149, 151–54, 161, 164, 221, 267, 289–91

dual, 58

spatial, 41

spectrum, 29, 61, 85, 90–91, 94, 117, 233, 241, 253, 276

spin, 15, 114, 118–19, 121–22, 124–27, 129, 132–33, 136, 142, 151–52, 154, 161, 234, 251, 279

isobaric, 162, 342, 345

total, 132–33

spinors, 120–21, 270, 300, 319, 321, 335

spin spaces, 120, 162–63

S-state, 133, 374–76

states, 38–39, 57, 84, 106–7, 139–40, 144, 149–50, 152–55, 157–60, 162, 234–36, 242, 252–54, 281, 376

bound, 215, 376–77

discrete, 199

final, 211, 251–52, 258

macroscopic, 155

microscopic, 155

W

Z